MOLECULAR
BIOLOGY
INTELLIGENCE
UNIT

YEAST STRESS RESPONSES

Stefan Hohmann

Göteborg University
Göteborg, Sweden and
Catholic University of Leuven
Leuven, Belgium

Willem H. Mager

Vrije Universiteit
Amsterdam, The Netherlands

Illustrations: Stefan Hohmann

מכון ויצמו למד ש
השטרית
גנטיקה מולקולרית

CHAPMAN & HALL
I(T)P An International Thomson Publishing Company

New York • Albany • Bonn • Boston • Cincinnati • Detroit • London • Madrid • Melbourne •
Mexico City • Pacific Grove • Paris • San Francisco • Singapore • Tokyo • Toronto • Washington

R.G. LANDES COMPANY
AUSTIN

SYSTEM NO. מס' מערכת
68216-1

MOLECULAR BIOLOGY INTELLIGENCE UNIT
YEAST STRESS RESPONSES

R.G. LANDES COMPANY
Austin, Texas, U.S.A.

Please address all inquiries to the Publishers:
R.G. Landes Company, 810 S. Church Street, Georgetown, Texas, U.S.A. 78626
Phone: 512/ 863 7762; FAX: 512/ 863 0081

North American distributor:
Chapman & Hall, 115 Fifth Avenue, New York, New York, U.S.A. 10003

CHAPMAN & HALL

U.S. and Canada ISBN: 0-412-13251-6

Library of Congress Cataloging-in-Publication Data
Hohmann, Stefan, 1956-
   Yeast stress responses / Stefan Hohmann, Willem H. Mager;
illustrations, Stefan Hohmann.
      p. cm. — (Molecular biology intelligence unit)
   Includes bibliographical references and index.
   ISBN 1-57059-421-X (alk. paper)
   1. Saccharomyces cerevisiae—Effect of stress on.
   2. Saccharomyces cerevisiae—Physiology.   3. Yeast fungi—Effect of stress on.
   4. Yeast fungi—Physiology. I. Mager, Willem H., 1947- . II. Title. III. Series.
QK623.S23H64 1997
571.l6'29563—dc21
                                                            96-51840
                                                               CIP

PUBLISHER'S NOTE

R.G. Landes Company publishes six book series: *Medical Intelligence Unit, Molecular Biology Intelligence Unit, Neuroscience Intelligence Unit, Tissue Engineering Intelligence Unit, Biotechnology Intelligence Unit* and *Environmental Intelligence Unit.* The authors of our books are acknowledged leaders in their fields and the topics are unique. Almost without exception, no other similar books exist on these topics.

Our goal is to publish books in important and rapidly changing areas of bioscience and environment for sophisticated researchers and clinicians. To achieve this goal, we have accelerated our publishing program to conform to the fast pace in which information grows in bioscience. Most of our books are published within 90 to 120 days of receipt of the manuscript. We would like to thank our readers for their continuing interest and welcome any comments or suggestions they may have for future books.

Shyamali Ghosh
Publications Director
R.G. Landes Company

CONTENTS

EDITORS

Stefan Hohmann
Department of General and Marine Microbiology
Göteborg, Sweden, and
Laboratorium voor Moleculaire Celbiologie
Katholieke Universiteit Leuven
Leuven, Belgium
Introduction, Chapter 4

Willem H. Mager
Department of Biochemistry and Molecular Biology
Biocentrum Amsterdam
Vrije Universiteit Amsterdam
Amsterdam, The Netherlands
Introduction, Chapter 7

CONTRIBUTORS

Johannes H. de Winde
Laboratorium voor Moleculaire
  Celbiologie
Katholieke Universiteit Leuven
Leuven, Belgium
Chapter 1

Edwina K. Fuge
Department of Biology
University of New Mexico
Albuquerque, New Mexico, U.S.A.
Chapter 2

José A. Márquez
Instituto de Biologia Molecular
  y Celular de Plantas
Universidad Politecnica de Valencia -
  C.S.I.C.
Valencia, Spain
Chapter 5

Peter Piper
Department of Biochemistry
  and Molecular Biology
University College London
London, U.K.
Chapter 3

Gabino Ríos
Instituto de Biologia Molecular
  y Celular de Plantas
Universidad Politecnica de Valencia -
  C.S.I.C.
Valencia, Spain
Chapter 5

Helmut Ruis
Institut für Biochemie und
   Molekulare Zellbiologie
   und Ludwig Boltzmann-
   Forschungsstelle für Biochemie
Universität Wien
Wien, Austria
Chapter 8

Nicholas Santoro
Department of Biological Chemistry
University of Michigan Medical
   School
Ann Arbor, Michigan, U.S.A.
Chapter 6

Ramón Serrano
Instituto de Biologia Molecular
   y Celular de Plantas
Universidad Politecnica de Valencia -
   C.S.I.C.
Valencia, Spain
Chapter 5

Marco Siderius
Department of Biochemistry
   and Molecular Biology
Biocentrum Amsterdam
Vrije Universiteit Amsterdam
Amsterdam, The Netherlands
Chapter 7

Johan M. Thevelein
Laboratorium voor Moleculaire
   Celbiologie
Katholieke Universiteit Leuven
Leuven, Belgium
Chapter 1

Dennis J. Thiele
Department of Biological Chemistry
University of Michigan Medical
   School
Ann Arbor, Michigan, U.S.A.
Chapter 6

Margaret Werner-Washburne
Department of Biology
University of New Mexico
Albuquerque, New Mexico, U.S.A.
Chapter 2

Joris Winderickx
Laboratorium voor Moleculaire
   Celbiologie
Katholieke Universiteit Leuven
Leuven, Belgium
Chapter 1

SYMBOLS LEGEND

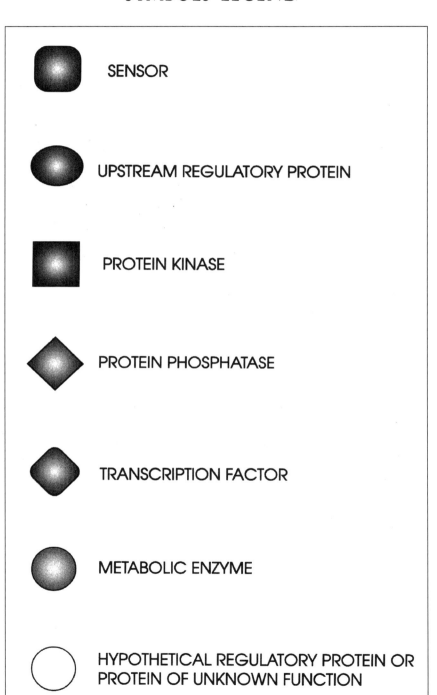

SENSOR

UPSTREAM REGULATORY PROTEIN

PROTEIN KINASE

PROTEIN PHOSPHATASE

TRANSCRIPTION FACTOR

METABOLIC ENZYME

HYPOTHETICAL REGULATORY PROTEIN OR
PROTEIN OF UNKNOWN FUNCTION

========= INTRODUCTION =========

STRESS RESPONSE MECHANISMS IN THE YEAST SACCHAROMYCES CEREVISIAE

Willem H. Mager and Stefan Hohmann

The survival of living cells is dependent on their ability to sense alterations in the environment and to appropriately respond to the new situation. The unicellular eukaryote *Saccharomyces cerevisiae* has to adapt to a steadily changing environment in order to maintain a high proliferation rate. Alterations in the chemical or physical conditions of the cell that impose a negative effect on growth demand rapid cellular responses which are essential for survival. The molecular mechanisms induced upon exposure of cells to such adverse conditions are commonly designated stress responses (illustrated in Fig. 1).

Stress response mechanisms aim to protect cells against potentially detrimental effects of stress challenges and to repair any molecular damage. Therefore the stress response leads to adjustment of metabolism and other cellular processes to the new status. According to their biological significance stress responses not only result in the repair of damage that has occurred but also lead to the acquisition of stress tolerance and thus to the establishment of mechanisms that prevent damage from occurring (Fig. 1). Indeed, exposure to a mild stress evokes improved resistance against severe stress. Thus as a result of the stress response the cell produces a number of proteins at a different level or with different activity than before stress exposure.

The best characterized stress response is that to heat shock. Shifting yeast cells from 23°C to 36°C results in the induced expression of a large set of so-called heat shock genes. Transcription of most heat shock genes in *S. cerevisiae* occurs via the activation of the heat shock-specific transcription factor Hsf1p which promotes transcription via so-called heat shock elements (HSEs) present in the promoter of target genes. Enhanced expression of the heat shock genes leads to the (transient) accumulation

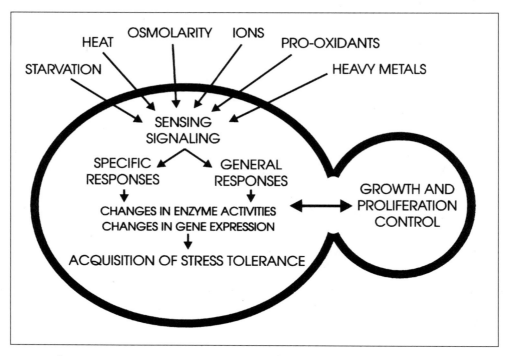

Fig. 1. Different types of stress conditions are sensed by the yeast cell and trigger both specific and general molecular responses. These stress responses result in changes at the level of enzyme activities and gene expression and lead to the acquisition of stress tolerance. Stress control plays an important role in the regulation of the cellular growth potential.

of heat shock proteins which play an important role in protecting other cellular proteins against thermal denaturation and in restoring their biological activity. Such heat shock proteins fulfill a similar role under stress as compared with normal growth conditions: they serve as molecular chaperones in protein folding and transport.

This feature holds true for probably all types of stress responses. Exposure of cells to a certain stress agent leads to the activation of molecular mechanisms normally implicated in homeostatic processes. Other clear examples are the responses of budding yeast to increased or decreased external osmolarity (osmotic stress response), to high ion concentrations (salt stress response) or to the presence or generation of reactive oxygen species (oxidative stress response). Most prominent among the events that are triggered by hyper-osmotic stress—i.e., by transfer of yeast to a high salt or sugar medium—is the elevated syn-

thesis and retainment of glycerol, the sole compatible solute of this organism. This part of the osmostress response thus makes use of the normal osmoregulation machinery which is expected to be involved in turgor maintenance and hence in growth processes under ambient environmental conditions. Exposing yeast cells to high NaCl in addition to an osmostress response also evokes specific Na^+-induced salt stress responses such as the elevated expression of the gene *ENA1* encoding the sodium pump and probably the stimulation of other ion transport processes. These transport mechanisms are normally involved in ion homeostasis and/or in the generation of gradients that drive active transport processes. Finally, yeast cells also contain enzymatic as well as non-enzymatic defense mechanisms against the harmful effects of reactive oxygen derivatives. The genes encoding components of this defense apparatus—such as those involved in glutathione or thioredoxin biosynthesis—display increased expression after an oxidative challenge. Glutathione and thioredoxin are normally involved in cellular redox-reactions in particular those involved in establishment and maintenance of the tertiary structure of proteins. In addition, redox active metals play a major part in the generation of reactive oxygen species in the cell. Therefore a strong connection exists between the homeostatic regulation of these metals and the occurrence of oxidative stress events.

Depletion of nutrients induces a starvation response in yeast cells which leads to an adaptation of cellular metabolism partially dependent on the nutritional component that is lacking. Probably most of the time yeast cells face limited nutrient availability in nature. Therefore mechanisms that allow the cells to survive such periods in a metabolic hold-on situation (with the option to rapidly regain full metabolic and proliferative activity as soon as nutrients are plentiful) are obviously highly important for any cell exposed to environmental changes. Upon long-term starvation yeast cells enter stationary phase in which they become highly stress tolerant. Stationary phase cells are clearly distinct from proliferating cells and they can survive as such for decades or even centuries and express a very different set of genes than proliferating and short-term starved cells do.

The stress-type specific responses mentioned above, however, only form part of the cellular response elicited by stress. For instance exposing yeast cells to high salt concentrations not only results in adaptation of the osmoregulatory mechanisms but also a transient increase in the synthesis of (at least some of the) heat shock proteins. These observations reflect the 'general stress' that cells experience by changing the external osmolarity. It is conceivable that osmostress may also affect the conformation of proteins since cells are shrinking at such condition and the intracellular macromolecular concentrations therefore transiently increase. The overlap between the

various stress responses may at least partly contribute to the phenomenon of cross protection: exposure to one stress condition in general leads to a certain tolerance against another type of stress. As mentioned above, nutrient starvation renders yeast cells resistant to various stress conditions and thus also induces a general stress response. This may reflect a general alarm system: starvation may be interpreted by the cell as a first sign for additional problems to come. In addition, metabolically less active cells might have less flexibility in their responses and therefore acquire stress tolerance at an early phase of adaptation to starvation.

Most likely the common basis for the general stress responsive gene expression resides in the occurrence in a large number of gene promoters of another stress-responsive cis-element—STRE—for (general) stress-responsive element. A typical example of such promoters are those of the genes encoding the small heat shock protein Hsp12 and catalase T. The multiple STREs in these promoters mediate not only activation after heat shock but also after high salt or peroxide exposures as well as upon nutrient starvation.

The common response to stress and nutrient starvation also exemplifies the important link between adaptation to stress and growth control. Transcription activation via STREs appears to be under negative control by cAMP-dependent protein kinase A whereas growth-related gene expression is stimulated by high protein kinase A activity. In this respect stress response and growth regulation therefore may be two sides of the same coin (Fig. 1).

The mechanisms of sensing of stress belong to the most intriguing but thus far ill-understood aspects of the stress response. Heat-stress might be sensed at the level of protein denaturation which may lead to the recruitment and activation of Hsf1p from its inactive complex with the heat shock protein Hsp70, but other mechanisms have also been proposed. External osmolarity is sensed by *S. cerevisiae* most probably in part by plasma membrane proteins resembling the two-component sensing and signaling systems in bacterial cells this, however, reveals little about their exact mechanism of function. It is completely unknown how reactive oxygen species or high salt concentrations evoke the corresponding specific response and how nutrient depletion is detected.

The underlying stress signaling pathways are beginning to be elucidated, however. It is clear by now that the High Osmolarity Glycerol (HOG) response pathway—which belongs to the family of Mitogen Activated Pathways (MAPs)—mediates at least part of the osmostress response and several components of this pathway have been identified. STREs represent targets of this pathway and putative corresponding transacting factors have recently been identified. On the other hand the HOG-responsive site in the promoter of the glycerol-3-phosphate dehydrogenase gene (*GPD1*)—which is specifically controlled by osmotic stress—has not been elucidated so far. Another MAP-kinase pathway—the PKC (for Protein Kinase C) pathway—can be stimulated specifically by hypo-osmotic stress (and probably by heat stress as well) but neither the sensor nor any targets of this response have been identified. Components of the salt-specific response may include the calcium-regulated protein phosphatase calcineurin. The target of the oxidative stress response as far as transcriptional activation is concerned turned out to be the transcription factor Yap1p, but is not known how activation of this factor takes place. The response to nutrient starvation and entry into stationary phase appears to be closely related to the activity of the yeast Ras-cAMP pathway which contains the yeast Ras proteins and cAMP-dependent protein kinase A as mentioned above. Also in this case the sensor and most of the targets of this pathway are unknown and the interplay of this signaling cascade of central importance to yeast growth control with other stress induced pathways is only very poorly understood.

In this book the molecular mechanisms underlying several stress responses in the budding yeast *Saccharomyces cerevisiae* will be reviewed and discussed. This organism has been a favorite eukaryotic species for investigation in the past years due to the wide range of genetic approaches available. Yeast genes can be isolated readily in many ways and the availability of the sequence of the entire yeast genome is presently revolutionizing yeast research once more. In addition, yeast genes can easily be deleted or specifically mutagenized and expressed at higher or lower levels or conditionally. There is no other eukaryotic organism where so many molecular details have been elucidated by biochemical and genetic analyses in different areas of molecular and cellular biology such as cell cycle control, signal transduction, gene expression, protein traffic and turnover, organelle biogenesis and many others. Most significantly, every new finding can be exploited by novel genetic approaches to find further genes/proteins operating in the same system and yeast geneticists are ever proving creativity in designing such screens for new mutations and genes. In addition,—as pointed out in most of the chapters—genes from higher eukaryotes can be functionally expressed in yeast, which has opened the possibility for a new era in yeast genetics: genetic analysis of higher eukaryotic systems by employing the powerful tools available for yeast studies.

The results obtained with this organism are indeed of a wider importance. As will be discussed in several chapters, different bodies of evidence exist that mechanisms operating in yeast also occur in complex eukaryotes. Several yeast mutations that affect different stress response mechanisms could be complemented by genes from higher eukaryotes, suggesting that structural and functional conservation during evolution has occurred. The presented data therefore have important impact for understanding stress response mechanisms in other organisms, such as the relationship of stress and certain pathological conditions and the engineering of crops towards improved stress tolerance.

This book will not only cover the current knowledge of the physiology of yeast stress responses but will also review the state of an important portion of the research on signal transduction pathways in this organism.

The different stress situation outlined above are covered in seven consecutive chapters. The first chapter describes the responses of yeast to nutrient starvation as well as the mechanisms re-adapting the cells once the full set of required nutrients becomes available. This chapter also introduces the universally important Ras-cAMP pathway controlling protein kinase A, which is mentioned in all chapters of the book, as well as the catabolite repression (Snf1p) pathway. In extension the second chapter describes the mechanisms by which yeast cells enter, maintain and leave the long-term starvation stage stationary phase. Chapter 3 describes the heat shock response with emphasis on the role of heat shock proteins both under stress and under normal growth conditions. The responses to

osmotic stress and salt stress are partially overlapping—as mentioned above—but in order to emphasize the distinct aspects of ion homeostasis and salt stress the responses to pure osmotic stress and to high ion concentrations are covered in separate chapters, 4 and 5. Chapter 6 extensively covers all aspects of the oxidative stress response and heavy metal homeostasis. The overlapping, general stress response is reviewed in chapter 7. In the final chapter conclusions will be drawn with respect to the choice of yeast as a model organism, approaches to study stress responses and the interplay between the various responses and underlying signaling pathways. Although difficult in such a rapidly developing field in molecular biology, in this book we will also try to predict some breakthroughs that might hopefully be achieved in the near future.

===================== CHAPTER 1 =====================

FROM FEAST TO FAMINE: ADAPTATION TO NUTRIENT DEPLETION IN YEAST

Johannes H. de Winde, Johan M. Thevelein and Joris Winderickx

1. FROM FEAST TO FAMINE: AN INTRODUCTION

Microorganisms share an unsurpassed ability to thrive within a seemingly barren environment. Of the eukaryotic kingdom only the species at the very bottom of the evolutionary ladder manage to make themselves a living on nothing but a simple carbon source, some elementary nitrogen-, phosphor- and sulfur compounds and some essential trace elements. Thus unicellular and simple multicellular fungi are not only at the bedrock of eukaryotic offspring, but are at the top of many food chains as well.

In this respect baker's yeast *Saccharomyces cerevisiae* may be regarded as a very useful example of a cultivated unicellular fungus. This budding yeast is widely used in beer brewery, winemaking, food production and synthesis of many useful compounds. In addition, its capacity to grow under a wide variety of culturing conditions has made this yeast a fruitful model to study various aspects of nutrient-induced metabolic phenomena and growth control.

As wild yeast growing in vineyards and apple trees, *Saccharomyces* sp. have to cope with long periods of nutritional shortage alternating with brief encounters of plentiful abundance (eating a grape takes only a small portion of the time to grow it). Therefore these yeasts have developed intricate ways to both profit from and survive on very low nutrient levels and to sense a sudden abundance of nutrients with efficient resetting of their metabolism and growth rate to a rich environment. When a single essential nutrient becomes limiting and eventually absent, recently the cellular proliferative machinery is efficiently shut down and a survival program is launched. In the absence of any one of the essential

Yeast Stress Responses, edited by Stefan Hohmann and Willem H. Mager.
© 1997 R.G. Landes Company.

nutrients yeast cells enter a specific, nonproliferative state known as stationary phase (chapter 2) with the ultimate aim of surviving the starvation period.[1]

2. SETTING THE STAGE: LIMITATION, STARVATION AND CELL CYCLE CHECK POINTS

In this chapter we will bring together current knowledge concerning the ways in which yeast cells respond to depletion of nutrients. An important discrimination is made between nutrient limitation and nutrient starvation. Since like all microorganisms yeast can use a wide variety of substances as nutrient sources, decreasing

availability of one substrate can in many instances be compensated by the utilization of another. Thus the response to nutrient limitation often encompasses a metabolic switch from the utilization of a richer to that of a poorer nutrient source. Specific signaling and metabolic pathways often have to be activated for this. For *S. cerevisiae* cells growing on rich, sugar-containing medium—be it in nature or under laboratory conditions—fermentable sugar will usually be the first limiting nutrient. Subsequently, metabolism and biosynthetic capacity are reprogrammed at the so-called 'diauxic shift' for the utilization of ethanol and acetate as carbon sources which have accumulated during previous fermen-

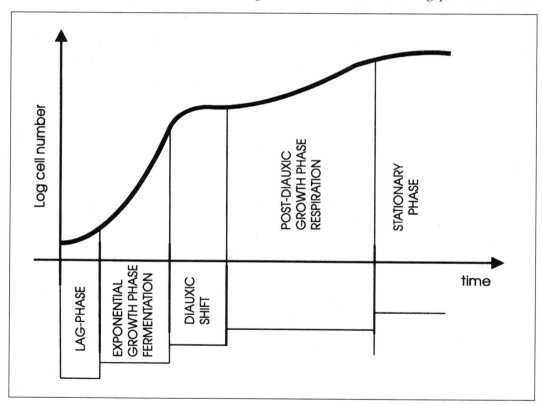

Fig. 1.1. Typical culture-density profile of a fermentative batch culture of Saccharomyces cerevisiae. A schematic representation of the increase in cell number and cell density of a batch culture of S. cerevisiae inoculated in complete medium with a rapidly fermentable sugar (glucose, fructose, mannose) as a carbon source. A relatively brief lag-period and a rapid, exponential fermentative growth phase are followed by a second lag-phase when the sugar has become limiting. During this 'diauxic shift' the cells reset their metabolic capacity from fermentation to respiration and subsequently resume growth using as a carbon source the ethanol, acetate and other products of the initial fermentative growth phase. When eventually the carbon source or another essential nutrient becomes exhausted the cells enter a resting state or stationary phase (chapter 2) with the ultimate goal to survive the starvation period.

tative growth. When eventually these compounds have been used up the cells will
enter stationary phase as a consequence of
carbon source starvation (Fig. 1.1).

In *S. cerevisiae* the availability of nutrients is checked within a narrow window
in the G1 phase of the cell division cycle[2]
(Fig. 1.2). The first events introducing a
new proliferation cycle will be executed
(Start A) when the nutrient supply is sufficient.[2] Limiting amounts of a certain nutrient compound causes the cells to stall
transiently in G1 while reprogramming
metabolism for the utilization of any available alternative nutrient. When the cells
experience definite starvation for one or
more nutrients they will enter stationary
phase which is distinct from G_1 in many
aspects and consequently named G_0.[1,3] It
is important to note that nutrient sensing
and growth regulation are confined to the

G_1 phase of the cell cycle also in other
yeast species with *Schizosaccharomyces pombe*
being a well-characterized example.[4-7]
Hence the general concept of eukaryotic
nutrient sensing and the signal transduction pathways and metabolic regulation
routes involved may be well conserved
between divergent eukaryotic species.

3. SPECIFIC RESPONSES TO NUTRIENT DEPLETION

3.1. CARBON SOURCE LIMITATION AND STARVATION

3.1.1. Responses operative in carbon source utilization

One must first understand how different carbon sources are used by the yeast
cells and get insight as to how particular
carbon sources affect yeast metabolism in

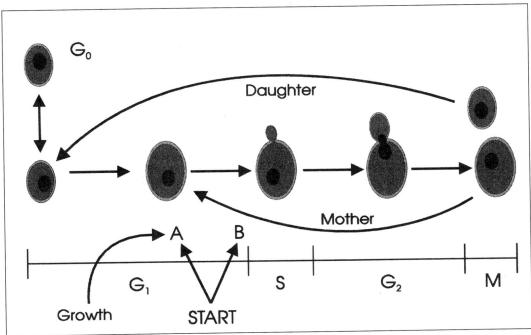

Fig. 1.2. The Saccharomyces cerevisiae *cell cycle. Proliferating budding yeast cells experience several distinct growth phases,[2,5] G_1 (gap1), S (DNA-Synthesis), G_2 (gap2) and M (Mitosis, cell division) phases each are readily definable when a mother cell starts a new round of proliferation resulting in the release of a new daughter. The latter will continue growing in G_1 until she reaches the critical cell size required to traverse START into a new proliferation cycle. The difference in size between mother and daughter cells is especially pronounced in rapidly growing, fermenting cultures. The START decision in G_1 is separated in two phases: START A being the nutrient and growth checkpoint and START B being the replication and proliferation checkpoint.[2] When one or more essential nutrients are depleted from the growth medium cells will enter an off-cycle quiescent state known as G_0.[1,3] When nutrients are available, cells will resume growth at START A.*

order to better understand the consequences of carbon source limitation and starvation in yeast.

Baker's yeast *S. cerevisiae* belongs to a group of so-called facultative anaerobic yeasts. These microorganisms will ferment hexose sugars like glucose and fructose under both aerobic and anaerobic growth conditions. In aerobic batch cultures of this yeast typically about 70% of the available glucose is fermented to ethanol and CO_2, 20% is incorporated into biomass, 8% is used in glycerol production and only 2% will yield CO_2 and H_2O via oxidative phosphorylation inside the mitochondria. Accordingly glycolytic flux is high and O_2 consumption is low.[8,9]

Saccharomyces sp. will utilize a wide variety of compounds as carbon and energy sources (Fig. 1.3). Glucose and fructose enter immediately into the glycolytic pathway where ATP is obtained from substrate phosphorylation and sugars are converted into pyruvate and then to ethanol and CO_2. Galactose and mannose are first converted to glucose-6-phosphate and fructose-6-

phosphate which will then enter glycolysis. Di- and trisaccharides that can be utilized as carbon sources are cleaved by specific glucosidases into their compound monosaccharides and in this way yield glucose, fructose and galactose.[10,11] As stated above, although only a small fraction of the fermentable carbon source is initially completely metabolized and much less ATP is produced than during respiratory growth, yeasts prefer fermentative growth. This may seem wasteful. However, since the ethanol as well as the acetate produced during fermentation will be metabolized in the postdiauxic growth phase (Fig. 1.1), as soon as the fermentable sugar is exhausted nearly all of the available carbon source will be used eventually. In addition, and perhaps more importantly the production of high ethanol concentrations inhibits the growth of most other microorganisms allowing *S. cerevisiae* to eventually dominate in spontaneous fermentation.

Nonfermentable (sometimes referred to as 'poor') carbon sources such as ethanol, acetate and lactate are metabolized via res-

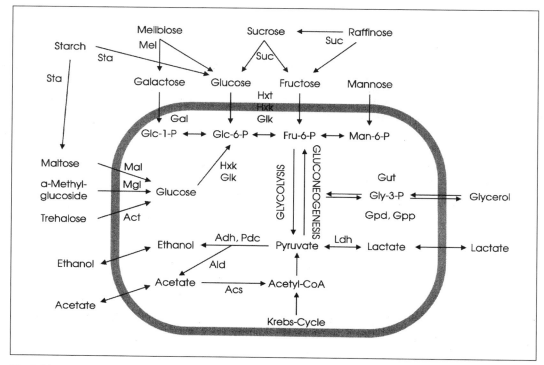

Fig. 1.3A.

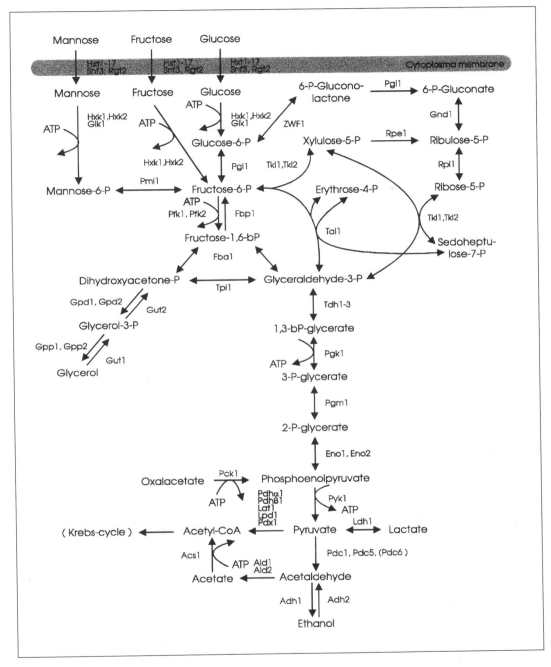

Fig. 1.3B.

Fig. 1.3. Carbon flow and intermediary metabolism in yeast. Overview of utilization (A) and intermediary metabolism (B) of diverse sugar and nonsugar carbon sources in Saccharomyces cerevisiae. Names of enzymes and enzyme families involved in the various utilization pathways are depicted in the usual abbreviations. In 'B' the flow of various carbon sources into the main metabolic routes—glycolysis, gluconeogenesis, pentose-phosphate cycle and Krebs cycle—is depicted in more detail. For more information, please refer to several excellent reviews in references 10-12.

piration in the TCA cycle with ATP being produced by oxidative phosphorylation.[12] Glycerol which enters glycolysis after its conversion to dihydroxyacetone phosphate (see chapter 4) also requires respiration in order to serve as an energy source because the excess NADH formed has to be re-oxidized. During respiratory growth the cells produce hexose-phosphates by gluconeogenesis for the biosynthesis of macromolecules. Most steps in the gluconeogenic pathway are catalyzed by enzymes that are also used in glycolysis. Hence in order to avoid futile cycling strict control is exerted at the level of two antagonistic enzyme pairs: fructose-1,6-bisphosphatase (Fbp1p) / phosphofructokinase (Pfk1p) and phosphoenolpyruvate carboxykinase (Pck1p) / pyruvate kinase (Pyk1p) that catalyze the two irreversible steps (Fig. 1.3).[10,11,13]

3.1.2. Carbon source-dependent growth control

Yeast batch cultures grown on glucose show several well-defined growth phases (Fig. 1.1) due to a multiple-level regulation of metabolism by the available carbon source. In the first phase characterized by rapid growth, glucose is fermented with the concomitant repression of genes required for respiratory growth. When glucose is exhausted the culture enters a short adaptive lag-phase known as diauxic shift. During this phase the glucose-repressed genes become derepressed and the culture adapts its metabolism for the subsequent utilization of ethanol and other byproducts of fermentation.[14] It is important to realize that derepression of the production of certain enzymes begins well before complete glucose depletion.[15] Enzymes of the gluconeogenic pathway, on the other hand, are very sensitive to glucose and remain repressed at glucose concentrations that are lower than the K_m of the high affinity glucose transport system.[16] This high sensitivity is likely required to avoid futile cycling and to enable a clear-cut shift from glycolysis to gluconeogenesis (see section 3.1.5). Interestingly, the *BCY1* transcript which encodes the regulatory subunit of protein

kinase A shows a transient 5-fold increase during the diauxic shift.[1] The increased expression of *BCY1* may lead to a sudden drop in the activity of protein kinase A which may be required to reset the regulatory machinery. Adaptation to respiratory growth is facilitated in this way.[17] Growth in the third, postdiauxic phase is much slower and ceases with the exhaustion of the available ethanol and acetate after which the culture enters stationary phase (chapter 2).

During the diauxic shift and postdiauxic growth phase many genes operative in stress response become derepressed. These include several heat shock proteins like *HSP12* and *SSA3*, *CTT1*, *DDR2* and *UBI4*. Furthermore, the cell starts to accumulate the reserve carbohydrates glycogen and trehalose (see section 3.1.3) just before glucose is exhausted.[18] The latter has been shown to be required for optimal resistance to starvation and general stress.[19-24]

S. cerevisiae prefers the utilization of fermentable sugars with glucose as the prime substrate of choice. Even when grown on a mixture of glucose and fructose or mannose, the glucose will be metabolized first.[11,25] The metabolic preference for fermentable sugar is mediated through posttranslational inactivation and degradation of several enzymes (section 3.1.6) and transcriptional repression of various genes involved in the utilization of alternative carbon sources (section 3.1.5). At the same time glucose induces the biosynthesis and/or the activation of enzymes required for its optimal metabolization (section 3.1.8) along with several components required to accomplish a faster growth rate.[11,26-29] Several signaling pathways are involved in this complex regulation. Of these the Ras-cAMP pathway (section 3.1.6) and the main glucose-repression pathway (section 3.1.5) have been studied in most detail.

3.1.3. Glycogen and trehalose metabolism

The intracellular concentrations of the storage carbohydrates glycogen and trehalose vary with the growth phase of the yeast cell.[18] In cells growing on a fermentable carbon source like glucose or fructose glycogen accumulates before the fermentable sugar is exhausted with peak concentrations occurring at the diauxic shift. In contrast, trehalose only begins to accumulate at the diauxic shift and peaks when cells enter stationary phase because of nutritional shortage. During stationary phase trehalose is slowly degraded and its disappearance is accompanied by loss of culture viability (chapter 2).[1,18] Hence the induction of glycogen and trehalose accumulation may be regarded as a nutrient limitation response.

Both glycogen and trehalose are synthesized from glucose-6-phosphate and UDP-glucose, however, biosynthetic control and the enzymes involved are completely different (Fig. 1.4). The onset of glycogen synthesis requires the presence of self-glucosylating initiator proteins encoded by *GLG1* and *GLG2*.[30] Glycogen synthase is encoded by the differentially regulated *GSY1* and *GSY2* genes[31,32] and the glycogen branching enzyme by *GLC3*.[33,34] A debranching enzyme has not yet been identified in yeast. Glycogen catabolism is achieved by glycogen phosphorylase (*GPH1*)[35] or by glucoamylase in sporulating cells.[29] Expression of *GSY2*, *GLC3* and *GPH1* is transcriptionally activated when the glucose concentration is decreasing.[32-35] Glycogen synthase is activated via dephosphorylation by protein phosphatase type 1 encoded by *GLC7*[36,37] which is targeted to glycogen particles and regulated by the *GAC1* and *GLC8* gene products.[36,38] A glycogen synthase kinase which would be required to inactivate glycogen synthase has

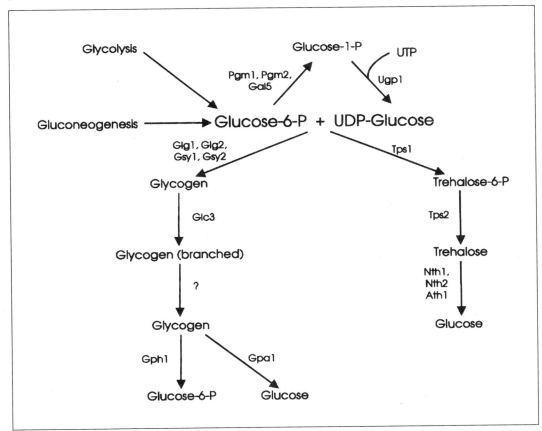

Fig. 1.4. Metabolism of the reserve carbohydrates glycogen and trehalose. Schematic depiction of synthesis and degradation routes for the storage carbohydrates glycogen and trehalose in Saccharomyces cerevisiae. *For abbreviated enzyme names please refer to paragraph 3.1.3 and refs. 10, 36, 41.*

not been conclusively identified in yeast (but see section 3.3.2). High activity of protein kinase A (see section 3.1.6) decreases glycogen synthesis, but the exact mechanism is unknown.[36,39] In addition, the Snf1p protein kinase—previously only implicated in carbon source-dependent control of gene expression (see section 3.1.5)—is required for normal accumulation of glycogen.[21,36,40] Hence Snf1p and protein kinase A appear to act antagonistically in separate regulatory pathways controlling nutritional responses in yeast.[20,21,40]

The concentration of trehalose in the yeast cell is the result of the synthetic activity of the trehalose synthase complex and the degradative activity of trehalase[41] (Fig. 1.4). The trehalose synthase complex consists of three subunits. Trehalose-6-phosphate synthase is encoded by TPS1[36,42-45] and trehalose-6-phosphate phosphatase by TPS2.[46] TSL1 and the related gene TPS3 both code for the largest and probably regulatory subunit of the complex[45] (Bell, Sun, Hohmann and Thevelein, unpublished results). Trehalose-6-phospate synthase activity is strongly enhanced by fructose-6-phosphate and inhibited by phosphate, whereas trehalose-6-phosphate phosphatase activity is enhanced by phosphate.[45,47] The biosynthesis of trehalose is in addition controlled via transcriptional and posttranslational regulation of the subunits of the synthase complex. The synthase subunit is proteolytically inactivated in the presence of glucose.[48] Transcription of the four subunit-encoding genes is coregulated by growth phase and stress conditions, possibly through cis-acting STRE-elements in their promoter regions.[49] These STRE-sequences have been implicated as central regulatory target sites in the control of various stress-related genes like CTT1, UBI4, SSA3, HSP12 and DDR2 and STRE-mediated transcription is negatively regulated by cAMP-dependent protein kinase A (chapter 7).[19,22-24,50-52] Interestingly, however, carbon source-dependent transcriptional conform to the trehalose synthase subunit genes does not confront to the typical STRE-mediated regulation. Rapid glucose-induced disappearance of the mRNA

has been observed for all four genes, however, only expression of TSL1 is permanently repressed under fermentative growth conditions. TPS1, TPS2 and TPS3 are expressed only a few-fold lower compared to postdiauxic or nonfermentative growth.[49] This probably relates to the observation that a certain minimal level of Tps1p (but not the other subunits of the trehalose synthase complex) is required even during fermentative growth.[53,54]

Trehalose is mobilized via hydrolysis by trehalase.[41,55] In *S. cerevisiae* neutral trehalase is encoded by NTH1 and presumably by a second redundant gene NTH2. The NTH1 gene appears to express the bulk of neutral trehalase under derepressing or stress conditions.[56-59] ATH1 encodes a vacuolar acid trehalase.[60] The latter has not been assigned a clear biological function as yet whereas neutral trehalase appears to constitute the main source of trehalose-degradative capacity in proliferating, stationary-phase or germinating yeast. Although some level of transcriptional regulation has been reported for NTH1 and NTH2[56,59] trehalase activity is apparently controlled mainly by phosphorylation and dephosphorylation. (Neutral) trehalase activity is stimulated by heat shock,[61] by readdition of nutrients in nutrient-starved cells[62,63] and by glucose or another fermentable sugar in glucose-depleted or postdiauxic cells.[64,65] An abrupt drop in activity is observed during the diauxic shift concomitantly with the start of trehalose accumulation in postdiauxic cells.[14] Activation of neutral trehalase is dependent on protein kinase A,[66] but cAMP is not essential as a second messenger (see also section 5.2).[62,63] This mode of trehalase activation has recently also been described for *S. pombe* and *Candida utilis*[67,68] and apparently represents a conserved signaling pathway operating in the eukaryotic nutrient response.

3.1.4. Trehalose metabolism and the control of sugar-induced signaling

The TPS1 gene encoding the trehalose-6-phosphate catalytic subunit has been cloned independently by different groups as complementing so-called glucose-negative mutations *byp1*, *fdp1* and *cif1*.[43,44] When yeast mutants carrying these alleles are grown on fermentable carbon sources they accumulate high levels of sugar-phosphates, thereby rapidly and fully depleting intracellular ATP and free phosphate and causing inhibition of growth. In addition, these mutants are deficient in various glucose-induced signaling phenomena such as a transient cAMP-increase (see section 3.1.6), inactivation of fructose-1,6-bisphosphatase Fbp1p (section 3.1.7) and activation of glycolytic enzymes (section 3.1.8). The apparent role in the control of sugar influx and of sugar-induced signaling suggested that the gene product involved were a component of a general glucose sensor complex and the gene was consequently renamed *GGS1*.[69-71] Recent evidence, however, indicates that the growth and signaling defects of *ggs1/tps1Δ* mutants can be suppressed by preventing ATP depletion through deletion of[72,73] or by mutations in *HXK2* encoding yeast hexokinase PII (Hohmann, Winderickx, de Winde, Thevelein, manuscript in preparation). The influx of fermentable sugar into glycolysis appears to be controlled at least in part via competitive inhibition of the hexokinases by trehalose-6-phosphate. Evidence for Tps1p-dependent but trehalose-6-phosphate-independent mechanisms especially during exponential growth on glucose has also been reported.[74,75] In conclusion it appears that the role of Ggs1p/Tps1p in sugar signaling is rather indirect. The strict control exerted on sugar influx into glycolysis ensures proper triggering of sugar-induced signaling by, among other proteins, the hexose kinases.[73]

3.1.5. Carbon catabolite repression

The term carbon catabolite repression implies that this regulatory circuit is activated by a preferred carbon source and controls gene expression when a rich fermentable carbon source is present. As will become clear from the following discussion this pathway can actually be regarded as a signaling cascade activated by nutrient depletion. Sugar limitation stimulates the

activity of the central protein kinase Snf1p and this in turn results in the relief of repression of the transcription of a large number of genes.

Several components have been identified that are involved in the pathway leading to transcriptional repression/derepression of genes encoding enzymes that are required for the utilization of nonfermentable carbon sources (Fig. 1.5). This pathway is known as the main glucose-repression or catabolite repression pathway and has been the topic of several detailed reviews.[11,26-29,76]

Glucose repression affects the synthesis of many enzymes which can be classified in three main groups.[76] The first group contains the specific gluconeogenic enzymes fructose-1,6-bisphosphatase (Fbp1p) and phosphoenolpyruvate carboxykinase (Pck1p), apparently together with glyoxylate cycle enzymes like isocitrate lyase (Icl1p). Synthesis of these enzymes is exquisitely sensitive to glucose[16,77] and their genes are strictly repressed in order to prevent simultaneous operation of gluconeogenesis and glycolysis. The second group is represented by the mitochondrial enzymes involved in the Krebs cycle and respiration. These are repressed because they are largely but not fully dispensable during fermentative growth.[78] Third, all proteins involved in the uptake and metabolization of alternative carbon sources like galactose and maltose are repressed as the cell switches to the utilization of a more preferred fermentable carbon source. The distinction between these three families of genes is reflected in the different sensitivities to and extent of repression exerted by fermentable sugars. As will be discussed below, transcriptional activation of these genes in addition to relief from catabolite repression requires a family-specific induction mechanism.

Although catabolite repression is often presented as one mechanism it does not affect all glucose-repressible genes in the same way. Increasing evidence indicates that various signal transduction routes are involved. Indeed, several mutants have been identified that only affect repression of a

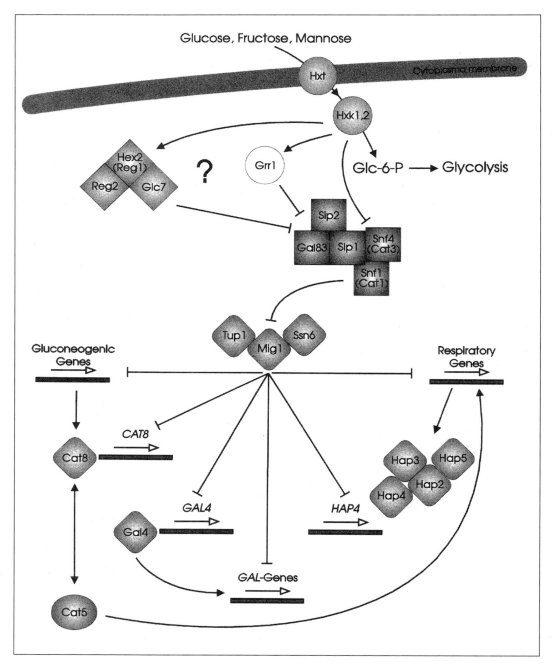

Fig. 1.5. Catabolite repression in Saccharomyces cerevisiae. Simplified schematic analysis of mediators and targets in the catabolite repression response in baker's yeast. The central glucose-dependent repressor complex Mig1p/Ssn6p/Tup1p exerts repression on diverse gene families including the family-specific transcriptional activator genes CAT8 (gluconeogenic genes), GAL4 (galactose utilization genes) and HAP4 (respiratory genes). Derepression is achieved through inhibition/inactivation of the repressor complex by a complex consisting of protein kinase Snf1p and several associated (regulatory) subunits. Under repressing conditions, Snf1p activity is inhibited by several 'upstream' mediators including Hxk1p and Hxk2p, Grr1p and protein phosphatase 1 Glc7p with associated regulatory subunits. For details please refer to the text.

subset of the glucose repressible genes. A clear example is presented by mutants in *HXK2* encoding the yeast hexokinase PII. This sugar kinase plays an important role in the mechanisms relating the presence of glucose to the repression machinery.[79,80] According to the classical glucose-repression model Hxk2p generates a signal that counteracts the activity of the serine-threonine kinase Snf1p/Cat1p[81-83] and its positive regulator Snf4p/Cat3p.[83,84] Snf1p is associated with Snf4p and interacts with and phosphorylates Gal83p, Sip1p, Sip2p, Sip3p and the transcription activator Sip4p,[85-87] but the exact function of the latter proteins remains to be established. The Snf1p kinase complex functions as a general activator required for derepression of all glucose-repressible genes. Its activity is believed to inhibit the function of a complex consisting of transcription factor Mig1p and corepressor proteins Ssn6p/Cyc8p and Tup1p/Cyc9p.[88,89] This repressor complex down-regulates transcription of several genes in the presence of glucose (Fig. 1.5).[89-94] Hence the regulatory phenomena generally referred to as glucose or catabolite repression might actually be regarded as a stimulation response induced by glucose depletion at the level of the central protein kinase Snf1p.

Mutations in or deletion of the *HXK2* gene abolish glucose repression of a large number of targets like the GAL and the SUC genes involved in galactose and sucrose/raffinose metabolism, respectively.[95-97] However, deletion of *HXK2* or *MIG1* does not affect repression of the gluconeogenic enzymes encoded by *FBP1* and *PCK1*, indicating the existence of a parallel control route.[98,99] Derepression of these genes requires the activation of the transcriptional inducer Cat8p, not only by Snf1p but also by Cat5p.[99-101] This is indicative of a specific induction mechanism operating for gluconeogenic genes, however, this has not yet been rigorously tested.

In addition, deletion of *HXK2* does not affect repression of the *ADH2* gene encoding repressible alcohol dehydrogenase which is required for metabolization of

ethanol as a carbon source. Consequently repression of this gene is also independent of the downstream effectors Ssn6p/Cyc8p, Tup1p/Cyc9p and Mig1p.[102] To regulate repression of *ADH2* in response to glucose yeast uses another mechanism that was initially thought to be based on protein kinase A-mediated inactivation of the transcription activator *ADR1*. However, several lines of evidence have questioned the role of protein kinase A in the *ADR1*-dependent transcription regulation of *ADH2*[102,103] and suggested a mechanism involving *HEX2/REG1*, a signal transducing protein previously characterized to influence the activity of Snf1p.[104-106] Combined with the Hxk2p-independence of *ADH2* regulation these data at least suggest that *HEX2/REG1* may be located at a branching point integrating different glucose-induced signals and transmitting the received information to different signaling pathways.

A central role in glucose-induced signaling for Hex2p/Reg1p has recently received new attention with the finding that this protein is a regulatory subunit of protein phosphatase type 1 (PP1).[107] *S. cerevisiae* PP1 is encoded by the essential *GLC7* gene, the product of which participates in the regulation of glycogen accumulation, sporulation, cell cycle progression and translation.[36,37,108-110] Hex2p/Reg1p is structurally and functionally distinct from the previously identified regulatory subunit Gac1p which targets the Glc7p catalytic subunit of PP1 to activate glycogen accumulation (see section 3.1.3).[38,111] Specificity of PP1 towards other substrates is achieved through interaction with various regulatory subunits.[36,109,112,113] Hex2p/Reg1p has a unique role in catabolite repression and together with the similar Reg2p regulatory subunit most likely directs Glc7p protein phosphatase PP1 activity towards substrates that are phosphorylated by the Snf1p protein kinase.[107,114]

Although protein kinase A does not appear to be involved in the establishment of glucose repression[21,73] it is participating in transcriptional regulation of glucose-repressible genes. Protein kinase A-dependent expression control of *SUC2* is mediated through the transcriptional repressor Sko1p/Acr1p.[115,116] Studies in fission yeast *Schizosaccharomyces pombe* suggest involvement of the cAMP pathway in glucose repression of *FBP1*.[117,118] Recent data indicate that in *S. cerevisiae* the gluconeogenic genes *FBP1* and *PCK1* are repressed by very low levels of glucose, but this regulation is again independent of the Ras-cAMP pathway and of protein kinase A.[16] At this point it should be mentioned that the expression of many other genes is regulated by protein kinase A. Many of these genes are involved in growth control and stress resistance and are not directly involved in the metabolic switch to utilization of the most preferred carbon source (see sections 3.1.2 and 4.2).

Catabolite repression influences the rate of transcription, but for several genes it also affects the stability of the corresponding mRNA. This has been demonstrated for iso-1 cytochrome c (*CYC1*)[119] and phosphoenolpyruvate carboxykinase (*PCK1*).[77] Recently glucose-induced mRNA turnover has been described for the invertase gene *SUC2* and *SDH1* encoding the iron protein subunit of the mitochondrial succinate dehydrogenase.[120,121] This fermentable sugar-stimulated mRNA turnover is likely to comprise the first phase of the biphasic repression response. This initial phase requires phosphorylation of the sugar by any of the sugar kinases Hxk1p, Hxk2p, or Glk1p.[73,120] Interestingly, of all the different components identified today in the glucose repression pathway only *HEX2/REG1* appears to be involved in accomplishing glucose-stimulated degradation of mRNA as well.[106,120]

Catabolite repression is not only induced by glucose and other rapidly fermented sugars, fructose and mannose, but also occurs with galactose[122] and maltose.[123] However, the mechanisms triggering repression with sugars other than glucose have not been studied in much detail. Recently we investigated fructose-induced repression and observed that both yeast hexokinase isoforms, Hxk1p and Hxk2p are

required.[73] In all cases where it has been studied fructose and mannose appear to generate the same signals and thus most probably affect the same targets as those affected by glucose. Repression by galactose or maltose, however, only affects a subset of the glucose-repressed genes. This nicely illustrates the above-mentioned subdivision of glucose-repressible genes in separate families that exhibit different repression sensitivities. Yeast cells growing on galactose or maltose have to repress the gluconeogenic enzymes in order to avoid futile cycling while at the same time the glucose-repressible GAL and MAL genes which encode proteins required for uptake and metabolism have to be expressed.

Galactose-induced repression of the cytoplasmic NAD-dependent glutamate dehydrogenase (Gdh2p; see section 3.2.1) and the mitochondrial L-lactate ferricytochrome c oxidoreductase Cyb2p is dependent on the transcriptional activator Gal4p of the *GAL* system.[124] Since Gal4p itself is glucose-repressible, this indicates that the mechanisms for galactose- and glucose-induced repression follow at least partially different signaling routes. Maltose does not activate the Ras-cAMP pathway, but at least partially activates the FGM-pathway (de Winde and Winderickx, unpublished data; see 5.2). Hence this sugar also apparently enhances protein kinase A activity. It remains to be established how galactose and maltose regulate expression of protein kinase A targets that are important in the switch from respiratory to fermentative growth, or vice-versa.

3.1.6. The Ras-cAMP pathway and protein kinase A

Addition of glucose or another easily fermentable sugar to yeast cells grown on poorer carbon sources triggers a rapid transient increase in the cellular cAMP concentration often referred to as the 'cAMP signal.[64,65,125-127] For this and other reasons—which will be discussed in section 4 on 'common responses to nutrient depletion'—cAMP has been implicated as an important secondary messenger in the nutrient response, accomplishing the metabolic adaptation to fermentative growth conditions through modulation of the activity of protein kinase A. Hence high intracellular levels of cAMP correlate with nutrient-rich growth conditions whereas low cAMP levels reflect nutrient limitation and poor growth conditions.

The concentration of cAMP in yeast cells is controlled by an elaborate regulatory pathway (Fig. 1.6; reviewed in refs. 69, 70, 128-130). Adenylate cyclase encoded by the *CYR1/CDC35* gene[131,132] is regulated by the products of the *RAS1* and *RAS2* genes which act as stimulatory G proteins. They are active in the GTP-bound state and inactive in the GDP-bound state.[133,134] Exchange of GDP for GTP is stimulated by the *CDC25* gene prod-

uct[135-137] and possibly also by the product of the homologous *SDC25* gene.[138] The Ras proteins possess intrinsic GTPase activity which causes self-inactivation, and this GTPase activity is greatly stimulated by the *IRA1* and *IRA2* gene products.[139,140] cAMP is hydrolyzed by a low-affinity phosphodiesterase (encoded by *PDE1*)[141] and a high-affinity phosphodiesterase (encoded by *PDE2*).[142] Like in other eukaryotic cells, cAMP activates protein kinase A by binding to the regulatory subunit (encoded by *BCY1*),[143] thereby causing the dissociation and concomitant activation of the catalytic subunits (encoded by *TPK1*, *TPK2* and *TPK3*).[144]

As will be discussed in more detail in sections 4 and 5 on 'common responses to nutrient depletion' the activity of protein

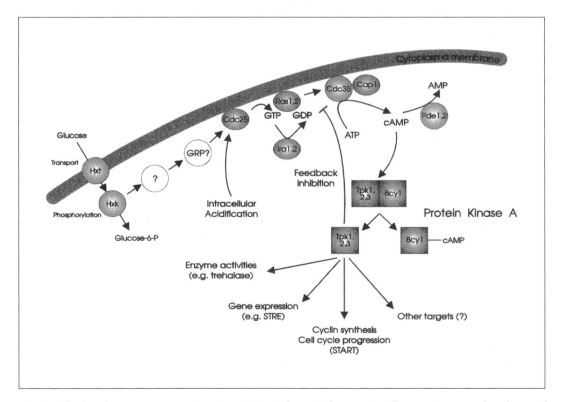

Fig. 1.6. The Saccharomyces cerevisiae Ras-cAMP pathway. When yeast cells experience an abundance of fermentable carbon source the Ras-cAMP signaling cascade is initiated at the level of hexose phosphorylation and relayed via a putative early glucose-repressible mediator Grp,[69] Cdc25p-activated Ras1p and Ras2p to adenylate cyclase Cdc35p. Subsequently synthesized cAMP causes the dissociation of the regulatory subunit Bcy1p from the catalytic Tpk subunit of protein kinase A. Activated protein kinase A mediates various regulatory processes leading to adaptation to fermentative metabolism. For details see section 3.1.6.

kinase A appears to be additionally controlled by a separate signaling route termed the Fermentable Growth Medium-induced (FGM) pathway.[49,63,69,70,145] Activation of the FGM-pathway is not only dependent on the availability of glucose or fructose but also on the availability of all other nutrients essential for growth. In contrast to the Ras-cAMP pathway only the presence of the sugar and not its phosphorylation is required for FGM-activation, indicating that this pathway acts as a specific nutrient sensing pathway.[145] The FGM-pathway has been shown to be independent of the second messenger cAMP and probably acts directly on the free catalytic subunits of protein kinase A.[62,63]

The activated catalytic subunits of protein kinase A phosphorylate a number of target proteins, some of which have been identified (reviewed in ref. 70). These targets are involved in different regulatory processes required to make the shift from gluconeogenic to fermentative growth. Other known targets of protein kinase A are involved in the breakdown of storage carbohydrates, in stress resistance and growth control.[19-24]

3.1.7. Catabolite inactivation

Addition of glucose to cell growing on nonfermentable carbon sources triggers the rapid inactivation of many enzymes. This process is called catabolite inactivation[146] and has been studied in most detail for the gluconeogenic enzymes PEP- carboxykinase (Pck1p) and fructose-1,6-bisphosphatase (Fbp1p).[147] Other enzymes that are known to be subject to catabolite inactivation include malate dehydrogenase and isocitrate lyase, which are required for the glyoxylate cycle,[148] proteins involved in the uptake and metabolism of maltose and galactose[149,150] and the high affinity transport systems for glucose.

For Fbp1p catabolite inactivation consists of two steps.[151,152] The enzyme is first reversibly inactivated by phosphorylation and then degraded by targeted proteolysis. The amino acid sequence that is phosphorylated in the first step has been identi-fied and shown to be identical to a site recognized by protein kinase A in vivo. Mutants deficient in adenylate cyclase do not show glucose-induced inactivation of Fbp1p.[153] Protein kinase A-mediated phosphorylation is regulated by cAMP and by the level fructose-2,6-bisphosphate via allosteric alteration of Fbp1p. Inactivation through phosphorylation is, however, not required for subsequent proteolytic degradation of the enzyme which despite some debate[154,155] appears to occur inside the vacuole.[156,157] Glucose-induced vacuolar targeting of Fbp1p appears to be mediated by small vesicles which are translocated to the vacuole presumably by microautophagy.[158,159]

A similar two-step inactivation pattern—a reversible followed by an irreversible inactivation—has also been reported for isocitrate lyase Icl1p.[148,160] Glucose-induced inactivation of malate dehydrogenase and phosphoenolpyruvate carboxy kinase cannot be reversed, however. Glucose-induced degradation of the maltose and galactose transporter proteins is initiated via translocation to the vacuole by the endocytic pathway.[149,150,158]

3.1.8. Catabolite activation and catabolite induction

Yeast cells experiencing an abundance of fermentable sugar exhibit increased expression of many genes and enhanced activity of various enzymes required for efficient transport and metabolism of the sugar. These include several glycolytic enzymes such as pyruvate decarboxylase (Pdc1p and Pdc5p), alcohol dehydrogenase I (Adh1p), enolase II (Eno2p), one of the isoenzymes of 6-phosphofructo-2-kinase (Pfk2p) and pyruvate kinase (Pyk1p).[161] The exact mechanism for their glucose-dependent induction is still not clarified. An increase in the concentration of different glycolytic metabolites may be the trigger for the induction of several genes (see ref. 162 and references therein).

Much attention has been focused on the transcription regulation of the glycolytic genes. Most of these contain in their promoters binding sites for multifunctional transcription factors like Rap1p, Gcr1p, Reb1p and Abf1p (reviewed in ref. 163). Interestingly, binding sites for Rap1p and Abf1p are also found in the promoters of various ribosomal protein genes (RPGs) that are also highly expressed in yeast cells growing on fermentable sugars (reviewed in ref. 164). For the RPGs it has been shown that expression induction is dependent on the protein kinase A activity but not on cAMP. Accordingly we and others have recently demonstrated that induction of the RPGs is largely regulated by the FGM-pathway, as will be discussed in more detail in section 5.2.[49,145,165,166]

Glucose induces components of its transport system. Enhanced transcription of various *HXT* genes is regulated by glucose in a concentration-dependent manner. Expression of *HXT1* is induced by high glucose concentrations, whereas transcription of *HXT2* and *HXT4* is activated by low levels of glucose and *HXT3* expression is induced independently of sugar concentration.[167] Low-glucose induction is mediated through the transporter homologue Snf3p[168] via activation of the early glucose-signaling component Grr1p[169] and inhibition of the repressor-corepressor complex Rgt1p/Htr1p-Ssn6p.[167,170-172] High-glucose induction is mediated through another transporter-homologue called Rgt2p which is very similar to Snf3p except for their extended C-terminal regions.[173] Hence Rgt2p and Snf3p are likely to act as receptors in transmitting the signal for glucose-induced gene expression.

Transcriptional induction of the *GAL*[29,97,170] and *MAL* genes[174-176] by galactose and maltose, respectively, is achieved through both relief of repression and activation by gene family-specific activators. The Gal4p activator[97] and the Mal-R activator[174] are themselves subject to catabolite repression. Upon derepression these activators induce transcription of their specific target genes which in several cases may additionally be direct targets of catabolite repression themselves. The same general mechanism holds true for induc-

tion of gluconeogenic genes[99,100] (see section 3.1.5) and genes encoding proteins for mitochondrial respiration and function[78,177] when glucose is limiting or absent. In this case the Cat8p activator[99] or the Hap4p activator subunit of the Hap2p/3p/4p/5p regulator complex[178,179] are derepressed and specifically activate their respective target gene families.

3.2. NITROGEN SOURCE LIMITATION AND STARVATION

3.2.1. Response mechanisms involved in nitrogen assimilation

S. cerevisiae can utilize a wide variety of nutritional compounds as nitrogen sources. In many instances the presence of a certain nitrogen compound in the growth medium triggers biosynthesis of the enzymes that are required for its metabolism. The expression of nitrogen catabolic genes is controlled in response to the quality of the nitrogen source available. That is the presence of preferred nitrogen sources that are readily transported and metabolized represses the expression of genes encoding transporters and catabolic enzymes necessary for the uptake and utilization of nitrogen sources whose utilization requires more energy. Thus growth of *Saccharomyces* sp. on media containing ammonia, glutamine, or asparagine as sole nitrogen source causes nitrogen catabolite repression of enzymes necessary for the utilization of—among others—proline, allantoin or γ-aminobutyrate.[180-182]

The use of ammonia, glutamine and glutamate as preferred nitrogen sources, together with asparagine which readily yields ammonia and glutamate,[180,181] is easily explained by the key steps of ammonia utilization (Table 1.1). The second reaction catalyzed by glutamine synthase (GS encoded by *GLN1*) is absolutely essential in cells grown on other nitrogen sources than glutamine, including ammonia.

Glutamine is required for the synthesis of nucleotides and several amino acids. It is therefore not surprising that both the synthesis and the activity of GS are tightly regulated in response to the availability of glutamine and—to a lesser extent—ammonia (Fig. 1.7). When intracellular glutamine levels decrease, *GLN1* expression is enhanced by the Gln3p transcriptional activator, a GATAA-binding protein.[183] In the presence of glutamine, both Gln3p and GS (Gln1p) are inhibited through covalent modification by Ure2p, a yeast prion-forming protein[184] with similarity to glutathione-S-transferases.[185]

In the case of GS inactivation it has been shown to be accompanied by a decrease in the number of free sulfhydryl groups,[186] but the anticipated attachment of glutathione has not been proven. The *URE2-GLN3* regulatory route mediates activation of expression of a variety of genes sensitive to nitrogen catabolite repression.[187] Upon depletion of glutamine or ammonia synthesis and activity of the low affinity, high capacity general amino acid permease Gap1p are enhanced,[188] as is expression of *GDH2*.[189] Expression of other genes for the utilization of less preferred nitrogen sources such as the gene for asparaginase, the *DAL* and *DUR* genes encoding allantoin and urea degrading enzymes and probably the *UGA* genes encoding enzymes for the utilization of γ-aminobutyrate are induced in a *URE2-GLN3* dependent way. In addition to the latter, arginine and proline are specific inducers of their catabolic enzymes. Production of arginase (Car1p) and ornithine transaminase (Car2p) is induced by argine or ornithine. Synthesis of proline oxidase (Put1p) and D-pyrroline-5-carboxylate (P5C) dehydrogenase (Put2p) are induced by proline through the action of the Put3p transcriptional activator[190] whereas nitrogen repression of *PUT1* and *PUT2* expression is mediated in part through *URE2*.[187,191] Hence during growth on 'rich' nitrogen sources the catabolic routes for those amino acids mentioned above are shut off via binary control; inducer exclusion through inactivation and repression of the general amino acid permease Gap1p and control of gene expression by Ure2p-Gln3p. Furthermore, recent evidence indicates the existence of an additional, Gln3p-independent nitrogen regulatory mechanism in *S. cerevisiae*[182] operating through transcription activators encoded by the *NIL1* and *GAT1* genes.[192,193]

3.2.2. General amino acid control and the stringent response

The requirement of glutamine for the biosynthesis of various amino acids is reflected by an increase in the level of GS in response to amino acid deprivation. Under these conditions the regulatory mechanism known as general amino acid control (abbreviated GCN) derepresses the synthesis of the specific transcriptional activator Gcn4p, causing increased transcription of a large number of amino acid biosynthetic genes (reviewed in ref. 194). The increased

---

Table 1.1. Key steps of ammonia utilization in yeast

1. $NH_3 + \alpha\text{-ketoglutarate} + NADPH/H^+ \xrightarrow{\text{Gdh1}} \text{glutamate} + NADP^+$

2. $NH_3 + \text{glutamate} + ATP \xrightarrow{\text{Gln1 (GS)}} \text{glutamine} + ADP/P_i$

3. $\text{glutamine} + \alpha\text{-ketoglutarate} + NADH/H^+ \xrightarrow{\text{GluS}} 2\ \text{glutamate} + NAD^+$

4. $\text{glutamate} + NAD^+ \xrightarrow{\text{Gdh2}} \alpha\text{-ketoglutarate} + NH_3 + NADH/H^+$

Adapted from ref. 181.

amino acid biosynthetic capacity reduces the intracellular glutamine pool and consequently causes Gln3p-dependent induction of *GLN1* transcription. In addition, *GLN1* expression is directly activated by Gcn4p.[195]

The triggering event for derepression of Gcn4p by the GCN-control in response to decreasing amino acid availability is the occurrence of uncharged tRNAs. The levels of various uncharged tRNAs are

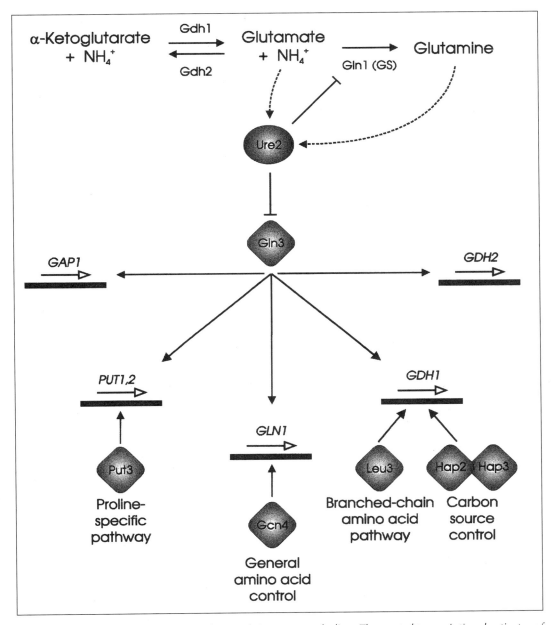

Fig. 1.7. Gln3p - Ure2p dependent regulation of nitrogen metabolism. The central transcriptional activator of the family of nitrogen metabolism genes Gln3p is inhibited by Ure2p when NH4+ and/or glutamine are abundant. When ammonia and glutamine are limiting Ure2p inhibition is relieved and Gln3p activated. Transcriptional regulation of several subfamilies of amino acid biosynthesis genes is additionally mediated by pathway-specific regulators ensuring tight coregulation of nitrogen control, carbon source control and general amino acid metabolism. For details see section 3.2.

monitored by the eIF-2a kinase Gcn2p through its C-terminal histidyl-tRNA synthetase-related domain.[196] Binding of any uncharged tRNA to this domain activates the protein kinase domain of Gcn2p which phosphorylates and thereby activates the α-subunit of the translation initiation factor eIF2. Subsequently, translation of the *GCN4* mRNA is specifically derepressed. Another important response to decreasing amino acid levels is the stringent control of ribosomal protein gene (*RPG*) transcription.[197] Upon amino acid starvation synthesis and accumulation of *RPG* mRNA and of rRNA are significantly reduced. In this way the protein synthetic capacity of the yeast cell is adjusted to the limitation of the protein building blocks. This stringent response is apparently also triggered by accumulating uncharged tRNAs. However, the control is exerted through a separate regulatory pathway since *gcn2* and *gcn4* mutations do not affect repression of *RPG* mRNA.[198] The downregulation of *RPG* mRNA is mediated through the transcriptional regulator Rap1p which is also required for normal and induced expression of ribosomal protein genes.[164,199-202] Rap1p-dependent control of *RPG* expression appears to be at least partially affected by cAMP-dependent protein kinase A.[203] However, this regulation is not directly mediated by cAMP but may be coupled to the rate of protein synthesis which is under protein kinase A control.[204] Protein synthesis is tightly coupled to the amino acid availability and intracellular amino acid pools appear to be controlled by the activity of the Ras-cAMP pathway.[205]

3.2.3. Integration of carbon and nitrogen control mechanisms

Strong evidence has been presented for cross-pathway regulation between nitrogen and carbon source signaling routes.[206] One of the important connection points between carbon and nitrogen metabolism is the reductive amination of α-ketoglutarate by NADP-linked glutamate dehydrogenase (NADP-GDH; Table 1.1) encoded by the *GDH1* gene. Probably for this reason the

regulation of *GDH1* appears to be quite complex. Nitrogen-dependent transcriptional control is mediated by Gln3p and by the branched-chain amino acid biosynthesis-specific activator Leu3p.[207] In addition, carbon source-dependent transcriptional control is mediated through the Hap2p/3p/4p/5p activator complex[206] previously implicated in carbon source dependent regulation of various genes required for mitochondrial biogenesis[178,179,208-210] (reviewed in refs. 78, 177). This finding substantiates the already long-standing observation[177] that yeast strains carrying *hap2* or *hap3* mutations display a severe defect in the utilization of ammonium sulfate as the sole nitrogen source.

3.2.4. Pseudohyphal differentiation: Integrated control exemplified

Another important response of *S. cerevisiae* to limitation of the available nitrogen source has recently been described. Heterozygous a/α diploids grown under conditions of limiting nitrogen source switch their budding pattern to polarized and elongated pseudohyphal growth.[211] Although originally observed under nitrogen limitation this morphological switch now is beginning to be recognized as a more general response to severe nutrient depletion.[212-214] Therefore as an illustration of the integration between signal transduction pathways operating in nutrient signaling and morphogenesis, pseudohyphal differentiation is presented in detail in the section on 'common responses to nutrient depletion' (section 4.3).

3.3. PHOSPHOR LIMITATION AND STARVATION

3.3.1. Response mechanisms involved in phosphate recruitment

S. cerevisiae like many other yeast species utilizes inorganic phosphate (P_i) as the preferred phosphor source. However, when P_i is limiting or absent a variety of organic compounds can serve as phosphate sources through the action of several acid phosphatases (reviewed in refs. 29, 215, 216).

The genes that are required for the utilization of these poorer phosphate sources are specifically repressed by P_i.

Upon P_i depletion expression of the *PHO5* gene encoding the major acid phosphatase with broad substrate specificity is activated by the Pho4p transcription factor.[217-219] Pho2p/Bas2p functions as an essential auxiliary transcriptional activator[217,220,221] and the interaction between Pho2p and Pho4p is absolutely necessary for normal transcriptional activation.[222] In the presence of high concentrations of P_i Pho4p is negatively regulated through binding of the specific inhibitors Pho80p and Pho85p.[223-225] Interestingly, *PHO80* and *PHO85* encode a cyclin and a cyclin-dependent kinase, respectively.[225,226] When P_i levels drop, Pho4p inhibition by Pho80p-Pho85p is alleviated by binding of the cyclin-dependent protein kinase inhibitor homologue Pho81p to the cyclin-cdk complex.[224,227] This specific response to phosphate depletion is schematically depicted in Figure 1.8 (see for a recent review ref. 228). Regulation of the Pho81p cyclin-dependent protein kinase inhibitor appears to be the critical phosphate switch,

but how this switch is operated and hence how intracellular phosphate levels control the *PHO* system is not yet clear. Evidence has been presented that Pho81p controls its own expression through a positive feedback loop with Pho4p again acting as transcriptional activator.[229,230] In addition, Pho81p may be regulated by posttranslational modification in response to low P_i levels.[227]

3.3.2. Integration of phosphate control in general metabolism and cell proliferation

The Pho85p cyclin-dependent protein kinase also associates with two other cyclin homologues, encoded by *HCS26/PCL1* and *ORFD/PCL2*.[225,231-233] The Pcl1p-Pho85p and Pcl2p-Pho85p cyclin-cdk complexes may have a definable albeit redundant role in the transition from G_1 to S-phase in the yeast cell cycle. In the absence of functional G_1 cyclins Cln1p and Cln2p, *PCL1* and *PCL2* as well as *PHO85* are required for viability.[232,233] Furthermore, Pcl1p and Pcl2p can activate the Pho85p kinase to inhibit Pho4p. Apparently the Pho85p kinase may operate in two regulatory

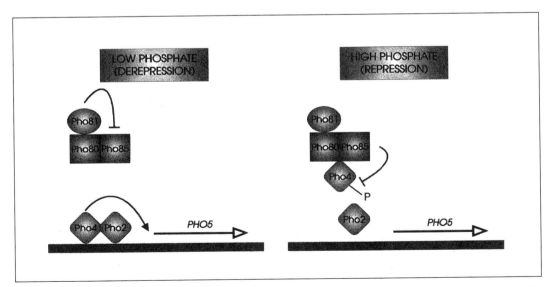

Fig. 1.8. Transcriptional regulation of the P_i-repressible acid phosphatase gene PHO5. The pathway-specific transcriptional regulator Pho4p is inhibited and inactivated by the cyclin-cdk complex Pho80p-Pho85p under conditions of high P_i concentration. Low P_i concentration causes Pho81p-mediated dissociation of the cyclin-cdk complex and activation of Pho4p which interacts with Pho2p and activates PHO5 expression. For details see section 3.3.1.

processes, involving phosphate signaling and cell cycle progression. This opens an exciting opportunity for regulatory cross-talk in the coordination of nutrient availability and cell cycle progression.[228]

Important progress concerning the relevance of this cross-talk has been made with the finding that the Pho85p cyclin-dependent protein kinase may act as a glycogen synthase kinase in yeast.[234,235] Disruption of and mutations in *PHO85* exhibit impaired glycogen synthase kinase activity, constitutive activity of glycogen synthase and consequently, hyperaccumulation of glycogen.[234] In addition, a *pho85* mutation causes increased *GSY2* expression[235] and suppresses the glycogen storage defect of *snf1* mutations.[234] A mutation in the protein phosphatase 1 gene *GLC7* only partially restores the glycogen accumulation phenotype.[235] Thus the identification of the Pho85p cyclin-dependent protein kinase as a physiological glycogen synthase kinase may reflect a close link between cell cycle decisions and the monitoring of the nutritional state and glycogen accumulation.

The Pho2p/Bas2p auxiliary transcription factor of the P_i-repressible *PHO* genes may be another candidate to operate in cross-talk between important regulatory mechanisms. This homeobox-containing protein regulates basal (Bas2p, together with Bas1p) expression of the *HIS4* gene encoding an enzyme in histidine biosynthesis.[236] In addition, Pho2p regulates expression of various genes required for biosynthesis of adenine[236] and of *TRP4* encoding the enzyme phosphoribosyl transferase which is central in tryptophane synthesis.[237] The Trp4p enzyme catalyzes a phosphoribosylpyrophosphate (PRPP) utilizing step. Interestingly, Pho2p inhibits Gcn4p-dependent transcriptional activation of *TRP4* when phosphate levels are low. Apparently utilization of phosphate-rich PRPP is disadvantageous to P_i-limited cells and therefore the Trp4p level has to be minimized. PRPP is also an important intermediate in purine and histidine biosynthesis. It is tempting to speculate that one important role of Pho2p is to coordinate

amino acid and purine metabolism with phosphate availability.

3.4. Sulfur Limitation and Starvation

3.4.1. Response mechanisms involved in sulfur assimilation

Sulfur is one of the low-abundance components in biomass since its requirement is largely restricted to synthesis of methionine and cysteine. Maybe this is why molecular details concerning the regulation of sulfur assimilation are only beginning to emerge. Sulphate is readily used by *Saccharomyces* sp. as a sulfur source. Following uptake through two sulfate-specific permeases the sulfate is reduced to sulfide which is immediately used in the biosynthesis of methionine and cysteine (Fig. 1.9).[238,239] All of the genes coding for enzymes of the sulfate reduction pathway are strongly repressed at the transcriptional level by S-adenosyl-methionine (AdoMet) or by methionine through its conversion to AdoMet.[240] Hence sulfate assimilation will be inhibited when methionine and/or cysteine are abundant and sulfate depletion will not be deleterious unless methionine becomes limiting. A specific transcriptional activator of the methionine biosynthesis genes has been identified[241] and trans-activating activity of this basic leucine-zipper (bZIP) activator Met4p is inhibited through an AdoMet-responsive domain.[242] This domain is likely to interact with the AdoMet-responsive transcriptional repressor Met30p.[243] The Met30p protein contains five consecutive repeats of the WD40 domain that is also present in the yeast transcriptional corepressors Tup1p[244-246] and Hir1p[247] and is presumably involved in protein-protein interactions.[248] *MET30* is an essential gene[243] indicating that its function is not restricted to the sulfate assimilation pathway.

The recent identification of a second bZIP protein involved in transcriptional activation of the *MET* genes and encoded by *MET28* sheds new light on the activating mechanism.[249] Transcriptional induc-

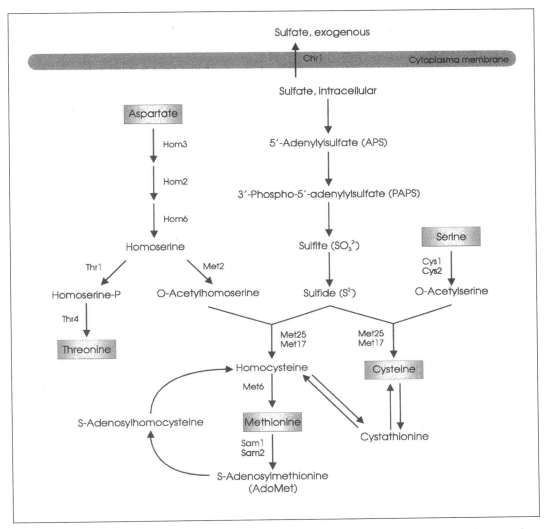

Fig. 1.9. *The sulfate assimilation pathway in* Saccharomyces cerevisiae. *Schematic depiction of intermediary sulfate metabolism in yeast. Adapted and modified from refs. 238, 239.*

tion is mediated by a complex consisting of the Met4p activator subunit, Met28p and the basic-helix-loop-helix (b/HLH) protein Cpf1p which is additionally involved in the regulation of other gene families and of chromosome segregation (see next paragraph for a more detailed description). The activator complex appears to be formed by Cpf1p-dependent tethering of Met4p and Met28p to the DNA.

3.4.2. Integration of sulfur control with carbon and phosphate utilization

The *MET19/ZWF1* gene, previously implicated in the sulfate assimilation pathway, has recently been shown to encode glucose-6-phosphate dehydrogenase (Zwf1p) which is the first enzyme of the pentose phosphate route.[250,250a] These authors conclusively show that the auxotrophy for organic sulfur of *met19/zwf1* mutants does not result from a depletion of NADPH by blocking the pentose phosphate pathway. Although the exact role of the pentose phosphate cycle in yeast metabolism is still a matter of debate, this finding sheds new light on the mechanisms that control anabolic carbon and sulfur fluxes. Indeed, in the biosynthesis of sulfur amino acids a strict coupling is to be expected between carbon flux yielding aspartate through the citric acid cycle and sulfate assimilation (Fig. 1.9).[238] Expression of *ZWF1* is strongly repressed by methionine and AdoMet as are most other *MET* genes. Expression of the genes encoding other enzymes of the pentose phosphate route is not controlled in this way.[250] This transcriptional repression of *ZWF1* is mediated through Met30p (cited in ref. 243), again indicating that the function of this repressor protein extends beyond the sulfate assimilation pathway.

A member of the family of global transcriptional regulators, Cpf1p[251] has been implicated in regulation of several genes of the methionine biosynthetic pathway.[241] This basic-helix-loop-helix and leucine zipper containing protein binds to the centromere core consensus sequence CDE1 and to diverse promoter regions, including those of several genes of the sulfate assimilation pathway[252-256] (reviewed in refs. 78, 164, 251). Cpf1p has been shown to be involved in transcriptional regulation of several genes[241,252,257,258] probably through modulation of chromatin structure in the promoter region surrounding its CDE1 consensus binding site.[259,260] Indeed, Cpf1p modulates gene activity through interaction with a family of chromatin-related proteins encoded by *SPT21*, *SIN3/RPD1*, *RPD3* and *CCR4*.[261] *CPF1* disruption mutants exhibit a severe methionine auxotrophy[253-255] and mRNA levels of *MET25* and *MET16* containing CDE1 consensus sites in their promoter regions are reduced.[241] However, site directed mutagenesis of the basic-helix-loop-helix region revealed that DNA binding, centromere functioning and chromatin structure modulation by Cpf1p are not related to and separable from its role in maintaining methionine prototrophy.[261,262] The recent finding that Cpf1p may tether the Met4p activator and Met28p to the DNA forming a MET-family specific activator complex[249] leads to the attractive hypothesis that Cpf1p functions in transcriptional control of sulfate assimilation genes—and probably also of other gene families—through heteromeric protein-protein interactions with other basic-leucine-zipper or basic-helix-loop-helix/zipper-containing regulators.

In this light it is interesting that an increased dosage of the phosphate control-specific basic-helix-loop-helix-activator Pho4p can relieve the methionine auxotrophy of a *cpf1* disruption mutant.[263] The Pho4p-dependent suppression requires the presence of *PHO2* and is enhanced by *pho80* mutations, which are known to derepress Pho4p trans-activating activity (see section 3.3.1). Hence, phosphate and sulfate metabolism may be cross-regulated at the level of two trans-activators that share many similarities in their DNA-binding and dimerization characterisitics.[218,255,264,265]

4. COMMON RESPONSES TO NUTRIENT DEPLETION

4.1. GENERAL CONCEPTS

As stated in the introduction to this chapter important distinctions should be made between either limitation or starvation for certain nutrient compounds when describing and investigating the cellular responses in yeast cells to gradual depletion of nutrients. Furthermore, we have to differentiate between specific responses to a certain depletion at stake and general responses to any nutritional depletion. In many instances the decreasing availability of one substrate induces or enhances specific metabolic pathways required for the utilization of another. When eventually one or more essential nutrient sources have been used up, the cells will enter stationary phase as a consequence of starvation (chapter 2).[1]

The common signaling cascades and metabolic changes which are triggered by starvation for different nutrients are closely related to the mechanisms operating in control of growth and proliferation. Yeast cells starved for a single essential nutrient will complete their current cell cycle and arrest in the next G_1 phase at 'Start A'[2] (Figs. 1.2 and 1.10). Subsequently they will progress into an 'off-cycle' G_0 or stationary phase in which they can survive nutrient starvation much longer than when arrested within the cell cycle (chapter 2).[1] This behavior is manifested with yeast cells growing on rapidly fermented sugars like glucose, fructose, mannose or maltose, but also with other, nonfermentable carbon sources like lactate, ethanol or acetate. When all essential nutrients are readily available the cells will exit G_0 and proceed with growth in the G_1 phase to reach their 'critical' cell size and traverse Start into S-phase, beginning a new proliferation round with DNA-synthesis.

Yeast cells that are starved for any essential nutrient and enter stationary phase display a number of specific characteristic properties. They accumulate elevated levels of the storage carbohydrates glycogen and trehalose[18] while expression of STRE-controlled genes like *CTT1*[19,266] and *SSA3*[50,267] is activated, transcription of

ribosomal protein genes is repressed, general stress resistance and in particular thermotolerance is greatly enhanced[268] and resistance to cell wall lytic enzymes is elevated[269] (for reviews see refs. 1 and 70 and chapters 2, 3 and 7). Interestingly, however, most of these characteristics are not very different from the properties of cells growing exponentially on a nonfermentable carbon source (Fig. 1.11). This is apparently the reason why general responses and metabolic consequences of nutrient starvation and nutritional resupplementation have not often been studied in yeast cells growing on a nonfermentable carbon source. This does not imply, however, that these responses do not exist. On the other hand differences between nutrient-deprived cells and exponentially growing cells on rapidly fermentable sugars are very pronounced for all the properties mentioned.

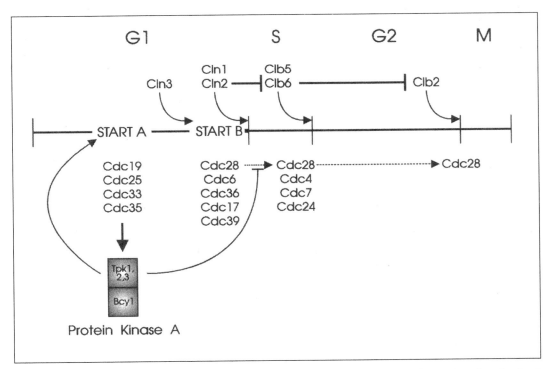

Fig. 1.10. The Start-concept of the yeast cell division cycle. Schematic depiction of genetic cell cycle check points in Saccharomyces cerevisiae. START A, START B and other proliferation-decision landmarks are genetically characterized by several CDC genes.[2] The Cdc28p protein kinase is the central mediator of phase-to-phase transitions. Phase-specific activation of Cdc28p is controlled by G_1 cyclins Cln1p, Cln2p and Cln3p, by S-phase cyclins Clb5p and Clb6p and by G_2 cyclins, including Clb2p (see refs. 271 and 272 and references therein).

4.2. Nutrient Starvation in Fermenting Yeast Cells

When yeast cells that are growing exponentially on a glucose-containing medium are starved for another essential nutrient like nitrogen, phosphate or sulfate they show the same metabolic response in all cases. The trehalose and glycogen level starts to increase, the expression of STRE-controlled genes is induced and the cells progressively acquire all characteristics of cells in stationary phase (chapter 2).[18,65] These metabolic changes could simply be the logical result of the growth arrest without being modulated themselves directly by the presence of the nutrients. However, the presence of the same characteristics in cells growing exponentially on nonfermentable carbon sources—albeit with a prolonged G_1 phase—already argues against this interpretation. Like the length of G_1, acquirement of these metabolic characteristics clearly can be modulated by nutrient availability alone, without being necessarily linked to growth arrest. In addition, the nutrient resupplementation response strongly argues for specific control of these metabolic responses by nutrients. The most telling example in this respect is the observation that re-addition of nitrogen, phosphate or sulfate to cells starved for these respective nutrients in each case causes within a few minutes a rapid, posttranslational activation of trehalase which triggers trehalose mobilization.[63] This activation is not dependent on protein synthesis. Hence this response is not a mere consequence of but rather precedes the induction of cell growth. The striking question arising from these observations is how depletion or resupplementation of nutrients so different as nitrogen, phosphate and sulfate can trigger such similar and rapid metabolic responses.

An important clue to the molecular mechanism underlying these common responses to the depletion or presence of different nutrients has come from studies on cAMP metabolism in yeast. The crucial observation was that yeast mutants devoid of cAMP also arrest at the same point in the cell cycle as wild type cells starved for nutrients.[2,39] In fact, for the temperature-sensitive cell cycle mutants *cdc35* and *cdc25* it was shown that they arrest at the restrictive temperature at the same point in the cell cycle as nutrient-starved cells before it was known that the gene products were implicated in cAMP metabolism.[2]

In some way protein kinase A activity is essential for the synthesis of cyclins and for progression over the start point of the cell cycle. Recently a connection has been revealed between cAMP and the critical cell size required for progression over Start. Elevated activity of protein kinase A causes repression of the G_1 cyclin genes *CLN1* and *CLN2* resulting in a larger critical cell size.[270,271] Presumably a counteracting growth related factor is establishing a balance between cyclin induction and repression resulting in a specific critical cell size.

Transcription factors or other downstream components controlled by protein kinase A and involved in control of G_1-cyclin expression have not been identified yet. Yak1p kinase and Sok1p appear to be components of a parallel proliferation-controlling pathway. Inactivation of *YAK1*[272,273] or overexpression of *SOK1*[274] suppresses the lethality of a complete inactivation of protein kinase A. The suppressive effect of a *YAK1* deletion depends on active Sok1p, suggesting a regulatory connection between the two. The *SOK2* gene product shares extensive similarity with (among others) the Phd1p transcriptional activator (see paragraph 4.2). *SOK2* has been identified as another potential target of protein kinase A acting in the control of genes important for growth regulation and morphological development.[275]

Two conditions have been identified which trigger a rapid increase in the cAMP level in yeast cells: (1) the addition of glucose to cells growing on a nonfermentable carbon source and (2) the addition of protonophores at low extracellular pH, causing internal acidification of the cells.[65] The stimulation of cAMP synthesis by glucose and the observation that nutrient-starved cells arrest at the same point in the cell cycle as cAMP-depleted cells led to the idea that cAMP would be a general second messenger for nutrient availability in yeast. Progression over the start point of the cell cycle would then be triggered by nutrient-induced elevation of the cAMP level. However, many arguments have been accumulating disfavoring this hypothesis. The glucose-induced stimulation of cAMP

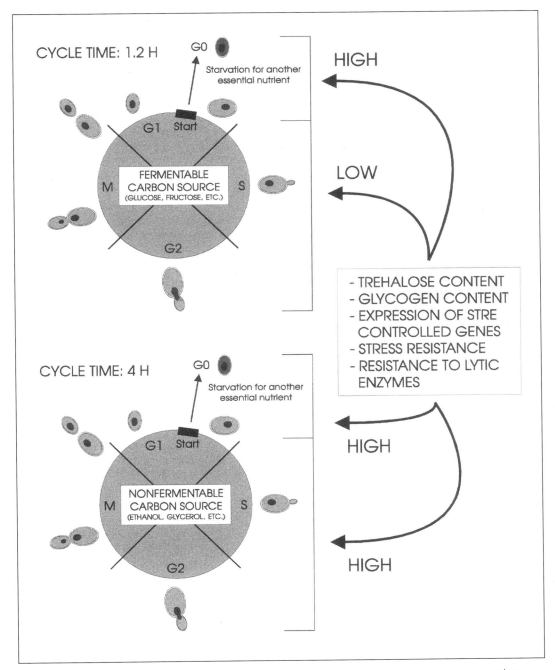

Fig. 1.11. Metabolic characteristics of yeast cells growing exponentially or entering stationary phase, on fermentable or nonfermentable carbon sources. Several starvation- and stress-linked characteristics of Saccharomyces cerevisiae as indicated in the box are absent only in cells growing exponentially on fermentable carbon sources.

synthesis is only observed in derepressed cells. This might be related to the necessity of downregulating protein kinase A for proper entrance into G_0 in cells that are starved for an essential nutrient on a glucose-containing medium.[276] In addition, no effect of other nutrients on the cAMP level has been found.[63] The latter is quite remarkable in view of the previously mentioned rapid changes in activity of enzymes controlled by protein kinase A, such as trehalase. However, it has been demonstrated conclusively using mutants lacking the regulatory subunit Bcy1p of protein kinase A that these rapid, posttranslational nutrient-induced activity changes do not require mediation by cAMP.[62] In addition, other nutrient-dependent regulation of known protein kinase A-controlled properties such as glycogen content, heat resistance, sporulation capacity[277] and *CTT1* expression[266] also are not mediated through cAMP. It has been proposed originally, that yeast cells with an overactive Ras-cAMP pathway are unable to properly arrest at Start in G_1 upon nutrient depletion.[134,142] This would fit with a nutrient sensing function for this pathway. When constitutively activated it would be unable to sense depletion and as a result cells would progress over Start regardless of the environmental conditions. More recently it has been shown that cells with an overactive Ras-cAMP pathway indeed do not arrest in G_1 upon nutrient starvation but rather arrest growth anywhere throughout the cell cycle.[205] Since these cells contain only small storage reserves (glycogen, trehalose, amino acids) they apparently are unable to complete their current cell cycle and therefore arrest randomly upon nutrient depletion.[130,205]

4.3. MORPHOLOGICAL DIFFERENTIATION AS A RESPONSE TO NUTRIENT LIMITATION

A drastic morphogenetic switch has been identified in *S. cerevisiae*, originally as an important response to limitation of the available nitrogen source but more recently also as a more general response to nutrient limitation. Heterozygous a/α diploids grown on medium containing low ammonia concentrations or proline as a sole nitrogen source switch their bipolar budding pattern to a strictly polarized and elongated morphology, leading to pseudohyphal growth[211] (Fig. 1.12, adapted from refs. 278-280). The unipolar pseudohyphal cell shape and the accompanying invasive growth resemble the polarized growth morphology of many filamentous fungi[281] and permits the cells to forage for limiting nutrients. The dimorphic switch is specifically induced by poor nitrogen availability. In addition, mutations in *SHR3* encoding a protein involved the uptake of many amino acids to enhance the morphological transition.[211] More recently a similar dimorphic shift and invasive growth pattern have been described for haploid cells although several differences with diploid pseudohyphal growth were observed.[282] In pseudohyphal-growing cells the short period of hyperpolarized growth normally observed in budding cells only in late G_1 is extended well into the G_2 phase. Mother and daughter cells divide when the daughter cell is full-grown and budding is synchronized.[278] This contrasts with the characteristic mother-daughter asynchrony normally observed in budding cells which is illustrated by the fact that the daughter cell requires significant growth during G_1 to reach the critical cell size to start a new proliferation cycle.[2] It is not clear how this alteration in cell cycle regulation is achieved. Interestingly, in yeast mutants defective in *GRR1* encoding a protein involved in several nutrient-linked signaling pathways[169,172,283] the G_1 cyclins Cln1p and Cln2p are stabilized.[284] This cyclin stabilization is likely to antagonize mitosis and to prolong polarized growth. As a consequence *grr1* mutants are constitutively pseudohyphal.

The signaling pathways operating in this morphogenic developmental switch in *Saccharomyces* are only beginning to be revealed. A direct involvement of the Ras-cAMP pathway has been suggested, since the dominant activated *RAS2*[Val19] mutation enhances pseudohyphal differentiation.[211] In addition, overexpression of *PDE2* encoding the most active phosphodiesterase inhibits pseudohyphal growth.[275] However, the in-volvement of the Ras-cAMP pathway in pseudohyphal development may be more indirect. The *RAS2*[Val19] mutation not only decreases storage carbohydrate levels but

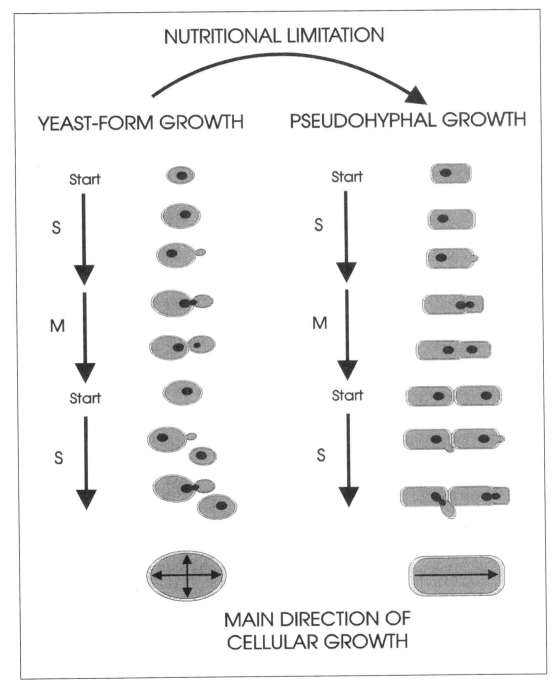

Fig. 1.12. Pseudohyphal differentiation in Saccharomyces cerevisiae. Nutritional limitation may cause a switch of the normal axial budding pattern to a strictly polarized and elongated morphology, termed pseudohyphal growth.[211] Adapted from refs. 279-281.

also severely diminishes intracellular amino acid pools.[205] The enhanced tendency of these mutants to switch to pseudohyphal growth in response to nitrogen limitation may well be related to these restricted amino acid stores, which will be more readily depleted than in wild type yeast. On the other hand *RAS2*[Val19]-stimulated pseudohyphal growth is dependent on the Rho-family GTPase Cdc42p[285] which has previously been implicated in the control of cell polarity and of the mating pheromone response.[286-288] Thus Ras2p may directly participate in dimorphic signaling by activating the upper part of the mating pheromone response pathway.

Indeed, several other components of the mating pheromone-induced MAPK cascade—Ste20p, Ste11p and Ste7p kinases and the Ste12p transcription factor (see chapter 4)—are controlling invasive growth in both diploids and haploids.[282,289] The MAP-kinases Fus3p and Kss1p of this pathway and the cascade-specific scaffolding protein Ste5p[290] are not required. Thus the coordinated regulation of polarized growth and specific transcriptional response in both mating pheromone signaling and pseudohyphal differentiation may be mediated through shared central signaling components and distinct, specialized MAP-kinases and scaffolds. Besides Ste12p several putative target transcription factors required for or involved in the dimorphic switch have been characterized. *PHD1* was identified by virtue of its ability to induce pseudohyphal growth when overexpressed.[291] Similarly, *MSS10*[214] which has also been isolated as the activator *MSN1* in the expression of the invertase gene *SUC2*,[292] as *FUP1* involved in iron homeostasis[293] and as *PHD2*[291] stimulates pseudopyphal development when overexpressed. *SOK2* encodes an inhibitor of pseudohyphal differentiation exhibiting a high degree of identity in its DNA-binding domain with the DNA-binding region of Phd1p.[275] Interestingly, the Sok2p protein appears to be a mediator of protein kinase A-dependent regulation of growth and development, whereas Phd1p does not. Several additional muta-

tions causing constitutive pseudohyphal growth have been described in an extensive genetic analysis of this morphological differentiation.[213,294,295] This study has thus far revealed the involvement of a novel protein kinase Elm1p and protein phosphatase 2A (PP2A)[294] and of phosphoribosylpyrophosphate (PRPP) synthase encoded by *PPS1*.[295] Pps1p is a central enzyme in nitrogen metabolism and PRPP is essential for the biosynthesis of nucleotides, histidine and tryptophane.[238,296]

Most interestingly, recent findings indicate that the onset of pseudohyphal differentiation in *S. cerevisiae* may not be restricted to conditions of nitrogen limitation, but rather appears to be a more general response to nutrient limitation[212-214] and is strictly dependent on the presence of oxygen.[297] Differences in susceptibility of the dimorphic switch-induction to various nutrient limitation conditions are likely to derive from differences in genetic background of the yeast strains under investigation. Haploid yeast cells grown on medium containing the poor carbon source ethanol and leucine as a nitrogen source develop large hyphal-like extensions.[298] This finding has been substantiated since 'fusel' alcohols like isoamylalcohol which are the natural catabolic products of several amino acids induce pseudohyphal growth.[212] Several starch degrading strains of *S. cerevisiae* display strong pseudohyphal differentiation and invasive growth when grown either on low ammonium medium or on medium with excess nitrogen but with maltotriose or starch as the sole—"poor" carbon source.[214] The induction of the expression of the genes encoding the starch degrading enzymes Sta1p-Sta3p is coregulated with expression of a mucin-like yeast protein, Muc1p, which is essential for pseudohyphal and invasive growth under every nutrient limiting condition. However, only the *MUC1* gene is activated by the Mss10p transcription factor mentioned above. The apparent strict coupling of pseudohyphal differentiation and nutrient depletion further substantiates the idea of extensive cross-talk between signal transduction cascades operating in nutrient signaling and morphogenic development.

5. RETURNING TO THE FEAST: SENSING NUTRITIONAL RESUPPLEMENTATION

5.1. RETURNING TO PROLIFERATION

When yeast cells starved for an essential nutrient are again replenished with the lacking nutrient they will exit stationary phase and progress through the G_1 phase of the cell cycle. On a fermentable medium this is associated with dramatic metabolic changes. Glycogen and trehalose levels drop, STRE-controlled genes are repressed and stress resistance drops rapidly. It is presumably because of these dramatic metabolic changes that this transition to proliferation under fermentable conditions has been studied in great detail, as opposed to characteristics of yeast cells resuming proliferation in media containing a nonfermentable carbon source.

5.2. THE FERMENTABLE GROWTH MEDIUM-INDUCED (FGM) PATHWAY

When a nitrogen source, phosphate or sulfate is added to yeast cells starved on a glucose-containing medium for the same nutrient a rapid activation of trehalase occurs which is responsible for the rapid mobilization of trehalose.[63] Similar rapid changes in activity have been described for other enzymes of glycogen and trehalose metabolism and glycolysis.[48,299] These rapid changes in enzyme activity are independent of protein synthesis and hence occur at the posttranslational level presumably by protein modifications. Under the same conditions ribosomal protein gene transcription is rapidly induced and mRNA levels of STRE-controlled genes rapidly decrease.[49,165] There is good evidence that these effects are controlled by protein kinase A. Notably, however, these effects triggered by re-addition of an essential nutrient in a sugar-containing medium do not appear to be triggered by cAMP. Under such conditions no transient increase in the cAMP level is generated[63] and the effects

mentioned are observed even in mutants lacking the regulatory subunit Bcy1p of protein kinase A.[62,266,277] Apparently the catalytic subunits of protein kinase A are activated in a novel cAMP-independent way. Since this activation is dependent on the presence of glucose or another rapidly-fermented sugar and concomitantly on all nutrients required for growth, we have called the pathway involved the 'Fermentable Growth Medium-induced (FGM) pathway'[70] (Fig. 1.13). Presumably, the pathway can also be activated by re-addition of other essential nutrients to cells starved for such a nutrient because the starvation logically will lead to the acquirement of all typical stationary-phase characteristics while the re-addition will lead to their disappearance.

At present the concept in which the catalytic subunits of protein kinase A are activated in a novel way by the FGM pathway is the simplest. It cannot be excluded, however, that protein kinase A controls the expression or the activity of another protein kinase which actually mediates the signaling by the FGM pathway. A putative candidate is the Sch9p protein kinase which shares extensive sequence similarity with the catalytic subunits of protein kinase A, Tpk1p, Tpk2p and Tpk3p[20,300] (de Winde, Thevelein and Winderickx, unpublished data). Recently the existence of this glucose sensing pathway acting in parallel to the glucose-inducible Ras-cAMP pathway has been underscored by the unexpected finding that activation of the FGM pathway does not require phosphorylation of the fermentable sugar.[145] In contrast, for activation of the Ras-adenylate cyclase pathway by glucose or fructose, phosphorylation of the sugar by any one of the known yeast sugar kinases is essential.[125] This finding now opens the interesting question as to whether for activation of the FGM pathway to occur the glucose is 'sensed' by a glucose carrier or by one or more of the many glucose carrier-like proteins present in yeast.[301] This glucose-sensing may even be dependent on the Snf3p and/or Rgt2p transporter homologues[173] discussed in section 3.1.8. The nutrient-

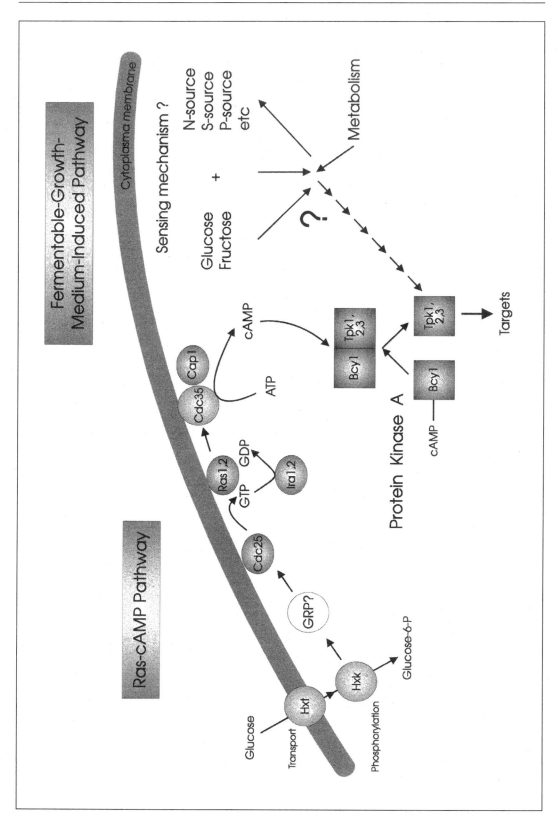

induced upshift of ribosomal protein gene expression on fermentable medium is also largely independent of phosphorylation of the available sugar.[145] This indicates that the boost of ribosomal biogenesis that accompanies the resumption of cellular proliferation is not a mere consequence of growth induction or any of its metabolic requirements. Rather it suggests the existence of specific nutrient sensing and signaling pathways that are independent of cellular metabolism.

6. PERSPECTIVES

Signal transduction mechanisms have become major research topics in molecular and cellular biology. Adequate perception of and efficient response to the environment is of vital importance to all living organisms ranging from simple prokaryotes to multicellular eukaryotes. As the number of identified components increases and the complexity of the signal transduction pathways involved is revealed, the most striking finding of all is the high degree of evolutionary conservation of the underlying basic mechanisms, especially within the eukaryotic kingdom. Since baker's yeast *S. cerevisiae* is a relatively simple, genetically and biochemically well characterized and easy-to-handle eukaryote it serves as a valuable tool in the scrutiny of eukaryotic signal transduction cascades and networks.

The way in which cells of *Saccharomyces* respond to nutritional stimuli more and more appears to present models for nutrient sensing and signaling in various specialized cell types of multicellular higher eukaryotes. An astonishing example is presented by the sugar-induced signaling mediated by Rgt2p, Snf3p or other sugar transporters (homologues?) and mediated by hexokinases and glucokinase in

yeast.[73,173] The initiation of this signaling cascade may be very similar to the glucose response in pancreatic β-cells of mammals. In these cells tissue-specific low-affinity transporter Glut2p and glucokinase—with high similarity to yeast hexokinase—are required for the glucose-induced insulin secretion response.

Another very interesting and well conserved eukaryotic nutrient sensing mechanism which is, however, at present only beginning to emerge may be the signaling pathway involving the phosphatidylinositol kinase family represented by Tor1p and Tor2p in *Saccharomyces*.[302] This pathway appears to integrate the cellular perception of nutrient availability with the control of protein synthesis (translational initiation) and progression through the G_1 cell cycle phase. Although mechanistically still quite elusive, the TOR pathway and the FGM pathway described above (section 5.2) may be linked quite closely. Future research will tell the tale.

REFERENCES

1. Werner-Washburne M, Braun E, Johnston GC, Singer RA. Stationary phase in the yeast *Saccharomyces cerevisiae*. Microbiol Rev 1993; 57:383-401.
2. Pringle JR, Hartwell LH. The *Saccharomyces cerevisiae* cell cycle. In: Strathern JN, Jones EW, Broach JR, eds. The Molecular Biology of the Yeast *Saccharomyces cerevisiae*; Life Cycle and Inheritance. Cold Spring Harbor, New York: Cold Spring Harbor Laboratory Press, 1981:97-142.
3. Drebot MA, Barnes CA, Singer RA, Johnston GC. Genetic assessment of stationary phase for cells of the yeast *Saccharomyces cerevisiae*. J Bacteriol 1990; 172:3584-3589.
4. Soto T, Fernandez J, Cansado J, Vicente-Soler J, Gacto M. Glucose-induced, cyclic-

Fig. 1.13 (opposite). Fermentable Growth Medium (FGM) control in Saccharomyces cerevisiae. A sugar phosphorylation-independent signaling cascade termed the FGM pathway is initiated by a complete fermentable growth medium and activates the free catalytic subunits of protein kinase A. The FGM pathway operates independently from and in parallel to the sugar phosphorylation-dependent Ras-cAMP cascade. In addition, several lines of evidence suggest the existence of a third control route measuring cellular metabolic capacity and modulating protein kinase A activity.[70]

AMP-independent signalling pathway for activation of neutral trehalase in the fission yeast *Schizosaccharomyces pombe*. Microbiol 1995; 141:2665-2671.

5. Nurse P. Eukaryotic cell cycle control. Biochem Soc Trans 1992; 20:239-242.

6. Byrne SM, Hoffman CS. Six *git* genes encode a glucose-induced adenylate cyclase activation pathway in the fission yeast *Schizosaccharomyces pombe*. J Cell Sci 1993; 105:1095-1100.

7. Woollard A, Nurse P. G(1) regulation and checkpoints operating around Start in fission yeast. Bioessays 1995; 17:481-490.

8. Oehlen LJWM, van Doorn J, Scholte ME, Postma PW, van Dam K. Changes in the incorporation of carbon derived from glucose into cellular pools during the cell cycle of *Saccharomyces cerevisiae*. J Gen Microbiol 1990; 136:413-418.

9. Oehlen LJWM, Scholte ME, de Koning W, van Dam K. Inactivation of the *CDC25* gene product in *Saccharomyces cerevisiae* leads to a decrease in glycolytic activity which is independent of cAMP levels. J Gen Microbiol 1993; 139:2091-2100.

10. Fraenkel D. Carbohydrate Metabolism. In: Strathern JN, Jones EW, Broach JR, eds. The Molecular Biology of the Yeast *Saccharomyces*: Metabolism and Gene Expression. Cold Spring Harbor, New York: Cold Spring Harbor Laboratory Press, 1982:1-37.

11. Wills C. Regulation of sugar and ethanol metabolism in *Saccharomyces cerevisiae*. Crit Rev Biochem Mol Biol 1990; 25:245-280.

12. De Vries S, Marres CAM. The mitochondrial respiratory chain of yeast structure and biosynthesis and the role in cellular metabolism. Biochem Biophys Acta 1988; 895:205-239.

13. Alonso AC, Pascual C, Herrera L, Gancedo JM, Gancedo C. Metabolic imbalance in a *Saccharomyces cerevisiae* mutant unable to grow on fermentable hexoses. Eur J Biochem 1984; 138:407-411.

14. Francois JM, Eraso P, Gancedo C. Changes in the concentration of cAMP, fructose-2,6-bisphosphate and related metabolites and enzymes in *Saccharomyces cerevisiae* during growth on glucose. Eur J Biochem 1987; 164:369-373.

15. Moehle CM, Jones EW. Consequences of growth media, gene copy number and regulatory mutations on the expression of the *PRB1* gene of *Saccharomyces cerevisiae*. Genetics 1990; 124:39-55.

16. Yin Z, Smith RJ, Brown AJP. Multiple signalling pathways trigger the exquisite sensitivity of yeast gluconeogenic mRNAs to glucose. Mol Microbiol 1996; 20:751-764.

17. Boy-Marcotte E, Tadi D, Perrot M, Boucherie H, Jaquet M. High cAMP levels antagonize the reprogramming of gene expression that occurs at the diauxic shift in *Saccharomyces cerevisiae*. Microbiol 1996; 142:459-467.

18. Lillie SH, Pringle JR. Reserve carbohydrate metabolism in *Saccharomyces cerevisiae*: responses to nutrient limitation. J Bacteriol 1980; 143:1384-1394.

19. Bissinger PH, Wieser R, Hamilton B, Ruis H. Control of *Saccharomyces cerevisiae* catalase T gene (*CTT1*) expression by nutrient supply via the Ras-cyclic AMP pathway. Mol Cell Biol 1989; 9:1309-1315.

20. Denis CL, Audino DC. The Ccr1 (Snf1) and Sch9 protein kinases act independently of cAMP-dependent protein kinase and the transcriptional activator Adr1 in controlling yeast *ADH2* expression. Mol Gen Genet 1991; 229:395-399.

21. Hubbard EJA, Yang X, Carlson M. Relationship of the cAMP-dependent protein kinase pathway to the SNF1 protein kinase and invertase expression in *Saccharomyces cerevisiae*. Genetics 1992; 130:71-80.

22. Werner-Washburne M, Becker J, Kosic-Smithers J, Craig EA. Yeast *Hsp70* RNA levels vary in response to the physiological status of the cell. J Bacteriol 1989; 171:2680-2688.

23. Tanaka K, Matsumoto K, Toh-e A. Dual regulation of the expression of the polyubiquitin gene by cyclic AMP and heat shock in yeast. EMBO J 1988; 7:495-502.

24. Praekelt UM, Maecock PA. *HSP12*, a new small heat shock gene of *Saccharomyces cerevisiae*: analysis of structure, regulation and function. Mol Gen Genet 1990; 223:97-106.

25. Suomalainen H, Oura E. Yeast nutrition and solute uptake. In: Rose AH, Harrison JH,

eds. The Yeasts. Vol. 2. New York, NY: Academic Press, 1971:3-74.

26. Gancedo JM. Carbon catabolite repression in yeast. Eur J Biochem 1992; 206:297-313.

27. Trumbly R. Glucose repression in the yeast *Saccharomyces cerevisiae*. Mol Microbiol 1992; 6:15-21.

28. Carlson M. Regulation of sugar utilization in *Saccharomyces* species. J Bacteriol 1987; 169:4873-4877.

29. Johnston M, Carlson M. Regulation of carbon and phosphate utilization. In: Jones EW, Pringle JR, Broach JR, eds. The Molecular and Cellular Biology of the Yeast *Saccharomyces*. Cold Spring Harbor: Cold Spring Harbor Laboratory Press, 1992: 193-281.

30. Cheng C, Mu J, Farkas I, Huang D, Goebl MG, Roach PJ. Requirement of self-glucosylating intiator proteins Glg1p and Glg2p for glycogen accumulation in *Saccharomyces cerevisiae*. Mol Cell Biol 1995; 15:6632-6640.

31. Farkas I, Hardy TA, De Paoli-Roach AA, Roach PJ. Isolation of the *GSY1* gene encoding yeast glycogen synthase and evidence for the existence of a second gene. J Biol Chem 1990; 265:20879-20886.

32. Farkas I, Hardy TA, Goebl MG, Roach PJ. Two glycogen synthase isoforms in *Saccharomyces cerevisiae* are coded by distinct genes that are differentially controlled. J Biol Chem 1991; 26:15601-15607.

33. Thon VJ, Vigneron-Lesens C, Marianne-Pepin T, Montreuil J, Decq A. Coordinate regulation of glycogen metabolsm in the yeast *Saccharomyces cerevisiae*: induction of glycogen branching enzyme. J Biol Chem 1992; 267:15224-15228.

34. Rowen DW, Meinke M, LaPorte DC. *GLC3* and *GHA1* of *Saccharomyces cerevisiae* are allelic and encode the glycogen branching enzyme. Mol Cell Biol 1992; 12:22-29.

35. Hwang PK, Tugendreich S, Fletterick RJ. Molecular analysis of *GPH1*, the gene encoding glycogen phosphorylase in *Saccharomyces cerevisiae*. Mol Cell Biol 1989; 9:1659-1666.

36. Cannon JF, Pringle JR, Fiechter A, Khalil M. Characterization of glycogen-deficient *GLC* mutants of *Saccharomyces cerevisiae*.

Genetics 1994; 136:485-503.

37. Feng Z, Wilson SE, Peng ZY, Schlender KK, Reiman EM, Trumbly RJ. The yeast *GLC7* gene required for glycogen accumulation encodes a type 1 protein phosphatase. J Biol Chem 1991; 266:23796-23801.

38. Francois JM, Thompson-Jaeger S, Skroch JZ U, Spevak Wea. *GAC1* may encode a regulatory subunit for protein phosphatase type 1 in *Saccharomyces cerevisiae*. EMBO J 1992; 11:87-96.

39. Matsumoto K, Uno I, Ishikawa T. Genetic analysis of the role of cAMP in yeast. Yeast 1985; 1:15-24.

40. Thompson-Jaeger S, Francois J, Gaughran JP, Tatchell K. Deletion of SNF1 affects the nutrient response of yeast and resembles mutations which activate the adenylate cyclase pathway. Genetics 1991; 129:697-706.

41. Thevelein JM. Regulation of trehalose metabolism and its relevance to cell growth and function. In: Brambl R, Marzluf GA, eds. The Mycota III; Biochemistry and Molecular Biology. Berlin-Heidelberg: Springer Verlag, 1996:395-420.

42. Bell W, Klaassen P, Ohnacker M, Boller T, Herweijer M, Schoppink P, van der Zee P, Wiemken A. Characterization of the 56 kDa subunit of the yeast trehalose-6-phosphate synthase and cloning of its gene reveal its identity with the product of *CIF1*, a regulator of carbon catabolite inactivation. Eur J Biochem 1992; 209:951-959.

43. Gonzalez MI, Stucka R, Blazquez MA, Feldmann H, Gançedo C. Molecular cloning of *CIF1*, a yeast gene necessary for growth on glucose. Yeast 1992; 8:183-192.

44. Van Aelst L, Hohmann S, Bulaya B, de Koning W, Sierkstra L, Neves MJ, Luyten K, Alijo R, Ramos J, Coccetti P, Martegani E, de Magalhaes-Rocha NM, Brandao RL, Van Dijck P, Vanhalewyn M, Durnez P, A.W.H. J, Thevelein JM. Molecular cloning of a gene involved in glucose sensing in the yeast *Saccharomyces cerevisiae*. Mol Microbiol 1993; 8:927-943.

45. Vuorio OE, Kalkkinen N, Londesborough J. Cloning of two related genes encoding the 56-kDa and 123-kDa subunits of trehalose synthase from the yeast *Saccharomyces cerevisiae*. Eur J Biochem 1993; 216:849-861.

46. De Virgilio C, Bürckert N, Bell W, Jeno P, Boller T, Wiemken A. Disruption of *TPS2*, the gene encoding the 100-kDa subunit of the trehalose-6-phosphate synthase/phosphatase complex in *Saccharomyces cerevisiae*, causes accumulation of trehalose-6-phosphate and loss of trehalose-6-phosphate phosphatase activity. Eur J Biochem 1993; 212:315-323.

47. Londesborough J, Vuorio OE. Purification of trehalose synthase from baker's yeast, its temperature-dependent activation by fructose-6-phosphate and inhibition by phosphate. Eur J Biochem 1993; 216:841-848.

48. François J, Neves MJ, Hers HG. The control of trehalose biosynthesis in *Saccharomyces cerevisiae*—evidence for a catabolite inactivation and repression of trehalose-6-phosphate synthase and trehalose-6-phosphate phosphatase. Yeast 1991; 7:575-587.

49. Winderickx J, de Winde JH, Crauwels M, Hino A, Hohmann S, Van Dijck P, Thevelein JM. Regulation of genes encoding subunits of the trehalose synthase complex in *Saccharomyces cerevisiae*. Novel variations of STRE-mediated transcriptional control? Mol Gen Genet 1996; 252:470-482.

50. Boorstein WR, Craig EA. Regulation of a yeast Hsp70 gene by a cAMP responsive transcriptional control element. EMBO J 1990; 9:2543-2553.

51. Marchler G, Schüller C, Adam G, Ruis H. A *Saccharomyces cerevisiae* UAS element controlled by protein kinase A activates transcription in response to a variety of stress conditions. EMBO J 1993; 12:1997-2003.

52. Varela JCS, Praekelt UM, Meacock PA, Planta RJ, Mager WH. The *Saccharomyces cerevisiae HSP12* gene is activated by the high-osmolarity glycerol pathway and negatively regulated by protein kinase A. Mol Cell Biol 1995; 15:6232-6245.

53. Hohmann S, Van Dijck P, Luyten K, Thevelein JM. The *byp1-3* allele of the *Saccharomyces cerevisiae GGS1/TPS1* gene and its multi-copy suppressor tRNA(GLN) (CAG): Ggs1/Tps1 protein levels restraining. Curr Genet 1994; 26:295-301.

54. Neves MJ, Hohmann S, Bell W, Dumortier F, Luyten K, Ramos J, Cobbaert P, De Koning W, Kaneva Z, Thevelein JM. Control of glucose influx into glycolysis and pleiotropic effects studied in different isogenic sets of *Saccharomyces cerevisiae* mutants in trehalose biosynthesis. Curr Genet 1995; 27:110-122.

55. Müller J, Boller T, Wiemken A. Trehalose and trehalase in plants: Recent developments. Plant Sci 1995; 112:1-9.

56. Nwaka S, Kopp M, Holzer H. Expression and function of the trehalase genes *NTH1* and *YBR0106* in *Saccharomyces cerevisiae*. J Biol Chem 1995; 270:10193-10198.

57. Kopp M, Müller H, Holzer H. Molecular analysis of the neutral trehalase gene from *Saccharomyces cerevisiae*. J Biol Chem 1993; 268:4766-4774.

58. Kopp M, Nwaka S, Holzer H. Corrected sequence of the yeast neutral trehalase-encoding gene (*NTH1*): biological implications. Gene 1994; 150:403-404.

59. Nwaka S, Mechler B, Destruelle M, Holzer H. Phenotypic features of trehalase mutants in *Saccharomyces cerevisiae*. FEBS Lett 1995; 360:286-290.

60. Destruelle M, Holzer H, Klionsky DJ. Isolation and characterization of a novel yeast gene, *ATH1*, that is required for vacuolar acid trehalase activity. Yeast 1995; 11:1015-1025.

61. De Virgilio C, Bürckert N, Boller T, Wiemken A. A method to study the rapid phosphorylation-related modulation of neutral trehalase activity by temperature shifts in yeast. FEBS Lett 1991; 291:355-358.

62. Durnez P, Pernambuco MB, Oris E, Argüelles JC, Mergelsberg H, Thevelein JM. Activation of trehalase during growth induction by nitrogen sources in the yeast *Saccharomyces cerevisiae* depends on the free catalytic subunits of cAMP-dependent protein kinase, but not on functional ras proteins. Yeast 1994; 10:1049-1064.

63. Hirimburegama K, Durnez P, Keleman J, Oris E, Vergauwen R, Mergelsberg H, Thevelein JM. Nutrient-induced activation of trehalase in nutrient-starved cells of the yeast *Saccharomyces cerevisiae*: cAMP is not involved as second messenger. J Gen Microbiol 1992; 138:2035-2043.

64. Van der Plaat JB. Cyclic 3',5'-adenosine monophosphate stimulates trehalose degra-

dation in baker's yeast. Biochem Biophys Res Comm 1974; 56:580-587.

65. Thevelein JM. Fermentable sugars and intracellular acidification as specific activators of the Ras-adenylate cyclase signalling pathway in yeast: the relationship to nutrient-induced cell cycle control. Mol Microbiol 1991; 5:1301-1307.

66. Uno I, Matsumoto K, Adachi K, Ishikawa T. Genetic and biochemical evidence that trehalase is a substrate of cAMP-dependent protein kinase in yeast. J Biol Chem 1983; 258:10867-10872.

67. Carrillo D, Vicente-Soler J, Fernandez J, Soto T, Cansado J, Gacto M. Activation of cytoplasmic trehalase by cyclic-AMP-dependent and cyclic-AMP-independent signalling pathways in the yeast *Candida utilis*. Microbiol 1995; 141:679-686.

68. Soto T, Fernandez J, Vicente-Soler J, Cansado J, Gacto M. Activation of neutral trehalase by glucose and nitrogen source in *Schizosaccharomyces pombe* strains deficient in cAMP-dependent protein kinase activity. FEBS Lett 1995; 367:263-266.

69. Thevelein JM. The Ras-Adenylate cyclase pathway and cell cycle control in *Saccharomyces cerevisiae*. Antoni van Leeuwenhoek Int J Gen Mol Microbiol 1992; 62:109-130.

70. Thevelein JM. Signal transduction in yeast. Yeast 1994; 10:1753-1790.

71. Thevelein JM, Hohmann S. Trehalose synthase: Guard to the gate of glycolysis in yeast? Trends Biochem Sci 1995; 20:3-10.

72. Hohmann S, Neves MJ, de Koning W, Alijo R, Ramos J, Thevelein JM. The growth and signalling defects of the *ggs1* (*fdp1/byp1*) deletion mutant on glucose are suppressed by a deletion of the gene encoding hexokinase PII. Curr Genet 1993; 23:281-289.

73. De Winde JH, Crauwels M, Hohmann S, Thevelein JM, Winderickx J. Differential requirement of the yeast sugar kinases for sugar sensing in the establisment of the catabolite-repressed state. Eur J Biochem 1996; 241:633-643.

74. Blazquez MA, Lagunas R, Gancedo C, Gancedo JM. Trehalose-6-phosphate, a new regulator of yeast glycolysis that inhibits hexokinases. FEBS Lett 1993; 329:51-54.

75. Hohmann S, Bell W, Neves MJ, Thevelein JM. Evidence for trehalose-6-phosphate dependent and independent mechanisms in the control of sugar influx into yeast glycolysis. Mol Microbiol 1996; 20:981-991.

76. Ronne H. Glucose repression in fungi. Trends Genet 1995; 11:12-17.

77. Mercado JJ, Smith R, Sagliocco FA, Brown AJP, Gancedo JM. The levels of yeast gluconeogenic mRNAs respond to environmental factors. Eur J Biochem 1994; 224:473-481.

78. De Winde JH, Grivell LA. Global regulation of mitochondrial biogenesis in *Saccharomyces cerevisiae*. Progr Nucl Acid Res Mol 1993; 46:51-91.

79. Entian KD. Genetic and biochemical evidence for hexokinase PII as a key enzyme involved in carbon catabolite repression in yeast. Mol Gen Genet 1980; 178:633-637.

80. Ma H, Botstein D. Effects of null mutations in the hexokinase genes of *Saccharomyces cerevisiae* on catabolite repression. Mol Cell Biol 1986; 6:4046-4052.

81. Celenza JL, Carlson M. A yeast gene that is essential for release from glucose repression encodes a protein kinase. Science 1986; 233:1175-1180.

82. Celenza JL, Carlson M. Cloning and genetic mapping of SNF1, a gene required for expression of glucose-repressible genes in *S.cerevisiae*. Mol Cell Biol 1984; 4:49-53.

83. Entian KD, Zimmermann FK. New genes involved in carbon catabolite repression and derepression in the yeast *Saccharomyces cerevisiae*. J Bacteriol 1982; 151:1123-1128.

84. Celenza JL, Carlson M. Mutational analysis of the *Saccharomyces cerevisiae* SNF1 protein kinase and evidence for functional interaction with SNF4 protein. Mol Cell Biol 1989; 9:5034-5044.

85. Yang XL, Jiang R, Carlson M. A family of proteins containing a conserved domain that mediates interaction with the yeast SNF1 protein kinase complex. EMBO J 1994; 13:5878-5886.

86. Lesage P, Yang XL, Carlson M. Analysis of the Sip3 protein identified in a Two-Hybrid screen for interaction with the Snf1 protein kinase. Nucl Acids Res 1994; 22:597-603.

87. Lesage P, Yang XL, Carlson M. Yeast Snf1 protein kinase interacts with Sip4, a C-6 zinc cluster transcriptional activator: a new role for Snf1 in the glucose reponse. Mol Cell Biol 1996; 16:1921-1928.

88. Vallier LG, Carlson M. Synergistic release from glucose repression by *mig1* and *ssn* mutations in *Saccharomyces cerevisiae*. Genetics 1994; 137:49-54.

89. Treitel MA, Carlson M. Repression by SSN6-TUP1 is directed by MIG1, a repressor activator protein. Proc Natl Acad Sci USA 1995; 92:3132-3136.

90. Lundin M, Nehlin JO, Ronne H. Importance of a flanking AT-rich region in target site recognition by the GC Box-binding zinc finger protein Mig1. Mol Cell Biol 1994; 14:1979-1985.

91. Nehlin JO, Ronne H. Yeast MIG1 repressor is related to the mammalian early growth response and Wilms' tumour finger proteins. EMBO J 1990; 9:2891-2898.

92. Nehlin JO, Carlberg M, Ronne H. Control of yeast *GAL* genes by MIG1 repressor: a transcriptional cascade in the glucose response. EMBO J 1991; 10:3373-3377.

93. Östling J, Carlberg M, Ronne H. Functional domains in the Mig1 repressor. Mol Cell Biol 1996; 16:753-761.

94. Keleher CA, Redd MJ, Schultz J, Carlson M, Johnson AD. Ssn6-Tup1 is a general repressor of transcription in yeast. Cell 1992; 68:709-719.

95. Ma H, Bloom LM, Walsh CT, Botstein D. The residual enzymatic phosphorylation activity of hexokinase II mutants is correlated with glucose repression in *Saccharomyces cerevisiae*. Mol Cell Biol 1989; 9:5643-5649.

96. Rose M, Albig W, Entian KD. Glucose repression in *Saccharomyces cerevisiae* is directly associated with hexose phosphorylation by hexokinase-PI and hexokinase-PII. Eur J Biochem 1991; 199:511-518.

97. Johnston M, Flick JS, Pexton T. Multiple mechanisms provide rapid and stringent glucose repression of *GAL* gene expression in *Saccharomyces cerevisiae*. Mol Cell Biol 1994; 14:3834-3841.

98. Niederacher D, Schüller HJ, Grzesitza D, Gutlich H, Hauser HP, Wagner T, Entian KD. Identification of UAS elements and binding proteins

necessary for derepression of *Saccharomyces cerevisiae* fructose-1,6-bisphosphatase. Curr Genet 1992; 22:363-370.

99. Hedges D, Proft M, Entian KD. *CAT8*, a new zinc cluster-encoding gene necessary for derepression of gluconeogenic enzymes in the yeast *Saccharomyces cerevisiae*. Mol Cell Biol 1995; 15:1915-1922.

100. Proft M, Kotter P, Hedges D, Bojunga N, Entian KD. *CAT5*, a new gene necessary for derepression of gluconeogenic enzymes in *Saccharomyces cerevisiae*. EMBO J 1995; 14:6116-6126.

101. Rahner A, Scholer A, Martens E, Gollwitzer B, Schüller HJ. Dual influence of the yeast Cat1p (Snf1p) protein kinase on carbon source-dependent transcriptional activation of gluconeogenic genes by the regulatory gene *CAT8*. Nucleic Acids Res 1996; 24:2331-2337.

102. Dombek KM, Camier S, Young ET. *ADH2* expression is repressed by REG1 independently of mutations that alter the phosphorylation of the yeast transcription factor ADR1. Mol Cell Biol 1993; 13:4391-4399.

103. Denis CL, Fontaine SC, Chase D, Kemp BE, Bemis LT. *ADR1*c mutations enhance the ability of ADR1 to activate transcription by a mechanism that is independent of effects on cyclic AMP-dependent protein kinase phosphorylation of Ser-230. Mol Cell Biol 1992; 12:1507-1514.

104. Niederacher D, Entian KD. Isolation and characterization of the regulatory *HEX2* gene necessary for glucose repression in yeast. Mol Gen Genet 1987; 206:505-509.

105. Niederacher D, Entian KD. Characterization of Hex2 protein, a negative regulatory element necessary for glucose repression in yeast. Eur J Biochem 1991; 200:311-319.

106. Tung KS, Norbeck LL, Nolan SL, Atkinson NS, Hopper AK. *SRN1*, a yeast gene involved in RNA processing, is identical to *HEX2/REG1*, a negative regulator in glucose repression. Mol Cell Biol 1992; 12:2673-2680.

107. Tu JL, Carlson M. REG1 binds to protein phosphatase type 1 and regulates glucose repression in *Saccharomyces cerevisiae*. EMBO J 1995; 14:5939-5946.

108. Francisco L, Wang W, Chan CSM. Type 1 protein phosphatase acts in opposition to Ipl1 protein kinase in regulating yeast chromosome segregation. Mol Cell Biol 1994; 14:4731-4740.

109. Hisamoto N, Frederick DL, Sugimoto K, Tatchell K, Matsumoto K. The *EGP1* gene may be a positive regulator of protein phosphatase type 1 in the growth control of *Saccharomyces cerevisiae*. Mol Cell Biol 1995; 15:3767-3776.

110. Wek RC, Cannon JF, Dever TE, Hinnebusch AG. Truncated protein phosphatase GLC7 restores translational activation of *GCN4* expression in yeast mutants defective for the eIF-2alpha kinase GCN2. Mol Cell Biol 1992; 12:5700-5710.

111. Stuart JS, Frederick DL, Varner CM, Tatchell K. The mutant type 1 protein phosphatase encoded by *glc7-1* from *Saccharomyces cerevisiae* fails to interact productively with the *GAC1*-encoded regulatory subunit. Mol Cell Biol 1994; 14:896-905.

112. Zhang SR, Guha S, Volkert FC. The *Saccharomyces SHP1* gene, which encodes a regulator of phosphoprotein phosphatase 1 with differential effects on glycogen metabolism, meiotic differentiation, and mitotic cell cycle progression. Mol Cell Biol 1995; 15:2037-2050.

113. Tu J, Song W, Carlson M. Protein phosphatase type 1 interacts with proteins required for meiosis and other cellular processes in *Saccharomyces cerevisiae*. Mol Cell Biol 1996; 16:4199-4206.

114. Frederick DL, Tatchell K. The *REG2* gene of *Saccharomyces cerevisiae* encodes a type 1 protein phosphatase-binding protein that functions with Reg1p and the Snf1 protein kinase to regulate growth. Mol Cell Biol 1996; 16:2922-2931.

115. Nehlin JO, Carlberg M, Ronne H. Yeast *SKO1* gene encodes a bZIP protein that binds to the CRE motif and acts as a repressor of transcription. Nucl Acids Res 1992; 20:5271-5278.

116. Vincent AC, Struhl K. ACR1, a yeast ATF/CREB repressor. Mol Cell Biol 1992; 12:5394-5405.

117. Nocero M, Isshiki T, Yamamoto M, Hoffman CS. Glucose repression of *fbp1* transcription in *Schizosaccharomyces pombe* is partially regulated by adenylate cyclase activation by a G protein alpha subunit encoded by *gpa2* (*git8*). Genetics 1994; 138:39-45.

118. Hoffman CS, Winston F. Glucose repression of transcription of the *Schizosaccharomyces pombe fbp1* gene occurs by a cAMP signaling pathway. Genes Dev 1991; 5:561-571.

119. Zitomer RS, Montgomery DL, Nichols DL, Hall BD. Transcriptional regulation of the yeast cytochrome *c* gene. Proc Natl Acad Sci USA 1979; 76:3627-3631.

120. Cereghino GP, Scheffler IE. Genetic analysis of glucose regulation in *Saccharomyces cerevisiae*: control of transcription versus mRNA turnover. EMBO J 1996; 15:363-374.

121. Cereghino GP, Atencio DP, Saghbini M, Beiner J, Scheffler IE. Glucose-dependent turnover of the mRNAs encoding succinate dehydrogenase peptides in *Saccharomyces cerevisiae*: Sequence elements in the 5' untranslated region of the Ip mRNA play a dominant role. Mol Biol Cell 1995; 6:1125-1143.

122. Polakis ES, Bartley W. Changes in the enzyme activities of *Saccharomyces cerevisiae* during aerobic growth on different carbon sources. Biochem J 1965; 97:284-297.

123. Eraso P, Gancedo JM. Catabolite repression in yeast is not associated with low levels of cAMP. Eur J Biochem 1984; 141: 195-198.

124. Lodi T, Donnini C, Ferrero I. Catabolite repression by galactose in overexpressed *GAL4* strains of *Saccharomyces cerevisiae*. J Gen Microbiol 1991; 137:1039-1044.

125. Beullens M, Mbonyi K, Geerts L, Gladines D, Detremerie K, Jans AWH, Thevelein JM. Studies on the mechanism of the glucose-induced cAMP signal in glycolysis and glucose repression mutants of the yeast *Saccharomyces cerevisiae*. Eur J Biochem 1988; 172:227-231.

126. Mbonyi K, Van Aelst L, Argüelles JC, Jans AWH, Thevelein JM. Glucose-induced hyperaccumlation of cyclic AMP and defective glucose repression in yeast strains with reduced activity of cyclic AMP-dependent protein kinase. Mol Cell Biol 1990; 10:4518-4523.

127. Mbonyi K, Beullens M, Detremerie K, Geerts L, Thevelein JM. Requirement of one

functional *RAS* gene and inability of an oncogenic ras-variant to mediate the glucose-induced cAMP signal in the yeast *Saccharomyces cerevisiae*. Mol Cell Biol 1988; 8:3051-3057.

128. Broach JR, Deschenes RJ. The function of *RAS* genes in *Saccharomyces cerevisiae*. Adv Cancer Res 1990; 54:79-139.

129. Broach JR. *RAS* Genes in *Saccharomyces cerevisiae*—Signal transduction in search of a pathway. Trends Genet 1991; 7:28-33.

130. Tatchell K. *RAS* genes in the budding yeast *Saccharomyces cerevisiae*. In: Kurjan J, Taylor BJ, eds. Signal Transduction. Prokaryotic and Simple Eukaryotic Systems. San Diego: Academic Press, 1993:147-188.

131. Kataoka T, Broek D, Wigler M. DNA sequence and characterization of the *S. cerevisiae* gene encoding adenylate cyclase. Cell 1985; 43:493-505.

132. Matsumoto K, Uno I, Ishikawa T. Identification of the structural gene and nonsense alleles for adenylate cyclase in *Saccharomyces cerevisiae*. J Bacteriol 1984; 157:277-282.

133. Broek D, Samiy N, Fasano O, Fujiyama A, Tamanoi F, Northup J, Wigler M. Differential activation of yeast adenylate cyclase by wild type and mutant Ras proteins. Cell 1985; 41:763-769.

134. Toda T, Uno I, Ishikawa T, Powers S, Kataoka T, Broek D, Cameron S, Broach J, Matsumoto K, Wigler M. In yeast, Ras proteins are controlling elements of adenylate cyclase. Cell 1985; 40:27-36.

135. Broek D, Toda T, Michaeli T, Levin L, Birchmeier C, Zoller M, Powers S, Wigler M. The *S. cerevisiae CDC25* gene product regulates the Ras-adenylate cyclase pathway. Cell 1987; 48:789-799.

136. Jones S, Vignais M-L, Broach JR. The CDC25 protein of *Saccharomyces cerevisiae* promotes exchange of guanine nucleotides bound to RAS. Mol Cell Biol 1991; 11:2641-2646.

137. Camonis JH, Kalékine M, Gondré B, Garreau H, Boy-Marcotte E, Jaquet M. Characterization, cloning and sequence analysis of the *CDC25* gene which controls the cyclic AMP level of *Saccharomyces cerevisiae*. EMBO J 1986; 5:375-380.

138. Camus C, Boy-Marcotte E, Jacquet M. Two subclasses of guanine exchange factor (GEF) domains revealed by comparison of activities of chimeric genes constructed from *CDC25, SDC25* and *BUD5* in *Saccharomyces cerevisiae*. Mol Gen Genet 1994; 245:167-176.

139. Tanaka K, Matsumoto K, Toh-e A. Ira1, an inhibitory regulator of the RAS-cyclic AMP pathway in *Saccharomyces cerevisiae*. Mol Cell Biol 1989; 9:757-768.

140. Tanaka K, Nakafuku M, Tamanoi F, Kaziro Y, Matsumoto K, Toh-e A. *IRA2*, a second gene in *Saccharomyces cerevisiae* that encodes a protein with a domain homologous to mammalian Ras GTP-ase activating protein. Mol Cell Biol 1990; 10:4303-4313.

141. Nikawa J, Sass P, Wigler M. Cloning and characterization of the low-affinity cyclic AMP phosphodiesterase gene of *Saccharomyces cerevisiae*. Mol Cell Biol 1987; 7:3629-3636.

142. Sass P, Field J, Nikawa J, Toda T, Wigler M. Cloning and characterization of the high-affinity cAMP phosphodiesterase of *S. cerevisiae*. Proc Natl Acad Sci USA 1986; 83:9393-9307.

143. Toda T, Cameron S, Sass P, Zoller M, Scott JD, McBullen B, Hurwitz M, Krebs EG, Wigler M. Cloning and characterization of *BCY1*, a locus encoding a regulatory subunit of the cyclic AMP-dependent protein kinase in *Saccharomyces cerevisiae*. Mol Cell Biol 1987; 7:1371-1377.

144. Toda T, S. C, Sass P, Zoller M, Wigler P. Three different genes in *Saccharomyces cerevisiae* encode the catalytic subunits of the cAMP-dependent protein kinase. Cell 1987; 50:277-287.

145. Pernambuco MB, Winderickx J, Crauwels M, Griffioen G, Mager WH, Thevelein JM. Differential requirement for sugar phosphorylation in cells of the yeast *Saccharomyces cerevisiae* grown on glucose or grown on non-fermentable carbon sources for glucose-triggered signalling phenomena. Microbiol 1996; 142: 1775-1782.

146. Holzer H. Catabolite inactivation in yeast. Trends Biochem Sci 1976; 1:178-181.

147. Gancedo JM, Gancedo C. Inactivation of gluconeogenic enzymes in glycolytic mutants of *Saccharomyces cerevisiae*. Eur J Biochem 1979; 101:455-460.

148. Ordiz I, Herrero P, Rodicio R, Moreno F. Glucose-induced inactivation of isocitrate lyase in *Saccharomyces cerevisiae* is mediated by an internal decapeptide sequence. FEBS Lett 1995; 367:219-222.

149. Riballo E, Herweijer M, Wolf DH, Lagunas R. Catabolite inactivation of the yeast maltose transporter occurs in the vacuole after internalization by endocytosis. J Bacteriol 1995; 177:5622-5627.

150. Lucero P, Herweijer M, Lagunas R. Catabolite inactivation of the yeast maltose transporter is due to proteolysis. FEBS Lett 1993; 333:165-168.

151. Gancedo C. Inactivation of fructose-1,6-bisphosphatase by glucose in yeast. J Bacteriol 1971; 107:401-405.

152. Holzer H. Mechanism and function of reversible phosphorylation of fructose-1,6-bisphosphatase in yeast. In: Cohen P, ed. Molecular Aspects of Cellular Regulation. Amsterdam: Elsevier, 1984:143-154.

153. Rose M, Entian KD, Hoffmann L, Vogel RF. Irreversible inactivation of *Saccharomyces cerevisiae* fructose-1,6-bisphosphatase independent of protein phosphorylation at Ser11. FEBS Lett 1988; 241:55-59.

154. Schork SM, Bee G, Thumm M, Wolf DH. Site of catabolite inactivation. Nature 1994; 369:283-284.

155. Schork SM, Bee G, Thumm M, Wolf DH. Catabolite inactivation of fructose-1,6-bisphosphatase in yeast is mediated by the proteasome. FEBS Lett 1994; 349:270-274.

156. Chiang H-L, Schekman R. Regulated import and degradation of a cytosolic protein in the yeast vacuole. Nature 1991; 350:313-318.

157. Chiang H-L, Schekman R. Site of catabolite inactivation. Nature 1994; 369:284.

158. Chiang H-L, Schekman R, Hamamoto S. Selective uptake of cytosolic, peroxisomal and plasma membrane proteins by the yeast vacuole. J Biol Chem 1996; 271:9934-9941.

159. Hoffman M, Chiang H-L. Isolation of degradation-deficient mutants defective in the targating of fructose-1,6-bisphosphatase into the vacuole for degradation in *Saccharomyces cerevisiae*. Genetics 1996; 143:1555-1566.

160. Lopez-Boado YS, Herrero P, Fernandez T, Fernandez R, Moreno F. Glucose-stimulated phosphorylation of yeast isocitrate lyase in vivo. J Gen Microbiol 1988; 134:2499-2505.

161. Entian KD, Barnett JA. Regulation of sugar utilization by *Saccharomyces cerevisiae*. Trends Biochem Sci 1992; 17:506-510.

162. Müller S, Boles E, May M, Zimmerman FK. Different internal metabolites trigger the induction of glycolytic gene expression in *Saccharomyces cerevisiae*. J Bacteriol 1995; 177:4517-4519.

163. Chambers A, Packham EA, Graham IR. Control of glycolytic gene expression in the budding yeast (*Saccharomyces cerevisiae*). Curr Genet 1995; 29:1-9.

164. Doorenbosch T, Mager WH, Planta RJ. Multifunctional DNA-binding proteins in yeast. Gene Expr 1992; 2:193-201.

165. Griffioen G, Mager WH, Planta RJ. Nutritional upshift response of ribosomal protein gene transcription in *Saccharomyces cerevisiae*. FEMS Microbiol Lett 1994; 123:137-144.

166. Griffioen G, Laan RJ, Mager WH, Planta RJ. Ribosomal protein gene transcription in *Saccharomyces cerevisiae* shows a biphasic response to nutritional changes. Microbiol 1996; 142:2279-2287.

167. Özcan S, Johnston M. Three different regulatory mechanisms enable yeast hexose transporter (*HXT*) genes to be induced by different levels of glucose. Mol Cell Biol 1995; 15:1564-1572.

168. Celenza JL, Marshall-Carlson L, Carlson M. The yeast *SNF3* gene encodes a glucose transporter homologous to the mammalian protein. Proc Natl Acad Sci 1988; 85:2130-2134.

169. Flick JS, Johnston M. *GRR1* of *Saccharomyces cerevisiae* is required for glucose repression and encodes a protein with leucine-rich repeats. Mol Cell Biol 1991; 11:5101-5112.

170. Erickson JR, Johnston M. Suppressors reveal two classes of glucose repression genes in the yeast *Saccharomyces cerevisiae*. Genetics 1994; 136:1271-1278.

171. Özcan S, Freidel K, Leuker A, Ciriacy M. Glucose uptake and catabolite repression in dominant *HTR1* mutants of *Saccharomyces cerevisiae*. J Bacteriol 1993; 175:5520-5528.

172. Vallier LG, Coons D, Bisson LF, Carlson M. Altered regulatory responses to glucose are associated with a glucose transport de-

fect in *grr1* mutants of *Saccharomyces cerevisiae*. Genetics 1994; 136:1279-1285.

173. Özcan S, Dover J, Rosenwald AG, Woelfl S, Johnston M. Two glucose transporters in *S. cerevisiae* are glucose sensors that generate a signal for induction of gene expresion. Proc Natl Acad Sci USA 1996; 93:12428-12432.

174. Hu Z, Nehlin JO, Ronne H, Michels CA. MIG1-dependent and MIG1-independent glucose regulation of *MAL* gene expression in *Saccharomyces cerevisiae*. Curr Genetics 1995; 28:258-266.

175. Needleman R. Control of maltase synthesis in yeast. Mol Microbiol 1991; 5:2079-2084.

176. Vanoni M, Sollitti P, Goldenthal M, Marmur J. Structure and regulation of the multigene family controlling maltose fermentation in budding yeast. Progress Nucl Acid Res Mol Biol 1989; 37:281-322.

177. Forsburg SL, Guarente L. Communication between mitochondria and the nucleus in regulation of cytochrome genes in the yeast *Saccharomyces cerevisiae*. Ann Rev Cell Biol 1989; 5:153-180.

178. McNabb DS, Xing YY, Guarente L. Cloning of yeast *HAP5*: A novel subunit of a heterotrimeric complex required for CCAAT binding. Genes Dev 1995; 9:47-58.

179. Forsburg SL, Guarente L. Identification and characterization of HAP4: a third component of the CCAAT-bound HAP2/HAP3 heteromer. Genes Dev 1989; 3:1166-1178.

180. Cooper TG. Nitrogen metabolism in Saccharomyces cerevisiae. In: Strathern JN, Jones EW, Broach J, eds. The Molecular Biology of the Yeast *Saccharomyces*: Metabolism and Gene Expression. Cold Spring Harbor, New York: Cold Spring Harbor Laboratory Press, 1982:39-99.

181. Magasanik B. Regulation of nitrogen utilization. In: Jones EW, Pringle JR, Broach J, eds. The Molecular and Cellular Biology of the Yeast *Saccharomyces*: Gene Expression. Cold Spring Harbor, New York: Cold Spring Harbor Laboratory Press, 1992: 283-317.

182. Coffman JA, Rai R, Cooper TG. Genetic evidence for Gln3p-independent, nitrogen catabolite repression-sensitive gene expression in *Saccharomyces cerevisiae*. J Bacteriol 1995; 177:6910-6918.

183. Minehart PL, Magasanik B. Sequence and expression of *GLN3*, a positive nitrogen regulatory gene of *Saccharomyces cerevisiae* encoding a protein with a putative zinc finger DNA-binding domain. Mol Cell Biol 1991; 11:6216-6228.

184. Wickner RB. [ure3] as an altered ure2 protein—evidence for a prion analog in *Saccharomyces cerevisiae*. Science 1994; 264:566-569.

185. Coschigano PW, Magasanik B. The *URE2* gene product of *Saccharomyces cerevisiae* plays an important role in in the cellular response to the nitrogen source and has homology to glutathione-S-transferases. Mol Cell Biol 1991; 11:822-832.

186. Kim KH, Rhee SG. Sequence of peptides from *Saccharomyces cerevisiae* glutamine synthetase. J Biol Chem 1988; 263:833-838.

187. Coffman JA, Elberry HM, Cooper TG. The Ure2 protein regulates nitrogen catabolic gene expression through the GATAA-containing UAS(NTR) element in *Saccharomyces cerevisiae*. J Bacteriol 1994; 176:7476-7483.

188. Jauniaux JC, Grenson M. Gap1, the general amino acid permease gene of *Saccharomyces cerevisiae*—Nucleotide sequence, protein similarity with the other bakers yeast amino acid permeases, and nitrogen catabolite repression. Eur J Biochem 1990; 190:39-44.

189. Miller SM, Magasanik B. Role of the complex upstream region of the *GDH2* gene in nitrogen regulation of the NAD-linked glutamate dehydrogenase in *Saccharomyces cerevisiae*. Mol Cell Biol 1991; 11:6229-6247.

190. Marczak JE, Brandriss MC. Analysis of constitutive and noninducible mutations of the PUT3 transcriptional activator. Mol Cell Biol 1991; 11:2609-2619.

191. Xu SW, Falvey DA, Brandriss MC. Roles of URE2 and GLN3 in the proline utilization pathway in *Saccharomyces cerevisiae*. Mol Cell Biol 1995; 15:2321-2330.

192. Stanbrough M, Magasanik B. Two transcription factors, Gln3p and Nil1p, use the same GATAAG sites to activate the expression of *GAP1* of *Saccharomyces cerevisiae*. J Bact 1996; 178:2465-2468.

193. Coffman JA, Rai R, Cunningham T, Svetlov V, Cooper TG. Gat1p, a GATA family protein whose production is sensitive to nitrogen catabolite repression, participates

in transcriptional activation of nitrogen-catabolic genes in *Saccharomyces cerevisiae*. Mol Cell Biol 1996; 16:847-858.

194. Hinnebusch AG. Transcriptional and translational regulation of gene expression in the general control of amino acid biosynthesis in *Saccharomyces cerevisiae*. Progr Nucl Acid Res Mol 1985; 38:195-240.

195. Minehart PL, Magasanik B. Sequence of the *GLN1* gene of *Saccharomyces cerevisiae*: role of the upstream region in regulation of glutamine synthetase expression. J Bacteriol 1992; 174:1828-1836.

196. Wek SA, Zhu S, Wek RC. The histidyl-tRNA synthetase-related sequence in the eIF-2a protein kinase Gcn2 interacts with tRNA and is required for activation in response to starvation for different amino acids. Mol Cell Biol 1995; 15:4497-4506.

197. Warner JR, Gorenstein C. Yeast has a true stringent response. Nature 1978; 275:338-339.

198. Moehle CM, Hinnebusch AG. Association of RAP1 binding sites with stringent control of ribosomal protein gene transcription in *Saccharomyces cerevisiae*. Mol Cell Biol 1991; 11:2723-2735.

199. Vignais ML, Woudt LP, Wassenaar GM, Mager WH, Sentenac A, Planta RJ. Specific binding of TUF factor to upstream activation sites of yeast ribosomal protein genes. Nucl Acids Res 1987; 1451-1457.

200. Kraakman LS, Mager WH, Maurer KTC, Nieuwint RTM, Planta RJ. The divergently transcribed genes encoding yeast ribosomal proteins L46 and S24 are activated by shared RPG-boxes. Nucl Acids Res 1991; 17:9693-9706.

201. Huet J, Sentenac A. TUF, the yeast DNA-binding factor specific for UAS$_{rpg}$ upstream activating sequences: Identification of the protein and its DNA-binding domain. Proc Natl Acad Sci USA 1987; 84:3648-3652.

202. Buchman AR, Kimmerley WJ, Rine J, Kornberg RD. Two DNA-binding factors recognizes specific sequences at silencers, upstream activating seqeunces, autonomously replicating seqeunces, and telomeres in *Saccharomyces cerevisiae*. Mol Cell Biol 1988; 8:210-225.

203. Klein C, Struhl K. Protein kinase A mediates growth-regulated expression of yeast ribosomal protein genes by modulating RAP1 transcriptional activity. Mol Cell Biol 1994; 14:1920-1928.

204. Kraakman LS, Griffioen G, Zerp S, Groeneveld P, Thevelein JM, Mager WH, Planta RJ. Growth-related expression of ribosomal protein genes in *Saccharomyces cerevisiae*. Mol Gen Genet 1993; 239:196-204.

205. Markwardt DD, Garrett JM, Eberhardy S, Heideman W. Activation of the Ras/cyclic AMP pathway in the yeast *Saccharomyces cerevisiae* does not prevent G(1) arrest in response to nitrogen starvation. J Bacteriol 1995; 177:6761-6765.

206. Dang V-D, Bohn C, Bolotin-Fukuhara M, Daignan-Fornier B. The CCAAT box-binding factor stimulates ammonium assimilation in *Saccharomyces cerevisiae*, defining a new cross-pathway regulation between nitrogen and carbon metabolism. J Bacteriol 1996; 178:1842-1849.

207. Hu YM, Cooper TG, Kohlhaw GB. The *Saccharomyces cerevisiae* Leu3 protein activates expression of *GDH1*, a key gene in nitrogen assimilation. Mol Cell Biol 1995; 15:52-57.

208. Pinkham JL, Olesen JT, Guarente LP. Sequence and nuclear localisation of the *S.cerevisiae* HAP2 protein, a transcriptional activator. Mol Cell Biol 1987; 7:578-585.

209. Hahn S, Pinkham J, Wei R, Miller R, Guarente L. The *HAP3* regulatory locus of *Saccharomyces cerevisiae* encodes divergent overlapping transcripts. Mol Cell Biol 1988; 8:655-663.

210. Olesen JT, Guarente L. The HAP2 subunit of yeast CCAAT transcriptional activator contains adjacent domains for subunit association and DNA recognition: model for the HAP2/3/4 complex. Genes Dev 1990; 4:1714-1729.

211. Gimeno CJ, Ljungdahl PO, Styles CA, Fink GR. Unipolar cell divisions in the yeast *S. cerevisiae* lead to filamentous growth: regulation by starvation and *RAS*. Cell 1992; 68:1077-1090.

212. Dickinson JR. 'Fusel' alcohols induce hyphal-like extensions and pseudohyphal formation in yeasts. Microbiol 1996; 142:1391-1397.

213. Blacketer MJ, Madaule P, Meyers AM. Mutational analysis of morphologic differ-

entiation in *Saccharomyces cerevisiae*. Genetics 1995; 140:1259-1275.

214. Lambrechts MG, Bauer FF, Marmur J, Pretorius IS. Muc1, a mucin-like protein that is regulated by Mss10, is critical for pseudohyphal differentiation in yeast. Proc Natl Acad Sci USA 1996; 93:8419-8424.

215. Oshima Y. Regulatory circuits for gene expression: the metabolism of galactose and phosphate. In: Strathern JN, Jones EW, Broach JR, eds. The Molecular Biology of the Yeast *Saccharomyces*; Metabolism and Gene Expression. Cold Spring Harbor, New York: Cold Spring Harbor Laboratory Press, 1982:159-180.

216. Vogel K, Hinnen A. The yeast phosphatase system. Mol Microbiol 1990; 4:2013-2017.

217. Vogel K, Hörz W, Hinnen A. The two positively acting regulatory proteins PHO2 and PHO4 physically interact with *PHO5* upstream activation regions. Mol Cell Biol 1989; 9:2050-2057.

218. Berben G, Legrain M, Gilliquet V, Hilger F. The yeast regulatory gene *PHO4* encodes a helix-loop-helix motif. Yeast 1990; 6:451-454.

219. Svaren J, Schmitz J, Horz W. The transactivation domain of Pho4 is required for nucleosome disruption at the *PHO5* promoter. EMBO J 1994; 13:4856-4862.

220. Sengstag C, Hinnen A. The sequence of the *Saccharomyces cerevisiae* gene *PHO2* codes for a regulatory protein with unusual amino acid composition. Nucl Acids Res 1988; 15: 233-246.

221. Bürglin TR. The yeast regulatory gene *PHO2* encodes a homeo box. Cell 1988; 53:339-340.

222. Shao D, Creasy CL, Bergmane LW. Interaction of *Saccharomyces cerevisiae* Pho2 with Pho4 increases the accessibility of the activation domain of Pho4. Mol Gen Genet 1996; 251:358-364.

223. Uesono Y, Tokai M, Tanaka K, Toh-e A. Negative regulators of the PHO system of *Saccharomyces cerevisiae*: characterization of *PHO80* and *PHO85*. Mol Gen Genet 1992; 231:426-432.

224. Hirst K, Fisher F, McAndrew PC, Goding CR. The transcription factor, the Cdk, its cyclin and their regulator: directing the transcriptional response to a nutritional signal. EMBO J 1994; 13:5410-5420.

225. Kaffman A, Herskowitz I, Tjian R, O'Shea EK. Phosphorylation of the transcription factor PHO4 by a cyclin-cdk complex, PHO80-PHO85. Science 1994; 263:1153-1156.

226. Toh-E A, Tanaka K, Uesono Y, Wickner R. *PHO85*, a negative regulator of the PHO system, is a homolog of the protein kinase gene, *CDC28*, of *Saccaromyces cerevisiae*. Mol Gen Genet 1988; 214:162-164.

227. Schneider KR, Smith RL, O'Shea EK. Phosphate-regulated inactivation of the kinase PHO80-PHO85 by the CDK inhibitor PHO81. Science 1994; 266:122-126.

228. Cross F. Transcriptional regulation by a cyclin-cdk. Trends Genet 1995; 11:209-211.

229. Creasy CL, Shao DL, Bergman LW. Negative transcriptional regulation of *PHO81* expression in *Saccharomyces cerevisiae*. Gene 1996; 168:23-29.

230. Yoshida K, Ogawa N, Oshima Y. Function of the *PHO* regulatory genes for repressible acid phosphatase synthesis in *S. cerevisiae*. Mol Gen Genet 1989; 217:40-46.

231. Ogas J, Andrews BJ, Herskowitz I. Transcription activation of CLN1, CLN2, and a putative new G1 cyclin (HCS26) by SWI4, a positive regulator of G1-specific transcription. Cell 1991; 66:1015-1026.

232. Espinoza FH, Ogas J, Herskowitz I, Morgan DO. Cell cycle control by a complex of the cyclin HCS26 (PCL1) and the kinase PHO85. Science 1994; 266:1388-1391.

233. Measday V, Moore L, Ogas J, Tyers M, Andrews B. The PCL2 (ORFD)-PHO85 cyclin-dependent kinase complex: a cell cycle regulator in yeast. Science 1994; 266:1391-1395.

234. Huang D, Farkas I, Roach PJ. Pho85p, a cyclin-dependent protein kinase, and the Snf1p protein kinase act antagonistically to control glycogen accumulation in *Saccharomyces cerevisiae*. Mol Cell Biol 1996; 16:4357-4365.

235. Timblin BK, Tatchell K, Bergman LW. Deletion of the gene encoding the cyclin-dependent protein kinase Pho85 alters glycogen metabolism in *Saccharomyces cerevisiae*. Genetics 1996; 143:57-66.

236. Arndt KT, Styles C, Fink GR. Multiple global regulators control *HIS4* transcription in yeast. Science 1987; 237:874-880.

237. Braus G, Mösch H-U, Vogel K, Hütter R. Interpathway regulation of the *TRP4* gene of yeast. EMBO J 1989; 8:939-945.

238. Jones EW, Fink GR. Regulation of amino acid and nucleotide biosynthesis in yeast. In: Strathern JN, Jones EW, Broach JR, eds. The Molecular Biology of the Yeast *Saccharomyces*; Metabolism and Gene Expression. Cold Spring Harbor, New York: Cold Spring Harbor Laboratory Press, 1982: 181-299.

239. Mountain HA, Byström HS, Tang Larsen J, Korch C. Four major transcriptional responses in the methionine/threonine biosynthetic pathway of *Saccharomyces cerevisiae*. Yeast 1991; 7:781-803.

240. Thomas D, Cherest H, Surdin-Kerjan Y. Elements involved in S-adenosyl methionine-mediated regulation of the *Saccharomyces cerevisiae MET25* gene. Mol Cell Biol 1989; 9:3292-3298.

241. Thomas D, Jaquemin I, Surdin-Kerjan Y. Met4, a leucine zipper protein, and centromere-binding factor 1 are both required for transcriptional activation of sulfur metabolism in *Saccharomyces cerevisiae*. Mol Cell Biol 1992; 12:1719-1727.

242. Kuras L, Thomas D. Functional analysis of Met4, a yeast transcriptional activator responsive to S-adenosylmethionine. Mol Cell Biol 1995; 15:208-216.

243. Thomas D, Kuras L, Barbey R, Cherest H, Blaiseau PL, Surdin-Kerjan Y. Met30p, a yeast transcriptional inhibitor that responds to S-adenosylmethionine, is an essential protein with WD40 repeats. Mol Cell Biol 1995; 15:6526-6534.

244. Williams FE, Trumbly RJ. Characterization of TUP1, a mediator of glucose repression in *Saccharomyces cerevisiae*. Mol Cell Biol 1990; 10:6500-6511.

245. Williams FE, Varanasi U, Trumbly RJ. The CYC8 and TUP1 proteins involved in glucose repression in *Saccharomyces cerevisiae* are associated in a protein complex. Mol Cell Biol 1991; 11:3307-3316.

246. Komachi K, Redd MJ, Johnson AD. The WD repeats of Tup1 interact with the homeo domain protein alpha 2. Genes Dev 1994; 8:2857-2867.

247. Sherwood PW, Tsang SVM, Osley MA. Characterization of *HIR1* and *HIR2*, two genes required for regulation of histone gene transcription in *Saccharomyces cerevisiae*. Mol Cell Biol 1993; 13:28-38.

248. Neer EJ, Schmidt CJ, Nambudripad R, Smith TF. The ancient regulatory protein family of WD-repeat proteins. Nature 1994; 371:297-300.

249. Kuras L, Cherest H, Surdin-Kerjan Y, Thomas D. A heteromeric complex containing the centromere binding factor 1 and two basic leucine zipper factors, Met4 and Met28, mediates the transcripton activation of yeast sulfur metabolism. EMBO J 1996; 15:2519-2529.

250. Thomas D, Cherest H, Surdin-Kerjan Y. Identification of the structural gene for glucose-6-phosphate dehydrogenase in yeast. Inactivation leads to a nutritional requirement for organic sulfur. EMBO J 1991; 10:547-553.

250a. Nogae I, Johnston M. Isolation and characterization of the *ZWF1* gene of *Saccharomyces cerevisiae*, encoding glucose-6-phosphate dehydrogenase. Gene 1990; 96: 161-169.

251. Diffley JFX. Global regulators of chromosome function in yeast. Anton van Leeuwenhoek Int J Gen Mol Microbiol 1992; 61:25-33.

252. Bram RJ, Kornberg RD. Isolation of an *S. cerevisiae* centromere DNA-binding protein, its human homolog and its possible role as a transcription factor. Mol Cell Biol 1987; 7:403-409.

253. Baker RE, Masison DC. Isolation of the gene encoding the *Saccharomyces cerevisiae* centromere-binding protein CP1. Mol Cell Biol 1990; 10:2458-2467.

254. Cai MJ, Davis RW. Yeast Centromere-Binding Protein CBF1, of the helix-loop-helix protein family, is required for chromosome stability and methionine prototrophy. Cell 1990; 61:437-446.

255. Mellor J, Jiang W, Funk M, Ratjen J, Barnes CA, Hinz T, Hegemann JH, Philippsen P. CPF1, a yeast protein which functions in centromeres and promoters. EMBO J 1990; 12:4017-4026.

256. Dorsman JC, Gozdzicka-Jozefiak A, van Heeswijk WC, de Winde JH, Grivell LA. Multifunctional DNA binding proteins in yeast: the factors GFI and GFII are identical to the ARS-binding factor ABF1 and the centromere-binding factor CPF1 respectively. Yeast 1991; 7:401-412.

257. De Winde JH, Grivell LA. Regulation of mitochondrial biogenesis in *Saccharomyces cerevisiae*: Intricate interplay between general and specific transcription factors in the promoter of the *QCR8* gene. Eur J Biochem 1995; 233:200-208.

258. De Winde JH, Grivell LA. Global regulation of mitochondrial biogenesis in yeast: ABF1 and CPF1 play opposite roles in regulating expression of the *QCR8* gene, encoding subunit VIII of the mitochondrial ubiquinol cytochrome *c* oxidoreductase. Mol Cell Biol 1992; 12:2872-2883.

259. De Winde JH, van Leeuwen HC, Grivell LA. The multifunctional regulatory proteins ABF1 and CPF1 are involved in the formation of a nuclease-hypersensitive region in the promoter of the *QCR8* gene. Yeast 1993; 9:847-857.

260. Kent NA, Tsang JSH, Crowther DJ, Mellor J. Chromatin structure modulation in *Saccharomyces cerevisiae* by Centromere and Promoter Factor 1. Mol Cell Biol 1994; 14:5229-5241.

261. McKenzie EA, Kent NA, Dowell SJ, Moreno F, Bird LE, Mellor J. The centromere and promoter factor-1, CPF1, of *Saccharomyces cerevisiae* modulates gene activity through a family of factors including SPT21, RPD1 (SIN3), RPD3 and CCR4. Mol Gen Genet 1993; 240:374-386.

262. Mellor J, Rathjen J, Jiang W, Dowell SJ. DNA binding of CPF1 is required for optimal centromere function but not for maintaining methionine prototrophy in yeast. Nucl Acids Res 1991; 19:2961-2969.

263. O'Connell KF, Baker RE. Possible cross-regulation of phosphate and sulphate metabolism in *Saccharomyces cerevisiae*. Genetics 1992; 132:63-73.

264. Dowell SJ, Tsang JSH, Mellor J. The centromere and promoter factor-1 of yeast contains a dimerisation domain located carboxy-terminal to the bHLH domain. Anal Biochem 1992; 20:4229-4236.

265. Fisher F, Goding CR. Single amino-acid substitutions alter helix-loop-helix protein specificity for bases flanking the core CANNTG motif. EMBO J 1992; 11:4103-4109.

266. Belazzi T, Wagner A, Wieser R, Schanz M, Adam G, Hartig A, Ruis H. Negative regulation of transcription of the *Saccharomyces cerevisiae* catalase-T (*CTT1*) gene by cAMP is mediated by a positive control element. EMBO J 1991; 10:585-592.

267. Boorstein WR, Craig EA. Transcriptional regulation of *SSA3*, an Hsp70 gene from *Saccharomyces cerevisiae*. Mol Cell Biol 1990; 10:3262-3267.

268. Plesset J, Ludwig J, Cox B, McLaughlin C. Effect of cell position on thermotolerance in *Saccharomyces cerevisiae*. J Bacteriol 1987; 169:779-784.

269. De Nobel JG, Klis FM, Priem J, Munnik T, van den Ende H. The glucanase-soluble mannoproteins limit cell wall porosity in *Saccharomyces cerevisiae*. Yeast 1990; 6:491-499.

270. Baroni MD, Monti P, Alberghina L. Repression of growth-regulated G1 cyclin expression by cyclic AMP in budding yeast. Nature 1994; 371:339-342.

271. Tokiwa G, Tyers M, Volpe T, Futcher B. Inhibition of G1 cyclin activity by the Ras-cAMP pathway in yeast. Nature 1994; 371:342-345.

272. Garret S, Broach J. Loss of Ras activity in *Saccharomyces cerevisiae* is suppressed by disruption of a new kinase gene, *YAK1*, whose product may act downstream of the cAMP-dependent protein kinase. Genes Dev 1989; 3:1336-1348.

273. Garrett S, Menold MM, Broach JR. The *Saccharomyces cerevisiae YAK1* gene encodes a protein kinase that is induced by arrest early in the cell cycle. Mol Cell Biol 1991; 11:4045-4052.

274. Ward MP, Garrett S. Suppression of a yeast cyclic AMP-dependent protein kinase defect by overexpression of *SOK1*, a yeast gene exhibiting sequence similarity to a developmentally regulated mouse gene. Mol Cell Biol 1994; 14:5619-5627.

275. Ward MP, Gimeno CJ, Fink GR, Garrett S. *SOK2* may regulate cyclic AMP-dependent pro-

tein kinase-stimulated growth and pseudohyphal development by repressing transcription. Mol Cell Biol 1995; 15:6854-6863.

276. Dumortier F, Argüelles JC, Thevelein JM. Constitutive glucose-induced activation of the Ras-cAMP pathway and aberrant stationary-phase entry on a glucose-containing medium in the *Saccharomyces cerevisiae* glucose-repression mutant *hex2*. Microbiology-Uk 1995; 141:1559-1566.

277. Cameron S, Levin L, Zoller M, Wigler M. cAMP-independent control of sporulation, glycogen metabolism and heat shock resistance in *S. cerevisiae*. Cell 1988; 53:555-566.

278. Kron SJ, Styles CA, Fink GR. Symmetric cell division in pseudohyphae of the yeast *Saccharomyces cerevisiae*. Mol Biol Cell 1994; 5:1003-1022.

279. Kron SJ, Gow NAR. Budding yeast morphogenesis: signalling cytoskeleton and cell cycle. Curr Op Cell Biol 1995; 7:845-855.

280. Wittenberg C, Reed SI. Plugging it in: signaling circuits and the yeast cell cycle. Curr Op Cell Biol 1996; 8:223-230.

281. Gow NAR. Growth and guidance of the fungal hypha. Microbiol 1994; 140:3193-3205.

282. Roberts RL, Fink GR. Elements of a single MAP kinase cascade in *Saccharomyces cerevisiae* mediate two developmental programs in the same cell type: Mating and invasive growth. Genes Dev 1994; 8:2974-2985.

283. Özcan S, Schulte F, Freidel K, Weber A, Ciriacy M. Glucose uptake and metabolism in *grr1/cat80* mutants of *Saccharomyces cerevisiae*. Eur J Biochem 1994; 224: 605-611.

284. Barral Y, Jentsch S, Mann C. G1 cyclin turnover and nutrient uptake are controlled by a common pathway in yeast. Genes Dev 1995; 9:399-409.

285. Mosch HU, Roberts RL, Fink GR. Ras2 signals vis the Cdc42/Ste20/mitogen-activated protein kinase module to induce filamentous growth in *Saccharomyces cerevisiae*. Proc Natl Acad Sci USA 1996; 93:5352-5356.

286. Ziman M, Preuss D, Mulholland J, O'Brien JM, Botstein D, Johnson DI. Subcellular localisation of Cdc42p, a *Saccharomyces cerevisiae* GTP-binding protein involved in the control of cel polarity. Mol Biol Cell 1993; 4:1307-1316.

287. Zhao ZS, Leung T, Manser E, Lim L. Pheromone signalling in *Saccharomyces cerevisiae* requires the small GTP-binding protein Cdc42p and its activator *CDC24*. Mol Cell Biol 1995; 15:5246-5257.

288. Simon MN, Devirgilio C, Souza B, Pringle JR, Abo A, Reed SI. Role for the Rho-family GTPase Cdc42 in yeast mating-pheromone signal pathway. Nature 1995; 376:702-705.

289. Liu HP, Styles CA, Fink GR. Elements of the yeast pheromone response pathway required for filamentous growth of diploids. Science 1993; 262:1741-1744.

290. Printen JA, Sprague GF. Protein-protein ineractions in the yeast pheromone response pathway: Ste5p interacts with all members of the MAP kinase cascade. Genetics 1994; 138:609-619.

291. Gimeno CJ, Fink GR. Induction of pseudohyphal growth by overexpression of *PHD1*, a *Saccharomyces cerevisiae* gene related to transcriptional regulators of fungal development. Mol Cell Biol 1994; 14:2100-2112.

292. Estruch F, Carlson M. Increased dosage of the *MSN1* gene restores invertase expression in yeast mutants defective in the Snf1 protein kinase. Nucl Acids Res 1990; 18:6959-6964.

293. Eide D, Guarente L. Increased dosage of a transcriptional activator gene enhances Iron-limited growth of *Saccharomyces cerevisiae*. J Gen Microbiol 1992; 138:347-354.

294. Blacketer MJ, Koehler CM, Coats SG, Meyers AM, Madaule P. Regulation of dimorphism in *Saccharomyces cerevisiae*: involvement of the novel protein kinase homolog Elm1p and protein phospatase 2A. Mol Cell Biol 1993; 13:5567-5581.

295. Blacketer MJ, Madaule P, Meyers AM. The *Saccharomyces cerevisiae* mutation *elm4-1* facilitates pseudohyphal differentiation and interacts with a deficiency in phosphoribosylpyrophosphate synthase activity to cause constitutive pseudohyphal growth. Mol Cell Biol 1994; 14:4671-4681.

296. Jones ME. Pyrimidine nucleotide biosynthesis in animals. Genes, enzymes and regulation of UMP biosynthesis. Ann Rev Biochem 1980; 49:253-279.

297. Wright RM, Repine T, Repine JE. Reversible pseudohyphal growth in haploid *Saccharomyces cerevisiae* is an aerobic process. Curr Genet 1993; 23:388-391.

298. Dickinson JR. Irreversible formation of pseudohyphae by haploid *Saccharomyces cerevisiae*. FEMS Microbiol Lett 1994; 119:99-104.

299. François J, Villanueva ME, Hers HG. The control of glycogen metabolism in yeast. 1. Interconversion in vivo of glycogen synthase and glycogen phosphorylase induced by glucose, a nitrogen source or uncouplers. Eur J Biochem 1988; 174:551-559.

300. Toda T, Cameron S, Sass P, Wigler M. *SCH9*, a gene of *Saccharomyces cerevisiae* that encodes a protein distinct from, but functionally and structurally related to cAMP-dependent protein kinase catalytic subunits. Genes Dev 1988; 2:517-527.

301. Andre B. An overview of membrane transport proteins in *Saccharomyces cerevisiae*. Yeast 1995; 11:1575-1611.

302. Hall MN. The TOR signalling pathway and growth control in yeast. Biochem Soc Trans 1996; 24:234-239.

======= CHAPTER 2 =======

STATIONARY PHASE IN THE YEAST *SACCHAROMYCES CEREVISIAE*

Edwina K. Fuge and Margaret Werner-Washburne

1. INTRODUCTION

Over millions of years of evolution there have been countless times in which yeast cells have experienced a slow or sudden depletion of nutrients from their environment. The mechanism by which microorganisms survive these repeated periods of starvation involves a dramatic decrease in metabolic activity, a cell cycle arrest and a variety of morphological and physiological changes that allow the cells to survive for long periods of time without added nutrients. This period in the life cycle is called stationary phase. Until recently, stationary phase was considered a relatively uninteresting time in the life of a cell. One reason may have been that man's traditional interest in yeast (i.e., brewing and baking) involves actively growing cultures. Another reason was that tools were not available to study these "quiet" cells and there was a lifetime of wonderful, dynamic things to study in yeast during exponential growth. With the development of techniques for studying life at the molecular level, the ability to examine the complexity and dynamics at this quiet but extremely critical time improved dramatically. Additionally, with the increasing awareness that by understanding yeast we can understand other eukaryotic cells, the study of stationary phase or G_0 became more compelling.

In natural environments the nutrient supply determines the state in which microorganisms exist. A plentiful nutrient supply permits microorganisms to grow and multiply rapidly, resulting in the fast consumption of available nutrients. As a consequence of their rapid growth, the resulting depletion of nutrients and their inability to propel themselves

Yeast Stress Responses, edited by Stefan Hohmann and Willem H. Mager.
© 1997 R.G. Landes Company.

from a nutrient-poor to a nutrient-rich environment, yeast like other microorganisms spend much of their lives in this quiescent state.[1] Microorganisms' ability to cope with repeated nutrient stresses by becoming quiescent and re-entering the mitotic cell cycle when nutrients become available provide a fertile area for research. This research also has significant agricultural, medical and environmental implications because the cells of most organisms are in some type of quiescent state.

Although our understanding of starvation-induced growth arrest and stationary phase in yeast has improved in the past several years, there are still many aspects of this important time that are not well understood or that have not been examined. For example, what is known about stationary-phase cells has been based on studies of yeast in the laboratory. In the laboratory the natural environment is mimicked by growing cells in batch culture until an essential nutrient is limiting. Morphological, physiological and biochemical characteristics are analyzed and frequently the assumption is made that these characteristics are shared by all stationary-phase cells. However, major differences may exist in the specific nutrient limitations that signal cells to enter stationary phase in the laboratory as compared with the natural environment. These differences in the environmental signals may result in subtle but important differences in the stationary-phase state attained by the cells. For this reason, when studies of stationary-phase cultures are carried out under conditions other than growth in YPD (yeast extract-peptone-glucose: complete medium), it is important to compare the responses of cells under the new conditions to those responses in YPD. In this way a connected body of knowledge about this important time in the yeast life cycle can be developed.

There are two relatively recent reviews on this topic discussing our understanding of stationary phase at the molecular level.[2,3] In this chapter it is our intention to give an overview of the study of stationary phase

in yeast, to identify some of the gaps in our knowledge and to discuss some of the techniques used to study stationary-phase cells.

2. STATIONARY PHASE, QUIESCENCE AND G_0

Strictly speaking, stationary phase describes a time in the growth of a microorganismal culture during which there is no increase or decrease in the cell number. Thus this term is used most correctly to describe a population phenomenon, i.e., a characteristic of the culture rather than an individual cell. Most scientists including ourselves also use this term more loosely to describe cells in a culture that are not dividing due to nutrient limitation: a stationary-phase cell.

The term quiescent describes a cell in a stationary-phase culture. In studies of mammalian cells quiescent cells are those cells that have stopped dividing but retain the ability to divide again (as opposed to senescent cells that cannot divide again).[4] Thus quiescence is an accurate term to describe the state of a nondividing cell in a stationary-phase yeast culture.[1]

G_0 describes a specific phase of the cell cycle in which cells are in an arrested state (see also chapter 1). In mammalian cells G_0 is induced by hormones and growth factors. In microorganisms G_0 is induced by nutrient limitation. It is not yet known whether G_0 in eukaryotic microorganisms is equivalent to G_0 in mammalian cells. One of the pieces of evidence that would relate G_0 in mammalian cells to yeast cells would be if genes induced or proteins expressed in G_0-arrested mammalian cells were also induced in G_0-arrested yeast cells. Growth-*arrest*-*s*pecific (gas) genes have been identified in G_0-arrested mammalian cells[4] most of which do not appear to have homologues in yeast. The exception is the gas gene statin.[5,6] Statin shares sequence identity with yeast elongation factor 1-α[7,8] that is encoded by the two genes *TEF1* and *TEF2*. It is not known whether either or both of these statin-related genes play a role during entry into stationary phase in yeast.[9] Nevertheless, the relationship between the G_0 states of mammalian and yeast cells will become clearer as more comparative studies are done and genes induced during quiescence are characterized.

A question that frequently arises about G_0 in yeast is whether cells actually arrest cell division in G_0 or whether they are just growing slowly and are therefore in a prolonged G_1 state. This question has not been completely resolved. The inability to detect cyclin mRNAs in stationary-phase cells suggests that at a population level the rate of cell division is extremely slow or occurs in only a small percent of the population.[10]

Genetic evidence that G_0 is separate from G_1 rests on two mutations, *gcs1*[11,12] and *grr1*,[13] that are associated with specific G_0 phenotypes. The phenotypes of these two mutants suggest that the genes play a role in reproliferation from stationary phase (see section 4.3). The absence of cyclin mRNA and genetic evidence provides the strongest support for there being a true G_0 nonproliferating state in yeast. As more attention is focused on the study of G_0

additional genes that are essential for entry into, survival during and exit from stationary phase are likely to be identified.

3. STATIONARY PHASE IS A RESPONSE TO NUTRIENT STARVATION

Nutrient starvation (chapter 1) is the fundamental condition that signals microorganisms to enter stationary phase. However, nutrient starvation may be perceived differently by cells depending on the strain of yeast that is studied, the culture conditions, the nutrient or nutrients that are limiting and the rate with which the nutrients become limiting. The following discussion emphasizes the potential importance of some of these factors and the potential problems of making broad generalizations from results obtained under specific conditions.

3.1. GETTING THERE

Yeast cultured in rich, glucose-based medium (YPD) exhibit several distinct phases of growth (Fig. 2.1). After a short

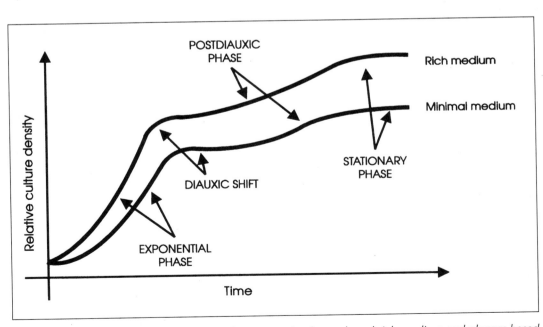

Fig. 2.1. Growth to stationary phase of wild-type yeast in glucose-based rich medium and glucose-based minimal medium. Rich medium (YPD): 1% yeast extract, 2% peptone, 2% glucose. Minimal medium: 0.67% yeast nitrogen base without amino acids, 2% glucose, 1% succinate, 6.7% NaOH and auxotrophic supplements.

period of adjustment to the medium (lag phase) cells divide exponentially (exponential phase) using energy derived primarily from fermentation. Following exhaustion of glucose (the diauxic shift) cells experience a transient cell-cycle arrest while they adjust to respiratory metabolism. Cell division resumes albeit at a reduced rate (the postdiauxic phase) after the diauxic shift, fueled by respiration of the end-products of fermentation and other available carbon sources. The postdiauxic phase may last as long as one week during which time the culture density usually doubles. When carbon sources are exhausted cells cease to divide (stationary phase).

The kinetics of entry into stationary phase are significantly affected by the medium in which the cells are grown. For example, cells cultured in minimal medium grow more slowly during exponential growth, exhibit a prolonged diauxic shift (10 hours in minimal medium compared with 1-2 hours in YPD) and enter stationary phase earlier and at a lower cell density than cells cultured in YPD (Fig. 2.1). It is not known whether this difference in the kinetics of entry into stationary phase affects the ultimate state of stationary-phase cells, i.e., whether cells grown to stationary phase in different media are physiologically different. However, some commonly used auxotrophic strains lose viability over time in certain minimal media (Fuge, unpublished results).

Some of the discrepancies in the literature concerning protein synthesis in stationary-phase cells can be attributed to the differences in culture conditions discussed above. For example, most two-dimensional-PAGE studies of protein synthesis after nutrient limitation have been done in minimal medium within a few hours of the limitation.[14,15] Some of the protein patterns observed in these studies appear to be due to the initial change in protein synthesis that occurs during glucose derepression after the diauxic shift.[14] We also have observed differences in the two-dimensional gel pattern of newly synthesized proteins from cells grown in YPD at the diauxic shift. However, in cultures grown 4 hours to 27 days past the diauxic shift the pattern of protein synthesis is remarkably similar to that observed in exponential-phase cells.[16]

A question that emerges from these studies is "when is a culture in stationary phase?" If the culture density doesn't change significantly in 5 hours, 12 hours or 24 hours, is that long enough? In minimal medium cells exhibit an arrest at the diauxic shift that can last 10 hours, after which growth resumes.[14] Certainly few people would wait more than 10 hours for cells to grow before deciding that the cells were in stationary phase. But if growth resumes in a culture to which no new nutrients have been added, can the culture truly be said to be in stationary phase? One of the more interesting growth curves we have seen involved cells grown in YEP 3% glycerol medium.[17] In this medium cells arrest growth after 2 days, remain in an arrested state for 2 days and then resume growth (4 days after inoculation).[17] This growth curve is especially interesting because there is no reason to expect a "diauxic shift" in cells growing in a nonfermentable carbon source because there is no need for them to switch from fermentative to respiratory metabolism. This suggests therefore that when cells exhaust one nutrient (possibly glycerol) in this medium they must adapt to utilizing another nutrient. Most of the studies of entry into stationary phase have been carried out in YPD and we are only beginning to compare characteristics of cells in different media. Therefore, there appears to be no absolute answer to the question of when cultures in different media are finally in stationary phase.

3.2. FAST VS. SLOW NUTRIENT LIMITATION

Starvation-induced arrest is achieved either by allowing cells to exhaust specific nutrients during the culture cycle or by transferring cells from nutrient-sufficient to nutrient-deficient medium. The initial responses of cells to slow vs. fast nutrient limitation are different and there are no published reports examining whether cells exposed to these distinct conditions are equally capable of entering stationary phase.

Differences in response to rapid vs. slow nutrient limitation at the level of gene expression are illustrated by two genes: *SNZ1* (Braun, unpublished results) and *YGP1*.[18] In cells grown to stationary phase in YPD, *SNZ1* mRNA accumulates 14 hours after glucose exhaustion at the diauxic shift, exhibits another accumulation 40 hours after glucose exhaustion (when carbon sources are limiting) and is maintained at a relatively high level in stationary phase. However, when exponentially growing cells are transferred to medium containing 0.05% glucose (glucose- and carbon-starvation conditions) *SNZ1* mRNA exhibits only a slight accumulation after 1 hour. *YGP1* is induced late during exponential growth when glucose is almost exhausted (<0.1% glucose) at the diauxic shift but is not induced by transferring exponentially growing cells to medium containing 0.05% glucose.[18]

These observations do not exclude the possibility that both rapid and slow nutrient limitation ultimately lead to the same stationary-phase arrest. When one considers the life cycle of yeast in nature it seems likely that cells can reach the stationary-phase state under both conditions. In fact exponential cultures or 7 day stationary-phase cultures can be transferred to water without measurable differences in their viabilities as long as 21 days after transfer (Weber, unpublished data). The differences observed in gene expression as a result of the kinetics of nutrient limitation may be due to different combinations of signals under the two conditions. For example, cells grown to stationary phase when exposed to slow nutrient limitation are likely to be able to maintain homeostasis more easily in the face of nutrient limitation than cells that have to respond very quickly to the absence of an essential nutrient. Nevertheless,, if both of these conditions can lead to the same stationary-phase state

then a careful comparison of entry into stationary phase under both conditions would be extremely valuable for identifying the essential components of this process.

3.3. LIMITATION FOR DIFFERENT NUTRIENTS

Limitation for nutrients other than glucose, including nitrogen, phosphorous and sulfur, can signal cells to arrest growth and develop some stationary-phase characteristics.[19-21] However, recent reports have suggested that only cells limited for glucose are able to survive long-term starvation.[22,23] This conclusion is based on studies of cells grown to stationary phase in YPD and transferred to water. When nitrogen supplements were added to these cultures viability was not affected, but when carbon was added the cells quickly lost viability. The authors hypothesized that the cells to which a carbon source was added died because they were then limited for other essential nutrients that could not act as signals for reentry into the stationary-phase state.

Other studies support the hypothesis that there are differences between carbon and nitrogen starvation as signals for stationary phase. For example, on more "native" media such as grape juice nitrogen is the initial limiting nutrient,[24] but it is not clear whether the cells in these conditions enter stationary phase as the result of the nitrogen limitation. With some yeast strains nitrogen limitation induces pseudohyphal growth instead of signaling cells to enter stationary phase (see chapter 1).[25] Pseudohyphal growth may be a potential mechanism for seeking richer food sources while carbon is still available. Ultimately pseudohyphae develop small, round cells perhaps when carbon is also limiting. Whether these small cells are in stationary phase is not known.

Metabolic differences between cells in these two starvation conditions have been demonstrated in studies of yeast grown in chemostat cultures.[26] These studies showed that cells limited for nitrogen exhibit much higher levels of respiration than cells limited for carbon, indicating that the cells were in distinct physiological states as a result of the different nutrient limitation. Morphological differences in both auxotrophic (strain W303-1A)[27] and prototrophic (strain S288C)[28] cells have also been observed in cultures limited for either carbon or nitrogen within 4 hours of the onset of starvation (Fig. 2.2) (see section 7.1). These differences persist at least 48 hours after the onset of starvation (Fuge, unpublished data).

3.4. CULTURE DENSITY

Little has been published on the effects of culture density on entry into stationary phase in yeast. Early work suggested that there were no soluble factors that signaled cells to enter stationary phase, i.e., there were no "stationary-phase pheromones" that would signal a density-dependent growth arrest.[21] In these experiments filtered medium from stationary-phase cultures was

added to exponentially growing cultures with no effect on the growth rate. It is still possible, however, that soluble stationary-phase signaling factors exist but that they are degraded by or are ineffective as signals for exponentially growing cells.

Culture density may have other, less immediate or apparent effects on stationary-phase survival. For example, stationary phase yeast cultures are dimorphic and composed of larger and smaller cells.[29,30] The smallest cells are the daughter cells from the last division, which occurs during the period of slow growth prior to entry into stationary phase. Normally cultures contain cells of different ages, with the oldest cells being the largest. It may be that over time as the oldest cells die and lyse they provide nutrients for cells in their proximity. This hypothesis suggests that viability of cells in dense cultures should remain higher for longer periods of time. In addition, flocculation,[31-34] which

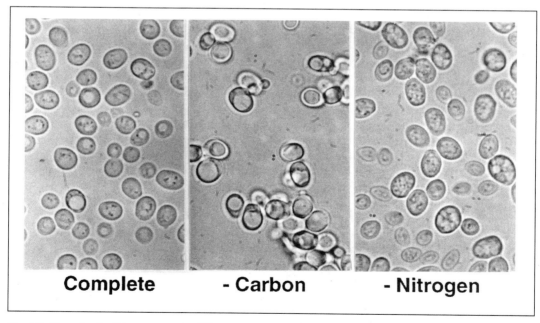

Fig. 2.2. Growth of wild-type yeast in different starvation media. Wild-type S288C cells were grown in YPD at 30°C. At OD$_{600}$ of 1 (approximately 2×10^7 cells/ml) cells were pelleted and washed at room temperature in distilled water. Cells were resuspended in either complete, carbon-minus or nitrogen-minus medium and incubated 4 hours at 30°C. Cells were visualized by light microscopy at 100X magnification. Complete medium: 0.67% yeast nitrogen base without amino acids, 2% glucose, 1% succinate, 6.7% NaOH and auxotrophic supplements. Carbon-minus medium was identical to complete medium except that glucose was excluded. Nitrogen-minus medium was identical to complete medium except that less yeast nitrogen base without amino acids (0.17%) was added.

occurs in many wild-type yeasts, may be another strategy for keeping older and younger cells in proximity as well as allowing faster colonization of an area once nutrients are available.

The effect of other microorganisms on stationary-phase survival in yeast has yet to be examined. In nature yeast is unlikely to be found in a pure culture. Thus the effect of other stationary-phase organisms on yeast and the effect of yeast on the growth of other organisms could be a potentially interesting and untapped area of research.

3.5. LIQUID VS. SOLID MEDIUM

Although cells appear to enter stationary phase both on solid and in liquid cultures, a comparison of stationary-phase yeast cells in colonies vs. cells grown in liquid cultures has not yet been published. Studies of *E. coli* colonial growth suggests that there may be communication between cells within colonies that allows colonies of single-cell organisms to respond differently to stimuli as a function of their position within the colony, much like cells in a multicellular organism.[35] In liquid medium stationary phase is determined when there is no further increase in colony-forming units per ml. Determining which cells in a colony are in stationary phase is a more difficult issue. Conventional wisdom holds that most cells in a colony on solid medium are in stationary phase, but this is difficult to prove in the absence of some measure of cell division within the colony. In liquid cultures all cells experience the same concentrations of nutrients, although oxygen concentrations may vary depending on the speed at which the culture is shaken. In a colony there are localized differences in nutrient availability as well as in pH and moisture. Colonies of cells carrying the *lacZ* gene fused to promoters that are induced during the post-diauxic phase accumulate blue pigment in the middle of the colony when grown on X-gal plates (M. Choder and M. Snyder, personal communications), suggesting that on the surface of the colony the cells im-

mediately adjacent to the solid medium are probably the last cells to enter stationary phase.

4. STATIONARY PHASE AS A DEVELOPMENTAL PROCESS

Stationary phase can be divided into three phases or stages (Fig. 2.3): entry into stationary phase, maintenance of viability in stationary phase and exit from stationary phase or reproliferation. Minimally, this process can be seen as requiring a cell-cycle arrest (G_0) and physiological changes that allow cells to survive for long periods of time without added nutrients and to respond quickly and appropriately when nutrients become available. The following sections provide an overview of the process of entry into stationary phase and the morphologic, biochemical and genetic alterations that accompany entry into the stationary-phase state.

4.1. ENTRY INTO STATIONARY PHASE

The process of entry into stationary phase may include a period of slow growth followed by a cell-cycle arrest. For cells growing in glucose-based rich medium or minimal medium this process can be viewed as including the diauxic shift (when glucose is exhausted and the cells undergo a transient arrest) and the postdiauxic, (respiratory or slow-growth) phase. For cells growing on carbon sources other than glucose, the process of entry into stationary phase has been less-well characterized. However, the growth curves of cultures grown on alternate carbon sources do include a period of slow growth prior to stationary phase. Since cells that have passed start when nutrients are limiting do not arrest until after the mitotic cycle is completed, it is likely that in most media this period of slow growth represents the final doubling of the culture.

Although nutrient limitation is the primary signal that causes cells to arrest growth and enter stationary phase, other secondary signals may be involved in the overall development of the stationary-phase cell. For example, in *E. coli* slow growth

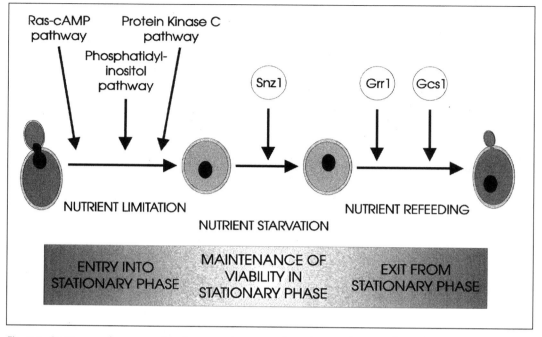

Fig. 2.3. Stationary phase as a developmental process. Entry into and transit through stationary phase is a continuum of events affected by the availability of nutrients. The points at which several signaling pathways and gene products discussed in the text are involved in this process are indicated.

has been suggested to activate growth-rate-sensitive or "gearbox" promoters.[36] Similarly, the induction of thermotolerance has been suggested to be a function of growth rate rather than simply a response to nutrient availability.[37] The fact that all stationary-phase cells are thermotolerant may be related to the slow-growth phase that all nutrient-limited cells experience prior to stationary-phase arrest. Because the experiments done to determine thermotolerance as a function of growth rate use different carbon sources to generate different growth rates, it is still possible that the induced thermotolerance is a function of the carbon source and not the growth rate.[38] Several signal transduction pathways are known or have recently been suggested to be involved in the cells' decision to enter stationary phase: the Ras-cAMP-protein kinase A pathway (chapter 1),[39-42] a protein kinase C mediated MAP (mitogen activated protein)-kinase cascade (chapter 4),[43] and a phosphatidylinositol kinase signaling pathway (Fig. 2.3).[44,45]

In yeast entering stationary phase protein kinase A activity decreases[39] in paral-

lel with the decrease in cAMP that occurs when nutrients are limiting.[46] Downregulation of protein kinase A is essential for survival during stationary phase and *bcy1* null mutants in which protein kinase A activity is not downregulated die at the diauxic shift (Werner-Washburne, unpublished results). This has been interpreted to mean that activation of the Ras-cAMP pathway prevents growth arrest in nutrient-limiting conditions. Recent studies suggest that loss of viability in *bcy* null mutants results from an inability to maintain adequate nutrient stores rather than an inability to arrest growth.[47] Analysis of other *bcy1* mutant alleles also suggests that downregulation of the protein kinase A is a progressive event and that different levels of kinase are required both as cells enter stationary phase and to maintain the stationary-phase state (V. Peck, personal communication).[48]

The protein kinase C (PKC) pathway is implicated in stationary phase through its interaction with Bro1 protein.[43] Strains lacking Bro1p die in the postdiauxic phase[43] at a stage after *bcy1* cells lose viability during growth to stationary phase. This suggests that Bro1p is required after the initial decrease in protein kinase A activity (based on the loss of viability of *bcy1* null strains.) The phenotype of *bro1* mutants further suggests that Bro1p plays a positive role in the progression of entry into stationary phase.

The phosphatidylinositol kinase pathway is implicated in stationary phase through identification of the two phosphatidylinositol kinase homologues Tor1p and Tor2p.[44] Phenotypic analysis of *tor1 tor2* temperature sensitive mutants suggests that the Tor proteins play a role in the entry of cells into stationary phase.[45] In these mutants even at the permissive temperature protein synthesis is about 10% of wild-type levels. At the nonpermissive temperature there is a severe reduction in the number of polysomes. Depletion of Tor2p experimentally achieved by using a *GAL1::TOR2* construct caused induction of *SSA3,UBI4* and other genes induced during the postdiauxic phase, i.e., during entry into stationary phase. The *TOR* function appears to be separate from the protein kinase A pathway and represents a novel, nutrient-sensing function that acts as a positive regulator of translation initiation and whose inactivation may be necessary for entry into stationary phase.

4.2. MAINTENANCE OF STATIONARY PHASE/QUIESCENCE

Previously, stationary-phase was considered simply a time when cells shut down and the processes critical for exponentially growing cells either did not occur or did so at negligible rates. Now, with the discovery of stationary phase sigma factors in *E. coli*[49] and G_0-specific (gas) genes in mammalian cells[4] there is a new appreciation of the positive changes that are likely to occur in all stationary-phase cells. Yeast genes known to be required for maintenance of stationary phase include *UBI4*, encoding polyubiquitin,[50] and a variety of genes involved in secretion, including *YPT1*.[51] Alleles of *BCY* encoding the regulatory subunit of protein kinase A, have been isolated that result in a loss of vi-

ability in stationary phase 5-8 days after the diauxic shift (Peck, personal communication). As mentioned earlier, the *BRO1* gene has recently been shown to play an important role in survival during the postdiauxic phase[43] and like the protein kinase A, may well play a role in maintenance of stationary phase.

Some published reports suggest that cells can remain viable in stationary phase for extremely long periods of time. Yeast cells have been cultured that have been in stationary phase (in the cold, in a bottle of beer at the bottom of the English Channel) for 175 years.[52] Stationary-phase bacterial cells have been reported to survive much longer.[53] We know very little about how cells arrested for so long maintain membrane integrity. What energy sources are used to maintain viability? Certainly trehalose, glycogen and lipids are potential energy sources and are known to accumulate in stationary phase cells.[21] But are other molecules such as amino acids used for energy? If so, how does the quiescent cell determine which proteins to degrade? What are the priorities for protein turnover and synthesis under these conditions?

We do know that within 4-7 days after culture inoculation the rates of protein synthesis decrease to 0.3% of that observed in exponentially growing cells. Interestingly, under the conditions used for protein isolation and separation few new, stationary-phase specific proteins were observed.[16] Nevertheless, most of the proteins that are synthesized in exponentially growing cells were also synthesized in stationary-phase cells and hence constitute housekeeping proteins. Within the constraints of this analysis (including proteins with isoelectric points from 4.5-7.5 and resolved by a 10% polyacrylamide gel) only one protein was observed that accumulated after cells were in stationary phase. This protein was named Snz1p and its function is unknown (Fig. 2.3). Phylogenetically Snz1p is one of the most highly conserved proteins yet seen (Braun, unpublished results).

For microorganisms the questions of maintenance of viability during stationary phase are perhaps some of the most interesting and the most difficult to answer. We know almost nothing about DNA repair during this time or whether there are "hot spots" for mutagenesis during stationary phase as might be implicated in *E. coli* by the frequency with which *rpo*[s] accumulates mutations in stationary-phase cells.[54] Clearer understanding of the levels of signaling that are possible in quiescent cells and the priorities that are established for long-term survival may give us insights not only into environmental issues concerning microorganisms but also into aging of cells in multicellular organisms.

4.3. REPROLIFERATION

The ability of cells to reenter the mitotic cell cycle is the hallmark of a successful stationary phase. In yeast only the reintroduction of nutrients is required for reproliferation. After reintroduction of nutrients cells in stationary phase exhibit a lag phase, which is believed to be required for the small daughter cells to attain the size required for cell division.[30]

A central question regarding stationary phase is whether specific genes might be required for reproliferation. In fact *GRR1* and *GCS1,* the only genes yet identified whose mutant phenotypes are stationary-phase specific, are required for reproliferation (Fig. 2.3). *GRR1* plays a role in glucose repression, divalent cation transport and cyclin turnover.[13,55,56] *Grr1* mutants are unable to reproliferate from stationary phase at 37°C.[13] *Gcs1* mutants are unable to reproliferate from stationary phase at cold temperatures (18°C).[11,12,20] *Gcs1* cells are viable when arrested as the result of nutrient limitation, become metabolically active when provided with fresh medium at 15°C but quickly lose viability and do not bud. The re-entry defect of *gcs1* strains is observed in cells limited for carbon and nitrogen as well as sulfur.[20] This suggests that despite the morphological differences observed in cells limited for carbon or nitrogen the Gcs1p-related arrest is common to cells under all these conditions.

Gcs1p has been identified as a Zn-finger protein[12] that is not a DNA-binding protein but appears to be related to GTPase activating proteins involved in secretion (G. Johnston, personal communication). The Zn-finger motif is essential for Gcs1p function.[12] As mentioned above, other proteins involved in secretion have been reported to be essential for stationary phase maintenance.[2,51] Some of these secretory mutants may actually be defective for reproliferation from stationary phase.

5. MORPHOLOGY AND PHYSIOLOGY OF STATIONARY-PHASE CELLS

Cells in stationary-phase cultures are morphologically distinct from cells in exponentially growing cultures.[57] By phase-contrast microscopy cells grown to stationary phase in YPD are unbudded and usually small, round and bright. Lipid vesicles and mitochondria, which become small and round in stationary-phase cells, are also more numerous.[57,58] The vacuole becomes large and prominent. Because stationary-phase cells are bright, organelles such as the vacuole and mitochondria, are not as readily seen as they are in exponentially growing cells. Visualization by electron microscopy of the vacuole of stationary-phase cells demonstrates that this organelle is filled with electron-dense material assumed to be polyphosphate.[57,59] The function of the polyphosphate is not known, but it may bind basic amino acids that also accumulate in the vacuole or be part of a physiological buffering system.[60]

With respect to the physiological characteristics of stationary-phase cells the ability to survive prolonged periods of time in the absence of added nutrients is of primary significance.[21-23] Using the common prototrophic lab strain S288C[28] we have observed that in cultures grown in YPD, colony forming units (CFU) per ml are sometimes reduced after exponential phase but then remain constant for at least 28 days.[16] Colony-forming units per ml of cultures of other lab strains (including those with auxotrophic markers) grown in YPD at 30°C also remain constant for extended periods of time. However, we have found that in defined minimal medium, prototrophic strains maintain viability but auxotrophic strains lose viability relatively quickly, suggesting that amino acid biosynthesis is crucial to maintenance of long-term viability.

Stationary-phase cells are thermotolerant,[61] are resistant to cell-wall degrading enzymes (primarily due to the increased thickness of their cell wall[57] but perhaps also to changes in the cell wall composition) and accumulate carbohydrates—specifically glycogen and trehalose.[21] These characteristics describe but do not define stationary-phase cells because cells in other phases of the life cycle may also exhibit these characteristics.[2] For example, thermotolerance and increased thickness of the cell wall may be induced as a result of slow growth rather than entry into stationary phase.[37,38]

A recent review suggests that the decreased metabolic rate is the most important characteristic of stationary-phase cells.[62] Of course stationary-phase cells have lowered metabolic rates, but cells incubated at low temperatures also have low metabolic rates. Yeast cells grown at low temperatures do not appear to be in stationary phase since *gcs1* mutants are able to reproliferate from exponentially growing cells but not from stationary-phase cells incubated at low temperatures.[11] Thus we believe that the best criteria for stationary phase remains the ability of cells to survive in the absence of nutrients.

6. MACROMOLECULAR FEATURES OF STATIONARY-PHASE CELLS

6.1. DNA

It is known that the genomic DNA persists as a single, unreplicated complement in stationary-phase cells,[63] but little is known about the chromatin of stationary-phase cells. Little work has been done since early studies showing that arrested cells have characteristically folded chromosomes.[64,65] It is not known whether this folded state occurs because of the synthe-

sis of new proteins, the modification of previously synthesized DNA-binding proteins or by some other mechanism. It is not known whether yeast have specific stationary-phase DNA-binding proteins as do *E. coli*.[66]

In *Escherichia coli* the novel DNA-binding protein Dps is synthesized at extremely high rates in stationary phase.[66] Dps synthesis depends on the starvation sigma factor δ^s and its binding to DNA renders the DNA resistant to DNase.[67] Resistance to DNase has not been thoroughly examined in stationary-phase yeast. And at least one study suggests that no new proteins are synthesized at levels comparable to that of Dps.[16] However, the failure to identify novel DNA-binding proteins in these experiments may have been due to the experimental methodology, which would have missed proteins that did not contain methionine or cysteine, small molecular weight proteins or proteins whose isoelectric points were outside of the range that was analyzed.

It seems reasonable that stationary-phase cells should have mechanisms to ensure the integrity of their DNA. Some recent experiments suggest small but reproducible increases in resistance to α- and γ-radiation-induced damage in stationary-phase yeast.[68] Other reports concerning UV-resistance suggest that either there is no difference between exponentially-growing and stationary-phase cells or that stationary-phase cells are actually less resistant to UV damage than are exponentially-growing cells.[69,70]

6.2. RNA

As in other stages of the yeast life-cycle regulation of gene expression is an integral and dynamic component of entry into stationary phase. Consistent with the observations related throughout this chapter, most studies pertaining to gene expression have been done during short-term limitation for specific nutrients. Less is known about the effects of long-term nutrient limitation, components of the transcription apparatus and the physical state of mRNA, such as polyadenylation and stability, on gene expression in stationary-phase cells.

In general, information about gene regulation in stationary-phase cells is derived from cells grown in YPD.[2,16,21,48,71] In YPD gene expression is dynamic as cells approach stationary phase.[72,73] After the diauxic shift most exponential mRNAs decrease in abundance a few mRNAs remain constant and a few increase in abundance.[72,73] In S288C cells at least one mRNA increases in abundance 40 hours after the diauxic shift (Braun, unpublished results) suggesting that yeast cells may be capable of responding transcriptionally to environmental signals even in stationary phase.

Regulation of transcription is an important control point during entry into and survival during stationary phase. The growth rate during the postdiauxic phase is known to be limited by *RPB4* (a subunit of RNA polymerase II). Cells lacking Rpb4p lose viability in stationary phase.[74-76] Transcription rates decrease about 3.7-fold and total mRNA decreases 2.7-fold on a per cell basis in cultures that have been in stationary phase for 1 day.[72] The decrease in both mRNA accumulation and transcription are dependent on topoisomerase I (*TOP1*). Because topoisomerase I is not thought to play a major role in transcription in log-phase cells,[77] it has been proposed that it acts indirectly to repress RNA polymerase II-mediated transcription during stationary phase.[72]

In yeast the interaction of subcomplexes with RNA polymerase II may influence survival during stationary phase analogous to the function of sigma factor σ^S in *E. coli*.[78] For example, the amount of Rpb4p that complexes with RNA polymerase II increases from 20% in exponential phase cultures to nearly 100% in stationary-phase cultures.[74,75] More recently another subunit of RNA polymerase II (Rpb7p) was observed to influence stationary-phase survival.[79] Rpb7p also forms a complex with Rpb4 which alters the initiation properties of RNA polymerase II.[74]

Because the changes in mRNA accumulation are dynamic as cells enter stationary phase few genes exist whose mRNA levels (as a function of total cellular RNA) are constant during this period and thus could be used as controls for RNA loading. When actin is used as a control, a mRNA that is present at detectable levels will appear to be highly induced in stationary phase simply because actin mRNA accumulation decreases so dramatically after the diauxic shift (Werner-Washburne, unpublished results).[72,80,81] Messenger RNAs encoded by *YAK1* and *RPB4* appear to be relatively constant as a function of total RNA when cells grow to stationary phase. *YAK1* encodes a kinase that suppresses the inviability of a strain lacking protein kinase A catalytic subunits.[82] The mRNA of the *BCY1* gene encoding the regulatory subunit of protein kinase A also remains relatively constant in some strains (Werner-Washburne, unpublished results). Frequently hybridization with one of the above genes and densitometric scans of the rRNA bands, which are also relatively constant, are used together as controls to determine whether the abundance of a particular mRNA changes as cells enter stationary phase.

Little is known about the state of rRNA and tRNA in stationary-phase cells. Yeast exhibit a stringent response (i.e., during amino-acid starvation rRNA synthesis decreases although tRNA is apparently unaffected).[83] An early report indicated that G_0-arrest is accompanied by a stringent response[15] and more recently it was reported that ribosomes decrease to less than 25% of normal (on a per cell basis) in stationary-phase cells.[84] However, one study reported little or no reduction in total RNA (rRNA and tRNA) from early stationary-phase cells grown in YPD.[72]

We have observed strain-dependent changes in ribosomal banding patterns in stationary-phase cells. An abundant stationary-phase-specific band that migrates between the 23S and 18S rRNA bands is observed when total RNA from W303-1A cells is separated on agarose gels (Padilla, unpublished observation). Similarly, a band having a higher molecular mass than the 23S rRNA appears when total RNA from stationary-phase S288C cells is separated on agarose gels (Fuge, unpublished observation). In *E. coli*[85,86] a protein associated with 100S stationary-phase ribosomes has been identified but no homologues are present in yeast (Werner-Washburne, unpublished observation). Although the origin of the RNAs we have observed is not known they are probably not RNA-protein conjugates and may reflect unprocessed pre-rRNAs.

6.3. PROTEINS

There are distinct changes in protein synthesis as cells grow to stationary phase in YPD. First, during glucose derepression at the diauxic shift there is a dramatic decrease in the rate of protein synthesis and in the pattern of proteins synthesized.[16] Second, during the postdiauxic phase the pattern of protein synthesis returns to that observed during exponential phase with the exception of several novel proteins whose synthesis is induced after glucose exhaustion at the diauxic shift.[16] Third, when cultures reach a point at which there is no further increase in cell number protein synthetic rates drop to about 0.3% of exponential phase rates, indicating a relatively dramatic decrease in metabolic activity as cells enter stationary phase.

As discussed earlier there are media-associated changes in protein synthesis patterns as cells approach stationary phase (see section 3.1). However, additional observations can be made from the analysis of stationary-phase proteins by two-dimensional PAGE. Few novel, abundant stationary-phase-specific proteins are synthesized in yeast. One of the most surprising observations in this study was that 1 month old stationary-phase cells exhibit the same protein synthesis patterns as exponentially growing cells.[16] Experiments using cycloheximide indicate that translation is essential for survival in stationary phase.[23] Thus the proteins synthesized at this time are

probably required for housekeeping functions. The fact that few novel proteins were identified in two-dimensional gels of stationary-phase proteins suggests that either few unique proteins are necessary for stationary-phase functions, they have isoelectric points outside of the range of the two-dimensional gel analysis or these stationary-phase-specific proteins do not contain methionine or cysteines or are synthesized very slowly and accumulate to very low levels. A comparison of the number of open reading frames and the number of protein spots visible on a two-dimensional gel indicates that only a small percentage of the proteins potentially synthesized by the cell are actually seen by two-dimensional gel analysis. This suggests that many proteins may in fact be below the level of detection by the current methods.

Comparative analysis of protein patterns from exponential-phase and stationary-phase cells also led to the conclusion that the most abundant proteins are not developmentally modified in a manner that changes their migration in isoelectric focusing or SDS-PAGE gels.[16] A second observation is relevant to the hypothesis that posttranslational modification may be much more significant in stationary-phase cells because the proteins could be activated or inactivated without the requirement of protein synthesis. As an example the regulatory subunit of protein kinase A Bcy1p has been shown by both one-dimensional SDS-PAGE and two-dimensional gel electrophoresis to be posttranslationally modified in stationary phase cells.[48]

There is no correlation between the abundance of specific mRNAs in stationary-phase cells and translation of the encoded proteins.[16] One example of this is found in the mRNA and protein encoded by the *HSP70*-related genes *SSA3* and *SSA1/2*. *SSA3* mRNA is relatively abundant in stationary-phase cells whereas the mRNA encoded by the *SSA1/2* genes is barely detectable by Northern analysis. Yet the rate of synthesis of the respective proteins is essentially the same. This suggests that translatability of specific mRNAs is differentially regulated in stationary-phase cells.

Degradation of stationary-phase proteins is an area of research that has not been rigorously investigated. In stationary-phase yeast there is a general increase in vacuolar proteases,[87,88] and two genes pertaining directly to protein degradation and that affect acquisition of stationary-phase characteristics have been identified. *UBI4*, encoding polyubiquitin, is required for survival in stationary phase.[50] *AAP1*, encoding an arginine-alanine peptidase, plays a role in glycogen accumulation.[89] Despite these observations it is unclear how increases in proteolytic capacity and rules for protein degradation (such as the N-end rule[90]) apply to stationary-phase proteins. In our studies of stationary-phase protein synthesis we observed that the amount of protein extracted per cell in stationary-phase cells was approximately 60% of the protein extracted from exponentially growing cells even though protein synthesis in stationary-phase cells was much less than 1% of that observed in exponentially growing cells.[16] This observation suggests that although the proteolytic capacity of the vacuole increases in stationary phase there is no dramatic or long-term increase in protein turnover in vivo in stationary-phase cells. Possibly the proteolytic capacity increases to insure that proteins are more quickly hydrolyzed into amino acid units for incorporation into new proteins or as a preparation for quick turnover of specific proteins when the cell begins to reproliferate.

Other studies demonstrate that proteolysis increases as a result of nutrient limitation.[91-93] Studies of nitrogen- and glucose-limited cells suggest that cells limited for nitrogen degrade proteins at the greatest rate (~2% per hour)[93] consistent with the high metabolic rates observed for nitrogen-limited cells in chemostat cultures. In glucose-limited cultures protein turnover was 0.5-1% per hour for 5 hours.[93] In phosphate-limited cultures protein degradation increased only marginally.[93] However, because these studies involved sudden nutrient limitation over a relatively short time period it should be emphasized that the results reflect proteolytic capacity rather than in vivo proteolysis.

Clearly protein turnover is important in stationary-phase cells because strains unable to synthesize polyubiquitin (*ubi4* strains) are unable to survive stationary phase and have about 50% loss of viability after 9 days in culture (Peck, personal communication). It has been argued that during exponential growth the lack of polyubiquitin is compensated for by ubiquitin donated or scavenged from proteins that are synthesized with ubiquitin termini. The necessity for a polyubiquitin in stationary phase may simply be that more ubiquitin is required in stationary phase or that different multiubiquitin isoforms may be involved in stationary-phase protein turnover.[94]

7. HOW TO STUDY STATIONARY PHASE YEAST CELLS

In this section we briefly describe a few experimental approaches to the study of stationary phase in yeast. Although clearly an interesting time in the life cycle of yeast, stationary phase has been difficult to study. Some of the difficulty is real and some of it is perceived.

7.1. CULTURE OF STATIONARY-PHASE CELLS

To obtain stationary-phase cells we presently inoculate a single colony into YPD and then grow cultures at 250 rpm in Erlenmeyer flasks at 30°C for 4-7 days, depending on the strain. Both auxotrophic and prototrophic strains enter stationary phase under these conditions and most auxotrophic strains exhibit very little loss of viability over a 1 month period of culture. Because contamination is a significant problem with long-term cultures, air shakers rather than water bath shakers are used for most of this work.

We have also recently begun to compare entry into stationary phase under different nutrient conditions. Examination of prototrophic cells (S288C)[28] grown overnight in YPD, washed and resuspended in defined, complete medium or medium lacking only a carbon source, only a nitrogen source or both carbon and nitrogen sources, revealed dramatic morphological differences under these conditions (Fig. 2.2). After 4 hours in complete medium S288C cells looked similar to cells grown in YPD (data not shown) and this persisted for the duration of the experiment (24 hours). After 4 hours in carbon-limited medium S288C cells developed large vacuoles surrounded by a thin band of cytoplasm. After the same time in nitrogen-limited medium, cells displayed a grainy or vesiculated appearance. Staining the cells with the dye Nile Red and visualization by epifluorescence microscopy indicated that the vesiculated appearance of nitrogen-limited cells was due to the presence of many lipid vesicles. Cells deprived of both carbon and nitrogen appeared morphologically similar to cells starved only for carbon (data not shown). The morphological differences observed under the different starvation conditions also persisted for 24 hours.

Two auxotrophic strains were also studied (data not shown). The S288C-derived auxotroph DS10 (*ura3-52 leu2-3-112 lys2 trp1Δ his3-11,15*) appeared healthy after 24 hours in complete medium (data not shown). In carbon-limiting and combined carbon- and nitrogen-limiting media DS10 cells developed large vacuoles surrounded by a thin band of cytoplasm, but in nitrogen-limiting medium they did not develop the grainy, vesiculated appearance observed in S288C wild-type cells. W303-1A[27] (*ura3-52 leu2-3-112 lys2 trp1Δ his3-11,15 ade2*), a different auxotrophic strain having the same amino acid auxotrophies as DS10 was also studied (data not shown). W303-1A cells die in complete medium lacking adenine which was absent in these experimental media. Within 1 hour of transfer to complete minimal medium W303-1A cells became very large, round and fragile. Like DS10 cells, W303-1A cells developed large vacuoles in carbon-

limited medium, however, W303-1A cells in nitrogen-limited medium were morphologically similar to S288C cells.

From these preliminary morphological studies it appears that rapid exposure to nitrogen-limiting conditions causes a distinct response from rapid exposure to carbon- or carbon- and nitrogen-limiting conditions. Additionally, cells limited for both carbon and nitrogen appear morphologically similar to cells limited only for carbon, suggesting that the carbon-limitation signal overrides nitrogen limitation. This is consistent with the observations described earlier that cells become able to withstand long periods of nutrient limitation only when carbon is the limiting nutrient.[22,23] Finally, the presence of auxotrophic mutations has a profound and often rapid effect on how cells respond to different nutrient limitations.

7.2. ASSESSMENT OF VIABILITY

Several methods are available for determining viability of stationary-phase cells. These include analysis of colony-forming units (CFUs) per ml, use of a vital dye such as phloxine B and use of a vital stain such as FUN1 (Molecular Bioprobes).[95-98] Each of these methods has advantages and disadvantages.

Assessment of colony-forming units is the traditional method to measure viability of stationary-phase cultures. Dilutions of the culture are made such that 100-300 colonies grow per plate. The advantage of this assay is the ability to quantitate the number of cells that can actually reproliferate. The disadvantages of the assay are that it requires at least 2 days to observe colonies and it does not allow a distinction to be made between mutants that die in stationary phase and those that are unable to reproliferate. The two mutants that are known to be defective for reproliferation are temperature sensitive, i.e., they can form colonies from stationary-phase cells at 30°C but not at the restrictive temperature (see section 4.3). Despite these disadvantages, measurement of CFUs remains an excellent way to identify mutants that lose viability in stationary phase or during reproliferation.

Phloxine B and erythrosine B are two vital dyes that are excluded from living cells.[95,99,100] These dyes are easily incorporated into solid medium allowing a visible assessment of the relative viability of a colony. The advantage of these dyes is that one can quickly determine which colonies contain dead cells and it is possible, by comparing colony colors over several days, to visually determine the order of death as mutants grow to stationary phase. This assay is not sensitive to mutants defective for reproliferation. The disadvantage of this assay is that it is not easily quantitatable, it is sensitive to local conditions on the plates and it is not sensitive to mutants that die after 9 days on solid medium (data not shown).

FUN1 (Molecular Bioprobes) is a vital stain that is taken up by living cells and accumulated in the vacuole where it can be seen by epifluorescence microscopy. The advantage of FUN1 staining is that it is a relatively quick measure of cell viability and allows quantitation of viable cells. In our hands percent viability determined by FUN1 is similar to results obtained by CFU analysis. Another advantage of this method is that the viability of different populations of cells is readily detectable. Because uptake of FUN1 is an active process cells must be incubated in glucose solution for a period of time. This is a disadvantage only if reproliferation mutants that do not respond to glucose are being studied since they will look the same as dead cells. In this specific case FUN1 analysis would indicate that the mutant cells were losing viability in stationary phase rather than during reproliferation.

A dye that we have not used but which can be used to identify viable cells in a culture is methylene blue. Use of this dye has been described elsewhere.[101]

7.3. TEMPORAL ORDERING OF STATIONARY-PHASE MUTANTS

If entry into stationary phase can be viewed as a developmental process then it

seems likely that mutants can be identified that arrest growth or lose viability at different stages of this process. To test the hypothesis several mutants identified in the literature as "stationary-phase" mutants were grown to stationary phase on solid and in liquid, rich, glucose-based medium. Viability was determined as discussed above using FUN1 (Molecular Bioprobes),[96] determining CFUs per ml and using phloxine B. The four strains carried different alleles: *bcy1-1* (a null allele of the gene encoding the regulatory subunit of protein kinase A),[40,41] *bcy1-65* (containing three point mutations in the 3' end of *BCY1*; Peck, unpublished data), *ubi4* (encoding the structural gene for polyubiquitin)[50] and *ard1* (encoding one subunit of N-acetyl transferase).[102-104] Our results showed that the mutants lost viability in the following order: *bcy1-1* at the diauxic shift, i.e., after 2 days; *ard1* at the onset of stationary phase, i.e., day 5; *bcy1-65* in stationary phase, i.e., day 6; *ubi4* in stationary phase, i.e., day 8. Having established a temporal ordering of viability for these mutants they can now be used for comparison with other stationary-phase mutants to develop a temporal map of the process of entry into and survival during stationary phase.

7.4. ANALYSIS OF STATIONARY-PHASE PROTEINS

We are presently using three methods to analyze stationary-phase proteins in yeast: one- and two-dimensional PAGE analysis, two-hybrid analysis and green-fluorescent protein tagging. Both labeled and unlabeled stationary-phase proteins have been analyzed by two-dimensional gel electrophoresis and one-dimensional SDS-PAGE and the results of this type of analysis have been discussed elsewhere in this chapter (see sections 3.1, 4.2 and 6.3). This approach is advantageous because global changes in protein synthesis and accumulation due to different nutritional regimes or stationary-phase mutants can be assessed. As an example, two-dimensional gel electrophoresis has been used recently to characterize the effect of high cAMP levels on protein synthesis at the diauxic shift.[105] It has also been a useful approach in establishing overall differences in protein synthesis between stationary-phase *E. coli* and stationary-phase yeast. Furthermore, in the absence of a genetic selection to identify genes involved in stationary-phase processes two-dimensional gel electrophoresis can be used comparatively to identify proteins (and their genes) affected by a particular regime. The proteins can be isolated from the gels, microsequenced and the genes subsequently identified. The major disadvantage of this approach is that only the most abundant proteins can be resolved. Only two newly-synthesized stationary-phase specific proteins were identified under the conditions we used.[16]

In the absence of readily detectable and specific phenotypes for identified stationary-phase genes, two-hybrid analysis[106,107] allows one to assess stationary-phase processes through protein-protein interactions. Further, these interactions do not necessarily have to be studied in stationary-phase cells, although the caveat that certain protein interactions occur because of protein modifications in stationary phase must be considered. Finally, the use of green fluorescent protein (GFP)-tagged proteins offers a new and promising approach to study the localization and ultimately provide insight into the function of stationary-phase proteins.[108,109]

8. SUMMARY

Yeast, like other microorganisms, enter stationary phase as a result of nutrient limitation and may survive in this state for long periods of time. One of the founding pieces of literature dealing with the biochemistry and physiology of stationary phase was written 16 years ago[21] and the cytological literature is even older.[57] Most researchers until relatively recently believed stationary phase was simply a time of decreased metabolic activity. Yet studies of human cells, bacteria and yeast indicate that this is a complex and well-coordinated time in the life of a cell. It is a paradoxical time when topoisomerases are involved

in repression of gene expression, when decreased protein kinase A activity is needed but more kinase is made and when mRNAs present in high concentrations are translated less efficiently than mRNAs present in low concentrations.

Methods now exist for analyzing the changes that occur in stationary-phase cells. To understand this time in the life of a yeast cell the experiments need to be well-designed and experimental controls have to take into consideration the role of specific auxotrophies and prototrophies in stationary-phase survival. To answer some of the very intriguing questions will require special creativity. For example, is there a level of altruism within a yeast colony such that the older cells prepare to sacrifice themselves for younger cells? Stationary phase is such a familiar and frequently overlooked phase of the life cycle. Who would have believed that this could be a new frontier?

NOTE IN PROOF:

Work pertaining to the *SNZ1* stationary-phase gene has been published recently. See: Braun EL, Fuge EK, Padilla AP, Werner-Washburne M. A stationary-phase gene in *Saccharomyces cerevisiae* is a member of a a novel, highly conserved gene family. J Bacteriol 1996; 178:6865-6872.

ACKNOWLEDGMENTS

We wish to thank the following members of our lab, V. Peck, E. Braun, P. Padilla, and J. Weber, who have generously allowed us to use unpublished data. We are also grateful to V. Peck for thoughtful discussions and M. Crawford and V. Peck for carefully reading the manuscript.

REFERENCES

1. Lewis DL, Gattie DK. The ecology of quiescent microbes. ASM News 1991; 57:27-32.
2. Werner-Washburne M, Braun E, Johnston GC, Singer RA. Stationary phase in the yeast *Saccharomyces cerevisiae*. Microbiological Reviews 1993; 57:383-401.
3. Werner-Washburne M, Braun EL, Crawford ME, Peck VM. Stationary phase in *Saccha-*

romyces cerevisiae. Molec Microbiol 1996; 19:1159-1166.

4. Schneider C, King RM, Philipson L. Genes specifically expressed at growth arrest of mammalian cells. Cell 1988; 54:787-793.

5. Wang E, Krueger JG. Statin, a nonprolif- eration-specific protein is associated with the nuclear envelope, and is heterogeneously distributed in cells leaving quiescent state. J Cell Physiol 1985; 140:418-426.

6. Sandig M, Bissonnette R, Liu CHL, Tomaszewski G, Wang E. Characterization of 57 kDa statin as a true marker for growth arrest in tissue by its disappearance from regenerating liver. J Cell Physiol 1994; 158:277-284.

7. Cottrelle P, Cool M, Thuriaux P et al. Ei- ther one of the two yeast EF-1 alpha genes is required for cell viability. Curr Genet 1985; 9:693-697.

8. Thiele D, Cottrelle P, Iborra F, Buhler JM, Sentenac A, Formageot P. Elongatin factor 1 alpha from Saccharomyces cerevisiae. Rapid large-scale purification and molecu- lar characterization. J Biol Chem 1985; 260:3084-3089.

9. Cavallius J, Zoll W, Chakraburtty K, Merrick WC. Characterization of yeast EF- 1 alpha: nonconservation of posttranslational modifications. Biochim Biophys Acta 1993; 1163(75-80).

10. Aparicio OM, Gottschling DE. Overcom- ing telomeric silencing: a *trans*-activator competes to establish gene expression in a cell cycle-dependent way. Genes & Devel- opment 1994; 8:1133-1146.

11. Drebot MA, Johnston GC, Singer RA. A yeast mutant conditionally defective only for reentry into the mitotic cell cycle from stationary phase. Proc Natl Acad Sci USA 1987; 84: 7948-7952.

12. Ireland LS, Johnston GC, Drebot MA et al. A member of a novel family of yeast 'Zn- finger' proteins mediates the transition from stationary phase to cell proliferation. EMBO 1994; 13:3812-3821.

13. Barral Y, Jentsch S, Mann C. G_1 cyclin turnover and nutrient uptake are controlled by a common pathway in yeast. Genes & Development 1995; 9:399-409.

14. Bataille N, Regnacq M, Boucherie H. In- duction of a heat-shock-type response in *S. cerevisiae* following glucose limitation. Yeast 1991; 7:367-378.

15. Iida H, Yahara I. Specific early-G_1 blocks accompanied with stringent response in *Saccharomyces cerevisiae* lead to growth arrest in resting state similar to the G_0 of higher cells. J Cell Biol 1984; 98:1185-1193.

16. Fuge EK, Braun EL, Werner-Washburne M. Protein synthesis in long-term stationary- phase cultures of *Saccharomyces cerevisiae.* J Bacteriol 1994; 176:5802-5813.

17. Bemis LT, Denis CL. Identification of func- tional regions in the yeast transcriptional activator *ADR1.* Mol Cell Biol 1988; 8:2125-2131.

18. Destruelle M, Holzer H, Klionsky DJ. Iden- tification and characterization of a novel yeast gene: the YGP1 gene product is a highly glycosylated secreted protein that is synthesized in response to nutrient limitation. Mol Cellular Biol 1994; 14:2740-2754.

19. Iida H, Yahara I. Durable synthesis of high molecular weight heat shock proteins in G_o cells of the yeast and other eucaryotes. J Cell Biol 1984; 99:199-207.

20. Drebot MA, Barnes CA, Singer RA, Johnston GC. Genetic assessment of station- ary phase for cells of the yeast *Saccharomyces cerevisiae.* J Bacteriol 1990; 172:3584-3589.

21. Lillie SH, Pringle JR. Reserve carbohydrate metabolism in *Saccharomyces cerevisiae:* re- sponses to nutrient limitation. J Bacteriol 1980; 143:1384-1394.

22. Granot D, Snyder M. Glucose induces cAMP-independent growth-related changes in stationary-phase cells of *Saccharomyces cerevisiae.* Proc Natl Acad Sci USA 1991; 88:5724-5728.

23. Granot D, Snyder M. Carbon source induces growth of stationary phase yeast cells, inde- pendent of carbon source metabolism. Yeast 1993; 9(5):465-479.

24. Bisson LF. Influence of nitrogen on yeast and fermentation of grapes. Proceedings of the International Symposium on Nitrogen in Grapes and Wine, June 18-19, 1991, Seattle, WA (American Society for Enology and Viticulture). 1991:78-89.

25. Gimeno CJ, Ljungdahl PO, Styles CA, Fink GR. Unipolar cell divisions in the yeast S cerevisiae lead to filamentous growth: regulation by starvation and *RAS*. Cell 1992; 68:1077-1090.

26. Larsson C, Stockar Uv, Marison I, Gustafssson L. Growth and metabolism of *Saccharomyces cerevisiae* in chemostat cultures under carbon-, nitrogen-, or carbon- and nitrogen-limiting conditions. J Bacteriol 1993; 175:4809-4816.

27. Thomas BJ, Rothstein R. Elevated recombination rates in transcriptionally active DNA. Cell 1989; 56:619-630.

28. Mortimer RK, Johnston JR. Genealogy of principal strains of the yeast genetic stock center. Genetics 1986; 113:35-43.

29. Johnston GC. Cell size and budding during starvation of the yeast *Saccharomyces cerevisiae*. J Bacteriol 1977; 132:738-739.

30. Johnston GC, Pringle JR, Hartwell LH. Coordination of growth with cell division in the yeast *Saccharomyces cerevisiae*. Exp Cell Res 1977; 105:79-98.

31. Mota M, Soares EV. Poplulation dynamics of flocculating yeasts. FEMS Microbiol Rev 1994; 14:45-52.

32. Soares EV, Mota M. Flocculation onset, growth phase, and genealogical age in *Saccharomyces cerevisiae*. Can J Microbiol 1996; 42:539-547.

33. Stratford M, Carter A-T. Yeast flocculation: Lectin synthesis and activation. Yeast 1993; 9:371-378.

34. Teunissen AWRH, Berg JAvd, Steensma HY. Transcriptional regulation of flocculation genes in *Saccharomyces cerevisiae*. Yeast 1995; 11:435-446.

35. Ben-Jacob E, Schochet O, Tenenbaum A, Cohen I, Czirók A, Vicsek T. Generic modelling of cooperative growth patterns in bacterial colonies. Nature 1994; 368:46-49.

36. Bohannon DE, Connell N, Keener J et al. Stationary-phase inducible "gearbox" promoters: differential effects of *katf* mutations and the role of s[70]. J Bacteriol 1991; 173:4482-4492.

37. Elliott B, Futcher B. Stress resistance of yeast cells is largely independent of cell cycle phase. Yeast 1993; 9:33-42.

38. Gross C, Watson K. Heat shock protein synthesis and trehalose accumulation are not required for induced thermotolerance in depressed *Saccharomyces cerevisiae*. BBRC 1996; 220:766-772.

39. Matsumoto K. Genetic analysis of the role of cAMP in yeast. Yeast 1985; 1:15-24.

40. Toda T, Cameron S, Sass P et al. Cloning and characterization of *BCY1*, a locus encoding a regulatory subunit of the cyclic AMP-dependent protein kinase in *Saccharomyces cerevisiae*. Mol Cell Biol 1987; 7:1371-1377.

41. Cannon JF, Tatchell K. Characterization of *Saccharomyces cerevisiae* genes encoding subunits of cyclic AMP-dependent protein kinase. Mol Cell Biol 1987; 7(8):2653-2663.

42. Cannon JF, Gitan R, Tatchell K. Yeast cAMP-dependent protein kinase regulatory subunit mutations display a variety of phenotypes. J Biol Chem 1990; 265:11897-11904.

43. Nickas ME, Yaffe MP. *BRO1*, a novel gene that interacts with the components of the Pkc1p-mitogen-activated protein kinase pathway in *Saccharomyces cerevisiae*. Mol Cell Biol 1996; 16:2585-2593.

44. Hall MN. The TOR signalling pathway and growth control in yeast. Biochem Soc Transactions 1996.

45. Barbet NC, Schneider U, Helliwell SB, Stansfield I, Tuite MF, Hall MN. TOR controls translation initiation and early G1 progression in yeast. Mol Biol of the Cell 1996; 7:25-42.

46. Francois J, Eraso P, Gancedo C. Changes in the concentration of cAMP, fructose 2,6-bisphosphate and related metabolites and enzymes in *Saccharomyces cerevisiae* during growth on glucose. Eur J Biochem 1987; 164:369-373.

47. Markwardt DD, Garrett JM, Eberhardy S, Heideman W. Activation of the Ras/cyclic AMP pathway in the yeast *Saccharomyces cerevisiae* does not prevent G1 arrest in response to nitrogen starvation. J Bacteriol 1995; 177:6761-6765.

48. Werner-Washburne M, Brown D, Braun E. Bcy1, the regulatory subunit of cAMP-dependent protein kinase in yeast, is differentially modified in response to the physi-

ological status of the cell. J Biol Chem 1991; 266(29):19704-19709.

49. Hengge-Aronis R, Klein W, Lange R, Rimmele M, Boos W. Trehalose synthesis genes are controlled by the putative sigma factor encoded by *rpoS* and are involved in stationary-phase thermotolerance in *Escherichia coli*. J Bacteriol 1991; 173:7918-7924.

50. Finley D, Özaynak E, Varshavsky A. The yeast polyubiquitin gene is essential for resistance to high temperatures, starvation, and other stresses. Cell 1987; 48:1035-1046.

51. Segev N, Botstein D. The ras-like yeast *YPT1* gene is itself essential for growth, sporulation, and starvation response. Mol Cell Biol 1987; 7:2367-2377.

52. Prokesch S. Small British brewers make a dent. *New York Times*, 1991 November 28:D-1.

53. Cano RJ, Borucki MK. Revival and identification of bacterial spores in 25- to 40-million-year-old Dominican amber. Science 1995; 268:1060-1064.

54. Zambrano MM, Siegele DA, Almirón M, Tormo A, Kolter R. Microbial competition: *Escherichia coli* mutants that take over stationary phase cultures. Science 1993; 259:1757-1760.

55. Flick JS, Johnston M. *GRR1* of *S. cerevisiae* is required for glucose repression and encodes a protein with Leucine-rich repeats. Mol Cell Biol 1991; 11(10):5101-5112.

56. Vallier LG, Carlson M. New SNF genes, *GAL11* and *GRR1* affect *SUC2* expression in *S. cerevisiae*. Genetics 1991; 129:675-684.

57. Matile P, Moor H, Robinow CF. Yeast cytology. In: Rose AH, Harrison JS, eds. The Yeasts. Biology of Yeasts. Vol 1. New York, NY: Academic Press, 1969:219-302.

58. Stevens B. Mitochondrial structure. In: Strathern J, Jones EW, Broach JR, eds. The Molecular Biology of the Yeast *Saccharomyces cerevisiae*: Life Cycle and Inheritance. Cold Spring Harbor, New York: Cold Spring Harbor Laboratory, 1981:471-504.

59. Cramer CL, Vaughn LE, Davis RH. Basic amino acids and inorganic polyphosphates in *Neurospora crassa*: Independent regulation of vacuolar pools. J Bacteriol 1980; 142:945-952.

60. Jennings DH. Some perspectives on nitro-

gen and phosphorus metabolism in fungi. In: Boddy L, Marchant R, Read DJ, eds. Nitrogen, Phosphorus and Sulphur Utilization by Fungi. Cambridge University Press, 1988:1-31.

61. Plesset J, Ludwig J, Cox B, McLaughin C. Effect of cell position on thermotolerance in *Saccharomyces cerevisiae*. J Bacteriol 1987; 169:779-784.

62. Hand SC, Hardewig I. Down regulation of cellular metabolism during environmental stress: mechanisms and implications. Ann Rev Physiol 1996; 58:539-563.

63. Pringle JR, Hartwell LH. The *Saccharomyces cerevisiae* cell cycle. In: Broach J, Strathern J, Jones E, eds. Molecular Biology of the Yeast *Saccharomyces*: Life Cycle and Inheritance. Cold Spring Harbor, New York: Cold Spring Harbor Laboratory, 1981:97-142.

64. Piñon R, Salts Y. Isolation of folded chromosomes from the yeast *Saccharomyces cerevisiae*. Proc Natl Acad Sci USA 1977; 74:2850-2854.

65. Piñon R. Folded chromosomes in noncycling yeast cells. Evidence for a characteristic G_0 form. Chromosoma 1978; 67:263-274.

66. Almirón M, Link D, Furlong D, Kolter R. A novel DNA binding protein with regulatory and protective roles in starved *E. coli*. Genes & Development 1992; 6:2646-2654.

67. Altuvia S, Almirón M, Huisman G, Kolter R, Storz G. The *dps* promoter is activated by OxyR during growth and by IHF and sˢ in stationary phase. Molec Microbiol 1994; 13:265-272.

68. Petin DV, Petin VG. Genetic control of RBE of a-particles for yeast cells irradiated in stationary and exponential phase of growth. Mutation Research 1995; 326:211-218.

69. Yoon H, Miller SP, Pabich EK, Donahue TF. *SSL1*, a suppressor of a *HIS4* 5'-UTR stem-loop mutation, is essential for translation initiation and affects UV resistance in yeast. Genes & Development 1992; 6:2463-2477.

70. Parry JM, Davies PJ, Evans WE. The effects of "cell age" upon the lethal effects of physical and chemical mutagens in the yeast *Saccharomyces cerevisiae*. Mol Gen Genet 1976; 146:27-35.

71. Werner-Washburne M, Stone DE, Craig EA. Complex interactions among members of an essential subfamily of *hsp70* genes in *Saccharomyces cerevisiae*. Mol Cell Biol 1987; 7(7):2568-2577.

72. Choder M. A general topoisomerase I-dependent transcriptional repression in the stationary phase of yeast. Genes & Development 1991; 5:2315-2326.

73. Werner-Washburne M, Becker J, Kosic-Smithers J, Craig EA. Yeast Hsp70 RNA levels vary in response to the physiological status of the cell. J Bacteriol 1989; 171(5):2680-2688.

74. Edwards AM, Kane CM, Young RA, Kornberg RD. Two dissociable subunits of yeast RNA polymerase II stimulate the initiation of transcription at a promoter in vitro. J Biol Chem 1991; 266:71-75.

75. Choder M, Young R. A portion of RNA polymerase II molecules has a component essential for stress responses and stress survival. Mol Cell Biol 1993; 13:6984-6991.

76. Choder M. A growth rate-limiting process in the last growth phase of the yeast life cycle involves RPB4, a subunit of RNA polymerase II. J Bacteriol 1993; 175:6358-6363.

77. Brill SJ, R. S. Transcription-dependent DNA supercoiling in yeast DNA topoisomerase mutants. Cell 1988; 54:403-411.

78. Hengge-Aronis R. Survival of hunger and stress: the role of rpoS in early stationary phase gene regulation in *E. coli*. Cell 1993; 72:165-168.

79. Khazak V, Sadhale PP, Woychik NA, Brent R, Golemis EA. Human RNA polymerase II subunit hsRPB7 functions in yeast and influences stress survival and cell morphology. Molec Biol of the Cell 1995; 6:759-775.

80. Papa FR, Hochstrasser M. The yeast *DOA4* gene encodes a deubiquitinating enzyme related to the human *tre-2* oncogene. Nature 1993; 366:313-319.

81. Praekelt UM, Meacock PA. *HSP12*, a new small heat shock gene of *Saccharomyces cerevisiae*: Analysis of structure, regulation and function. Mol Gen Genet 1990; 223:97-106.

82. Garrett S, Broach J. Loss of Ras activity in *Saccharomyces cerevisiae* is suppressed by disruptions of a new kinase gene, *YAK1*, whose product may act downstream of the cAMP-dependent protein kinase. Gen and Dev 1989; 3:1336-1348.

83. Warner JR, Gorenstein G. Yeast has a true stringent response. Nature 1978; 275:338.

84. Ju Q, Warner J. Ribosome synthesis during the growth cycle of *Saccharomyces cerevisiae*. Yeast 1994; 10:151-157.

85. Wada A, Yamazaki Y, Fujita N, Ishihama A. Structure and probable genetic location of a "ribosome modulation factor" associated with the 100S ribosomes in stationary phase *Escherichia coli* cells. Proc Natl Acad Sci 1990; 87:2657-2661.

86. Yamagishi et al. Regulation of the *Escherichia coli rmf* gene encoding the ribosome modulation factor: growth phase- and growth rate-dependent control. EMBO J 1993; 12:625-630.

87. Moehle CM, Aynardi MW, Kolodny MR, Park FJ, Jones EW. Protease B of *Saccharomyces cerevisiae*: isolation and regulation of the *PRB1* structural gene. Genetics 1987; 115:255-263.

88. Jones EW, Moehle C, Kolodny M et al. Genetics of vacuolar proteases. Yeast Cell Biol 1986:505-518.

89. Caprioglio DR, Padilla C, Werner-Washburne M. Isolation and characterization of *AAP1*: A gene encoding an alanine/arginine aminopeptidase in yeast. J Biol Chem 1993; 268:14310-14315.

90. Varshavsky A. The N-end rule. Cell 1992; 69:725-735.

91. Bakalkin GY, Kalnov SL, Zubatov AS, Luzikov VN. Degradation of total cell protein at different stages of *Saccharomyces cerevisiae* yeast growth. FEBS Lett 1976; 63:218.

92. Sumrada R, Cooper TG. Control of vacuole permeability and protein degradation by the cell cycle arrest signal in *Saccharomyces cerevisiae*. J Bacteriol 1978; 136:234-244.

93. López S, Gancedo JM. Effect of metabolic conditions on protein turnover in yeast. Biochem J 1979; 178:769-776.

94. Arnason T, Ellison MJ. Stress resistance in *Saccharomyces cerevisiae* is strongly correlated with assembly of a novel type of multi-

ubiquitin chain. Mol Cell Biol 1994; 14:7876-7883.

95. Bonneu M, Crouzet M, Urdaci M, Aigle M. Direct detection of yeast mutants with reduced viability on plates by erythrosine B staining. Anal Biochem 1991; 193:225-230.

96. Combs N, Hatzis C. Development of an epi-fluorescence assay for monitoring yeast viability and pretreatment hydrolysate toxicity in the presence of lignocellulosic solids. Applied Biochem & Biotech 1996; 57/58:649-657.

97. Crouzet M, Urdaci M, Dulau L, Aigle M. Yeast mutant affected for viability upon nutrient starvation: characterization and cloning of the *RVS161* gene. Yeast 1991; 7:727-743.

98. Sano A, Kurita N, Iabuki K et al. A comparative study of four different staining methods for estimation of live yeast from cells of *Paracoccidioides brasiliensis*. Mycopathologia 1993; 124:157-161.

99. Troshin AS. In: Widdas WF, ed. Problems of Cell Permeability. New York: Pergamon, 1966:181-217.

100. Gurr E. The Rational Use of Dyes. Leonard Hill, 1965.

101. Iida H, Yagawa Y, Anraku U. Essential role for induced Ca^{2+} influx followed by $[Ca^{2+}]i$ rise in maintaining viability for yeast cells late in the mating pheromone response pathway. A study of $[Ca^{2+}]$ in single *Saccharomyces cerevisiae* cells with imaging of fura-2. J Biol Chem 1990; 265:13391-13399.

102. Park E-C, Szostak JW. *ARD1* and *NAT1* proteins form a complex that has N-terminal acetyltransferase activity. EMBO J 1992; 11:2087-2093.

103. Mullen JR, Kayne PS, Moerschell RP et al. Identification and characterization of genes and mutants for an *N*-terminal acetyltransferase from yeast. EMBO J 1989; 8:2067-2075.

104. Whiteway M, Szostak JW. The *ARD1* gene of yeast functions in the switch between the mitotic cell cycle and alternative developmental pathways. Cell 1985; 43:483-492.

105. Boy-Marcotte E, Tadi D, Perrot M, Boucherie H, Jacquet M. High cAMP levels antagonize the reprogramming of gene expression that occurs at the diauxic shift in *Saccharomyces cerevisiae*. Microbiology 1996; 142:459-467.

106. Fields S, Song O. A novel genetic system to detect protein-protein interaction. Nature 1989; 340:245-246.

107. Phizicky EM, Fields S. Protein-protein interactions: methods for detection and analysis. Microbiological Reviews 1995; 59:94-123.

108. Prasher DC. Using GFP to see the light. Trends in Genetics 1995; 11:320-323.

109. Niedenthal RK, Riles L, Johnston M, Hegemann JH. Green fluorescent protein as a marker for gene expression and subcellular localization in budding yeast. Yeast 1996; 12:773-786.

CHAPTER 3

THE YEAST HEAT SHOCK RESPONSE

Peter Piper

1. INTRODUCTION

Disruption of a large number of cellular assemblies and processes, an increased protein unfolding and aggregation, and membrane structure alterations are paramount in cells exposed to supraoptimal temperatures. The heat shock response serves to counteract these events. Through it cells increase their thermotolerance or ability to withstand heat stress.[1-3]

Heat killing of a population of *S. cerevisiae* cells at extreme temperatures (48-52°C) is essentially a first order process (Fig. 3.1) provided that these cells are homogeneous with regard to their initial physiological state.[4] Cells that are in rapid exponential growth are most rapidly killed by heat, whereas nongrowing stationary cells are generally much more resistant ("thermotolerant"). Increases in thermotolerance are readily measured as an increase in the capacity to survive brief exposure at such temperature extremes. Induction of the heat shock response leads to rapid and dramatic induction of greater thermotolerance, but it is only one of several processes that can affect thermotolerance levels. The heat shock response also increases tolerances of certain other stresses (e.g., oxidative stress and osmostress (see ref. 3 for literature and chapters 4, 6 and 7). It therefore seems most probable that its major purpose is to increase cellular stress tolerances and especially tolerance of heat stress.

The most dramatic manifestation of triggering of the heat shock response is strong induction of heat shock proteins (Hsps) and—in yeast cells—accumulation of a large cytoplasmic pool of trehalose. Several physiological changes also occur at the same time. This chapter summarizes our current knowledge of the mechanisms whereby Hsps are synthesized in yeast through induced expression of the genes encoding these Hsps. It also briefly overviews the known functions of Hsps and discusses the less-well-documented physiological effects of heat stress. As will become apparent increases in thermotolerance in cells exposed to a heat shock

Yeast Stress Responses, edited by Stefan Hohmann and Willem H. Mager.
© 1997 R.G. Landes Company.

Fig. 3.1. The rates of killing of yeast cells by different temperatures. Triangles: 52°C, circles: 50°C and squares: 48°C. Reproduced with permission from Van Uden N, Adv Microb Physiol 1984; 25:195-251.

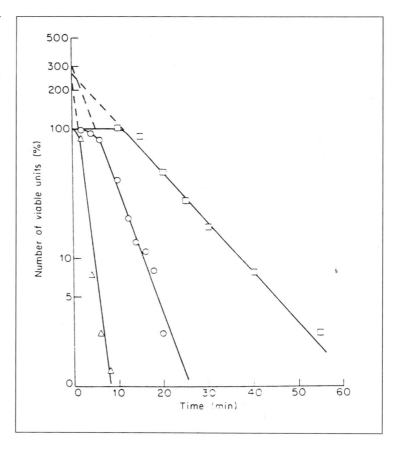

are not just due to induction of *HSP* gene expression but intimately involve several of these physiological changes caused by heat stress.

1.1. THE SENSOR OF THE HEAT SHOCK RESPONSE

Certain preconditions need to be met for *S. cerevisiae* cells to display the heat shock response. The response is essentially one of actively-growing cells. Stationary or quiescent yeast cells do not show a rapid response (as mentioned above they are already intrinsically thermotolerant).[5] Also the response is induced only over a comparatively narrow temperature range. In those laboratory *S. cerevisiae* strains that we have studied it is induced most strongly by temperature upshift to the maximum temperature at which the cells can grow (37-39.5°C) or upshift to 1-2°C above this maximum temperature of growth. The

lower the initial temperature of the yeast culture and thus the larger the temperature upshift to 37-41°C the more the cells seem to respond.[5] At temperatures above about 42-43°C *S. cerevisiae* seems incapable of appreciable protein synthesis, since this is the upper limit for pulse-labeling of proteins in this species. Remarkably though thermotolerance is induced most rapidly by brief exposure to temperatures as high as 45°C.[6] It would appear therefore that induction of thermotolerance at these temperatures must be by Hsp-independent mechanisms.

The heat shock response is triggered not just by heat shock but also by exposing cells to a wide variety of chemical agents. These chemicals include many that are disruptive to protein structure (e.g., alcohols and arsenite).[1] Other inducers (e.g., the antibiotic paromomycin) increase the frequency of protein synthesis errors.[7] This suggests that the response trigger might be a sudden increase in the intracellular levels of aberrant or misfolded protein. Protease inhibitors can also act as response inducers,[8] possibly because they inhibit removal of damaged proteins by intracellular degradation. In addition, the response can be induced by sphaeroplasting (removal of cell walls by β1-3 glucanase digestion of yeast cells in isotonic suspension) or by treatment with the RNA polymerase inhibitor thiolutin.[9] Ethanol is a potent inducer of the yeast heat shock response and the responses of yeast to ethanol and to heat shock are extremely similar.[10,11] However, it only induces a strong response when the ethanol addition to cultures exceeds a threshold of about 4% (v/v).[10] Ethanol is of course a natural product of yeast growth on fermentable sugars yet the extent to which cultures used in ethanol production are displaying a constitutive heat shock response seems not to have been studied in detail.

Even though it has been suspected for a number of years that one trigger for the heat shock response might be a sudden increase in the level of aberrant, misfolded or aggregated protein in the cell,[1,12,13] evidence is now emerging to suggest a very different sensory mechanism in which the cell membrane plays an important role. The proton motive force at the plasma membrane, or the fluidity of this membrane are both potentially very sensitive monitors of environmental change. This may make the plasma membrane the ideal location for stress-sensors. It is already known to be the location of the transmembrane osmostress-sensing histidine kinase Sln1p.[14] We noticed that reduced plasma membrane H+-ATPase activity lowers the ability of yeast to respond to heat shock.[15] In addition, treatments that depolarize the electrochemical potential gradient that yeast maintains at this membrane rapidly inhibit the heat shock response without abolishing the capacity of the cells for protein synthesis.[16] Such treatments include treatment with the plasma membrane H+-ATPase inhibitor diethylstilbestrol and exposure

(at low pH) to weak organic acids or uncouplers.[16] Furthermore, the groups of Vigh and Maresca have recently shown that the degree of fatty acid saturation in membranes strongly influences sensing of heat stress both in *Histoplasma capsulatum* and in *S. cerevisiae*. In *H. capsulatum* addition of palmitic acid (a 16:0 saturated fatty acid) to the growth medium strongly increased heat shock gene expression when cells were subsequently heat shocked to 37°C. On the other hand growth in the presence of oleic acid (18:1 unsaturated) reduced or totally eliminated heat shock gene induction by shift to 37°C and rendered cells intrinsically thermotolerant.[17] It was also shown that levels of *S. cerevisiae* heat shock gene transcription with stress depend on the degree of expression of the Δ^9-desaturase, an enzyme whose expression can be influenced by supplementation of the medium with unsaturated fatty acids (see ref. 17 for literature). Thus the ratio of saturated/unsaturated membrane lipid is involved in the perception of rapid temperature changes. This suggests that it is heat shock-induced disturbance of membrane physical state that leads to the signal that results in increased heat shock gene expression. It is one of the most exciting prospects for future work to unravel how this signal is propagated. The ability of alcohols to lower the threshold temperature for heat activation of the heat shock response[18] may reflect their tendency to partition into the lipid bilayer and so affect such signaling. Equally the ability of alcohols to trigger the heat shock response (which increases with their propensity to partition into the lipid bilayer[11,18]) and the tendency of weak lipophilic acids to inhibit the response in their uncharged protonated state[14] may reflect their effects on a sensory system responsive to membrane physical state.

2. EVENTS IN YEAST CELLS DISPLAYING THE HEAT SHOCK RESPONSE: AN OVERVIEW

Figure 3.2 summarizes some of the major events known to occur in *S. cerevisiae*

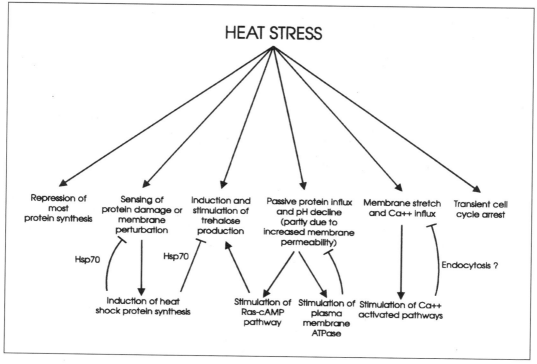

Fig. 3.2. Events in yeast cells during the heat shock response.

cultures initially growing below 30°C that are suddenly heat shocked to 36-41°C. The mechanisms of heat shock gene induction leading to stimulated Hsp production are discussed later. Here I will briefly overview other notable events.

2.1. TREHALOSE ACCUMULATION

Heat shock causes the extremely rapid accumulation of a large cytoplasmic pool of trehalose, a disaccharide composed of two α-α-linked glucose units. Trehalose is one of the most effective substances known for in vitro preservation of membranous structures and enzyme activities during desiccation, freezing or heating.[19,20] It accumulates in yeast under basically all conditions that lower growth rate and only in cells in rapid growth on glucose are trehalose levels very low. It is thought that its major role may be to act as a protectant against stresses such as desiccation, freezing and high temperatures rather than as a storage carbohydrate.[21] Consistent with this there is a good correlation between trehalose levels and thermotolerance in sta-

tionary phase and nonfermentative yeast cultures.[22] However, there is no such correlation between the trehalose and thermotolerance of fermentative yeast.[22] Evidence recently obtained with the *tps1Δ*, *tps1Δ hxk2Δ* and *nth1Δ* mutants indicates that the trehalose accumulated with heat shock of yeast probably contributes to the simultaneous rapid increase in thermotolerance, but that is by no means the only factor operating to establish of the normal increase in thermotolerance levels with mild heat shock.[22-24]

The very rapid induction of trehalose synthesis with heat shock of yeast reflects primarily alteration of activity of the enzymes of trehalose synthesis (trehalose synthase) and breakdown (trehalase) respectively.[3,24-27] There is some confusion in the literature as to whether it is due primarily to increases in intracellular pools of the precursors for trehalose synthesis (UDP-glucose and glucose-6-phosphate)[25] or is a reflection of activation of preexisting trehalose synthase and inactivation of trehalase at higher temperatures.[26,27] However, all these events may be contributing to the induction of trehalose together with fructose-6-phosphate (freely interconvertible with glucose-6-phosphate) acting as a potent activator of trehalose synthase.[27] Whatever the mechanism of this heat-induction of trehalose it is readily reversible, since the trehalose accumulated with heat shock is mobilized very rapidly with a subsequent temperature downshift.[3,21-26] This mobilization is considerably slower in the *nth1* mutant lacking neutral trehalase.[22,23] Conditions of extended stress are likely to result in altered levels of trehalose synthase, since all the genes for trehalose synthase subunits (*GGS1/TPS1*, *TPS2*, *TPS3* and *TSL1*) exhibit stress regulation, being induced by nutrient starvation, heat shock and osmotic stress of nonfermentative cultures.[28] However, heat shock of glucose-grown cultures is reported to cause only transient induction of these same genes.[28]

There are preliminary indications that the induction of Hsps and trehalose synthesis in the heat shock response may somehow be linked. Mobilization of trehalose with temperature shift-down is defective in strains with low Hsp70p levels, indicating that the induction of Hsp70p in heat shocked cells downregulates the simultaneous trehalose induction in these cells.[29] Even more tantalizing is the report that the *tps1Δ* mutant (defective in trehalose-6-phosphate synthase) does not show normal levels of Hsp synthesis with heat shock.[30] Unfortunately, the pleiotropic nature of the *tps1Δ* phenotype (see ref. 22 for literature) highlights the need for further work before this result can be satisfactorily explained.

2.2. CHANGES TO PLASMA MEMBRANE LIPIDS, PROTEINS AND ENZYMIC ACTIVITIES

Heat shock has both *immediate* and *delayed* effects on the cell membrane. It can be expected to produce immediate increases in membrane fluidity and in the permeability of membranes to protons and ions. An important consequence of increasing the proton permeability of the plasma membrane is the dissipation of the proton motive force maintained at this membrane mainly though the action of plasma membrane H^+-ATPase. This in turn leads to a reduction in intracellular pH and stimulation of the H^+-ATPase. The increase in *catalyzed* proton extrusion due to this stimulated H^+-ATPase activity may partly counteract the *passive* proton influx, thereby helping to restore homeostasis (as discussed in more detail in section 5).

More delayed effects of heat stress on the plasma membrane are: (i) alterations to protein composition, notably a considerable reduction in levels of plasma membrane H^+-ATPase and the appearance of the heat shock protein Hsp30p;[31] (ii) alterations to the saturation of membrane lipids which probably serve to reduce membrane fluidity (reviewed in ref. 32); and (iii) a stretch of this membrane during growth at high temperature that opens Ca^{2+} channels, this causing the activation of a number of Ca^{2+}-stimulated enzymes (also discussed later in section 5).

2.3. TRANSIENT CELL CYCLE ARREST

Vegetative yeast cultures that are up-shifted to stressful temperatures that still permit growth display a biphasic reduction in their growth rate over several hours.[4] Over a shorter time interval they are observed to transiently accumulate as unbudded cells, due to temporary arrest in the G1 phase of the cell cycle (at Start).[33] Resumption of growth then occurs provided that the temperature is not more than about 38-39°C. The normal transition over the Start step of the cell cycle requires activation of the p34[CDC28] protein kinase (see also chapter 1). This activation occurs when p34[CDC28] complexes with unstable proteins termed the G1 cyclins (products of a family of functionally-redundant genes; *CLN1*, *CLN2* and *CLN3* (*WHI1*)).[34-36] Heat shock decreases the abundance of the *CLN1* and *CLN2* (but not *CLN3*) transcripts.[35,36] On the other hand the continuous expression of cyclin from a heterologous promoter or the synthesis of more stable cyclins both eliminate the arrest at Start when cells are heat shocked without affecting other aspects of the heat shock response.[36] One reason for the transient arrest of yeast growth with mild heat shock therefore appears to be the temporary lack of cyclins causing a failure of p34[CDC28] protein kinase activation.[36] However, this may not be the complete explanation; heat shock-induced arrest at Start is prevented by the constitutive high cAMP-independent protein kinase A activity of *bcy1* point mutants even though *CLN1* and *CLN2* transcripts still decrease with heat shock of *bcy1* strains.[36]

3. GENE EXPRESSION CHANGES WITH HEAT SHOCK

3.1. THE HEAT SHOCK RESPONSE STIMULATES PRODUCTION OF HSPs BUT REDUCES SYNTHESIS OF MOST OTHER PROTEINS

Mild heat shock can have dramatically different effects on the expression of individual genes. For the majority of yeast genes it causes expression to be shut-off or lowered. Only for a minority (mostly heat shock genes) is expression strongly enhanced. In one study that analyzed the proteins labeled in unstressed and heat shocked *S. cerevisiae* on two-dimensional protein gels, synthesis of the vast majority of proteins was found to be reduced by mild (36°C) heat shock while more than 80 out of 500 proteins were transiently induced.[37] Both transcriptional and translational mechanisms are thought to operate to ensure that heat-induced proteins bypass the more general inhibition of synthesis of most proteins in heat shocked cells, but these mechanisms are still poorly understood.

Heat shocked cells do not generally maintain high Hsp synthesis for more than 20-40 minutes after a temperature upshift to 37-41°C. Instead, in cells heat shocked and then maintained at these temperatures Hsp synthesis becomes steadily reduced to a new steady-state level that is still considerably higher than the low levels of Hsp synthesis before stress was applied. Cells with very low Hsp70p levels display a constitutive heat shock response even at low

temperatures.[38] This indicates that heat elevation of Hsp70p (with induction of the *SSA1, SSA3* and *SSA4* genes) may contribute to downregulation of the response in normal cells that are shifted to 37-41°C and then maintained at these temperatures.

3.2. TWO DISTINCT PROMOTER ELEMENTS CAN DIRECT GENE INDUCTION BY HEAT SHOCK

Two quite distinct promoter sequences can direct the activation of a *S. cerevisiae* gene with heat shock. These are the Heat Shock Element (HSE) and the general STress Responsive Element (STRE). HSEs are the binding sites for heat shock factor (HSF). HSF is a transcriptional trans-activator that directs gene induction in response to heat stress and possibly those chemical agents mentioned above that cause membrane purturbation or intracellular accumulation of abnormal proteins. It has been proposed that increased association of chaperones (notably Hsp70p) with partially thermally-denatured proteins causes depletion of the "free" chaperone pool within cells and that this loss of free chaperone may be the actual signal for activation of HSF.[12,13,31] Strong heat-inducibility of HSE sequences is only seen with actively proliferating *S. cerevisiae* cells, since it is largely lost in cells that have entered the G_0 state (stationary phase).[5]

STREs—in contrast to HSEs—are activated not just by heat shock but also by a wide range of other stress conditions including osmostress, oxidative stress, nitrogen starvation, exposure to weak organic acids, low external pH or ethanol[13,39-42] (see also chapter 7). In addition, STREs differ from HSEs in that their heat induction is very strongly influenced by protein kinase A activity. These differences, together with the probable dissimilar functions of HSE-mediated and STRE-mediated gene inductions, are discussed in more detail in sections 3.3 and 3.4.

A single yeast heat shock gene can apparently be either under HSE control, under STRE control, or controlled by both the HSE and STRE. Studies on several different yeast heat shock gene promoters have revealed the heat induction of a few (e.g., *SSA4*, encoding a heat-inducible form of Hsp70p[43]) is regulated by the HSE alone. Still others are regulated by the STRE alone and sometimes multiple synergistically-acting STREs as in *HSP12*.[44] Genes of yet a third category (e.g., *SSA3*[45]) seem to be regulated by both HSE and STRE sequences. In these latter promoters HSEs and STREs may operate independently, although *SSA3* contains a STRE-like element that seems to be dependent on a downstream HSE for its activity.[45] Whether a heat shock gene is under HSE or STRE control in turn influences its stress-induction pattern. Thus STRE-regulated genes (e.g., *SSA3, HSP12*) are induced by several stresses and low protein kinase A activity whereas the HSE-regulated *SSA4* is induced mainly by heat shock alone. The situation is further complicated by recent findings that not all STRE-like promoter sequences direct identical expression patterns. Thus only certain STRE-like sequences seem to be heat-inducible while others such as the STRE-like element in *SSA3* are not.[39,45] Also not all of the STREs controlling genes for different subunits of the trehalose synthase complex direct the same patterns of regulation.[28]

In addition to having heat-inducible elements some yeast heat shock genes also have *cis*-acting promoter elements subject to other controls. For example, the promoter of the *UBI4* gene (encoding polyubiquitin) has a functional -542 element for binding of the yeast CCAATT box-binding factor (HAP2/3/4), an element that causes this gene to be expressed at higher levels in respiratory as compared to fermentative cultures.[46] *CYC7* and *HSP104* are two other heat shock genes that are also induced by respiratory growth[47,48] and a Hap1p control of the former gene has been demonstrated.[47]

3.3. HSF AND THE HEAT INDUCTION OF HSE-DIRECTED GENE EXPRESSION

The HSE—the DNA binding sequence of HSF—comprises at least three alternating repeats of the 5 bp sequence nGAAn (i.e., nGAAnnTTCnnGAAn or nTTCnnGAAnnTTCn), each repeat comprising one half-turn of the DNA double helix.[49] Studies on the *CUP1* promoter have shown that its two repeats of nGAAn are only sufficient to give slight thermal induction (see also chapter 6).[50] However, this small induction was increased by a point mutation in the DNA binding domain of HSF that presumably increased the strength of HSF binding to these two repeats.[50]

In unstressed cells of most eukaryotic organisms HSF exists as an inactive monomer. Only after stress induction does it trimerize, acquire high DNA-binding affinity and become an active transcription factor (see refs. 13, 49, 51 for literature). Since recombinant metazoan HSF expressed in bacteria constitutively forms trimers even at nonheat shock temperatures negative regulators appear normally to prevent the trimerization of this HSF which leads to its strong HSE binding. However, when this protein is expressed in eukaryotic systems (frog oocytes, cultured mammalian cells or reticulocyte lysates) latent heat shock-inducible HSF forms accumulate. Other indirect evidence that factors present in higher eukaryotes suppress the assembly of HSF trimers under normal conditions comes from expression studies of a human HSF in different cell types. These show that the temperature threshold for heat shock-dependent activation of human HSF is affected by its intracellular environment (see ref. 45 for literature). One possible means of HSF suppression is through interactions with Hsp70p. An association between metazoan HSF and Hsp70p has been observed and the overexpression of Hsps facilitates the dissociation of the metazoan HSF trimer.

The HSF of the budding yeasts *S. cerevisiae* and *Klyveromyces lactis* is fundamentally different from the HSF of higher eukaryotes (reviewed in refs. 13, 45). Firstly, regulation of its DNA binding activity is bypassed: yeast HSF is constitutively present as trimers bound to heat shock promoters.[49,51-55] Secondly, these yeast HSFs have a basal activity as transcriptional trans-activators even in unstressed cells.

Heat shock merely increases the trans-activation activity of these promoter-bound HSF trimers and causes them to become hyperphosphorylated.[55] Despite these differences HSFs from widely divergent eukaryotic sources share two regions of homology: a trimerization domain and a DNA binding region.[56] The latter comprises an 89 amino acid DNA-binding domain linked to 21 amino acids which may form a linker to the trimerization domain.[56] The crystal structure of the DNA binding domain of HSF from *K. lactis* shows this to be a variant of the helix-turn-helix family of DNA-binding domains.[57] The trans-activation ability of budding yeast HSF is localized in two separate domains broadly located near the N-terminus and C-terminus, respectively.[58-60] The C-terminal trans-activation domain was shown to overlap a leucine zipper-like motif[61] and its activity is influenced by other parts of the protein including a conserved heptapeptide (CE2[62]), the trimerization domain as well as the DNA binding domain.[58-63] These regions suppress the heat-induced trans-activation activity of yeast HSF under nonstress conditions, since their mutation causes constitutive high activity of heat shock promoters. Free Hsp70p levels may be involved—either directly or indirectly—in maintaining the promoter-bound yeast HSF in a state of basal transcriptional activity since yeast cells with greatly-reduced Hsp70p (due to loss of two Hsp70p genes) show constitutive high expression of heat shock genes.[38,45,64] Even though HSF also becomes hyperphosphorylated with heat shock this seems not to be essential for its increased activity.[63] Instead hyperphosphorylation may serve to deactivate the factor and thereby contribute to the transience of the heat shock response.[56]

Unlike most metazoans *S. cerevisiae* and *K. lactis* have only a single gene for HSF, a gene essential for the growth of even unstressed *S. cerevisiae*.[65] This may be because these yeast HSFs have both basal and heat-induced activities as transcriptional trans-activators. Thus their ability to direct low levels of expression of heat shock genes

even in unstressed cells may serve to supply Hsp chaperones that are needed under all conditions of yeast growth.[59] For instance, HSF is known to be essential in unstressed cells for the low basal-level expression of the *HSP82* gene, this being one of the two genes for the essential Hsp90p protein.[66] However, the requirement for HSF in unstressed cells might also reflect a multifunctional role for HSF protein. This possibility is indicated by the temperature-sensitive hsf1-m3 mutant, a strain that produces a defective nonheat-inducible HSF. This mutant exhibits defects in both mitochondrial protein import and the progression through the G2 phase of the cell division cycle when shifted to 37°C.[67]

3.4. HEAT SHOCK AND STRE-DIRECTED GENE EXPRESSION

A number of observations indicate that HSF is not the only factor contributing to heat stress resistance in yeast. The *hsf1-m3* mutation mentioned above causes loss of the heat-inducibility of HSE sequences and temperature-sensitivity, but remarkably has no effect on the acquisition of tolerance to extreme temperatures (50°C). This suggests HSF is not needed for the induction of resistance to severe stress.[67,68] In addition, a number of the genes subject to HSF control are needed for high temperature (37-39°C) growth but not for resistance to more severe heat stress (reviewed in refs. 1-3, 39). Also in *S. cerevisiae* there is a good inverse correlation between levels of 50°C thermotolerance and levels of protein kinase A activity (for literature, see refs. 3, 39). However, protein kinase A has very little influence over HSE induction levels.[41] Tolerance to extreme temperatures is also induced by treatments which cause no HSE induction, such as osmotic dehydration or exposure to weak organic acids.[3] It seems therefore that HSE-directed gene expression—although essential for *growth* under conditions of moderate heat stress—is not a major factor influencing resistance to more severe heat stress.

HSF-independent heat stress-activated promoter elements were originally identified in the *CTT1* and *DDR2* genes.[39-42,69] These STRE sequences (consensus AGGGG or CCCCT) are activated—as indicated above—not just by heat shock but also by nitrogen starvation, osmotic and oxidative stresses, low external pH, weak organic acids and ethanol. Very recently the high level expression of genes activated by STREs has been shown to require the transcription trans-activators Msn2p and Msn4p, proteins that bind STRE sequences in vitro (discussed in detail in chapters 7 and 8).[70]

The great variety of different stresses inducing STREs has led to the suggestion that STRE-directed gene expression may program the cell for the survival of several quite diverse conditions of stress, including extreme conditions.[39] STRE-mediated transcription has all the properties of a system—activated by diverse types of sublethal stress—whereby cells are re-programmed for the survival of more severe stress (e.g., desiccation, prolonged starvation, freezing).[39] In contrast, HSE sequences may primarily direct gene expression that will assist growth at high temperatures. Consistent with this mutants defective in the transcription activators Msn2p and Msn4p needed for STRE action exhibit hypersensitivity to several diverse stresses.[70] In addition, STRE-inducing conditions induce cross-resistance against several other types of stress (see also chapters 7 and 8).[39-42]

In its broadest sense a stress condition of a unicellular microbe can be defined as anything that lowers growth rate. It therefore includes any nutrient limitation (see chapter 1 and especially chapters 7 and 8). Expression of STRE-regulated stress genes is generally minimal in rapidly-growing cells but enhanced under growth-limiting conditions and starvation. The STRE seems to respond to diverse growth-reducing factors through a number of signal transduction pathways. One of these stress-sensing systems is the high osmolarity-activated HOG pathway (see chapter 4). Another is protein kinase A activity, which is closely related to levels of several stress toler-

ances[39-42] including thermotolerance.[3] Lowered activity of protein kinase A causes both slower growth and increased responsiveness of STRE sequences to stress (discussed further in chapter 7).[38-41]

Although STRE-like sequences are found in a large number of *S. cerevisiae* genes, it is surprising how few of these genes have actually been shown to contribute to the capacity to survive severe stress. They include *HSP104*,[48,71] *CTT1*[42,72] and—in *hsp104* mutant strains—the *SSA* genes for Hsp70p proteins.[73] Of these *CTT1* and *SSA3* have clearly been shown to be controlled by STRE-like elements,[39-42,45] while *HSP104* contains three STRE-like sequences in its promoter.[44] Genes of trehalose biosynthesis are also under STRE control[28,74] although—as mentioned earlier—the rapid heat shock-induction of trehalose is due mainly to alteration of the activities of preexisting trehalose synthase and trehalase, probably not alterations to de novo synthesis of these enzymes.

4. STRESS PROTECTIVE AND NORMAL ROLES OF HSPs

In this section I will briefly overview the roles of those yeast Hsps for which specific functions are known. One major class of *S. cerevisiae* Hsp has still not been shown to have any function. This is the class that comprises the small Hsps (Hsp12p, Hsp26p). Even though both of these proteins are induced under many conditions of stress[1,2,75,76] loss of both Hsp12p and Hsp26p in *S. cerevisiae* produces no apparent phenotype. However, an Hsp12p homologue of *S. pombe* suppresses a mutational defect in one of the late septation genes—*cdc4*—suggesting that this protein might participate in formation of the F-actin contractile ring at cytokinesis.[77] *S. cerevisiae* Hsp26p has near its C-terminus homology to the small Hsps of other species and to crystallin proteins. It also assembles into large 20S multimers of uniform size and molecular weight.[75] The heat shock-induced Hsp150p—a secretory glycoprotein—is another *S. cerevisiae* Hsp without any known function.[78]

4.1. Hsp60p, Hsp70p and Hsp90p Act as Molecular Chaperones

The proteins within cells are normally present in their native completely folded forms. However, during their biogenesis and their translocation into intracellular compartments such proteins temporarily exist in a totally unfolded or a partially-folded conformation. In such conformational states regions of the protein that are normally buried become exposed. Hsps—notably forms of Hsp70p and Hsp60p—interact with these normally buried surfaces. This binding is not just to prevent these unfolded or partially folded proteins from aggregating with each other. It also serves to facilitate proper protein folding, formation of correct protein-protein assemblies and translocation of polypeptides across membranes. These Hsps transiently associating with proteins during their biogenesis have therefore been termed "molecular chaperones."[79-81] In yeast as in other species Hsp70p levels probably increase with heat shock so that Hsp70p can sequester proteins that become partially unfolded though limited thermal denaturation. This stress-protective action of Hsp70p is therefore just a logical extension of the normal function of Hsp70p proteins as molecular chaperones.

Hsp70p of yeast and all other eukaryotes investigated to date occurs in many different forms encoded by a multigene family.[1,2,79-81] Based on genetic studies the genes for cytoplasmic Hsp70ps in *S. cerevisiae* have been divided into two groups: *SSA* (*SSA1-4*) and *SSB* (*SSB1,2*).[1,2,43,45,79-81] The *SSA* and *SSB* gene products are thought to be functionally distinct. *SSA* products are important for protein translocation across the ER and mitochondrial membranes. *SSB* products on the other hand are largely associated with polysome-associated primary translation products.[79-81] The expression of the *SSA* and SSB genes is also regulated differently. Three of the four *SSA* genes are induced by heat shock whereas the *SSB* genes are repressed by heat shock.[1,2,79-81]

All Hsp70ps bind ATP with high affinity and possess a weak ATPase activity. This ATPase is stimulated by binding to either unfolded proteins, synthetic peptides or Ydj1p. Ydj1p—a homologue of the DnaJ protein of *E. coli*—is partially localized to the cytoplasmic face of the ER and mitochondrial membranes in yeast.[82] The ATPase site of Hsp70p resides in the more conserved N-terminal 44 kDa fragment—a fragment for which a crystal structure is available[83]—while the peptide-binding domain is thought to be localized near the C-terminus (see refs. 79-81 for literature).

Figure 3.3 summarizes the roles of Hsp70p proteins in normal cells and under conditions of stress. The Hsp70p induced by heat shock is thought to mainly function in sequestering partially heat-damaged protein until this protein can be either degraded or reactivated (Fig. 3.2). In unstressed cells a major role of the different cytoplasmic forms of Hsp70p protein (mainly the constitutively-made Hsc70p) is to associate with newly synthesized polypeptide chains. This association keeps these precursors unfolded until they can be properly folded in the cytoplasm or translocated across the ER or mitochondrial membranes. This translocation across membranes is energy-dependent and may require Ydj1p.[79-82] Meanwhile in the lumen of the ER another form of Hsp70p (Bip in mammalian cells, the essential Kar2p in yeast) associates with newly-translocated proteins so as to maintain their unfolded conformation until the correct folding pathway can be adopted (Fig. 3.3). Also in the mitochondrial matrix yet another Hsp70p form (mt Hsp70p or Ssc1p in *S. cerevisiae*) is essential for protein import (Fig. 3.3), binding the extended precursor protein with high affinity as it appears in the matrix. Then, in an ATP-dependent step, Ssc1p releases this protein for Hsp60p-catalyzed folding (Fig. 3.3).

Although structurally distinct from Hsp70ps, Hsp60ps share functional features with them in that they also bind unfolded polypeptides. Hsp60ps play a crucial role

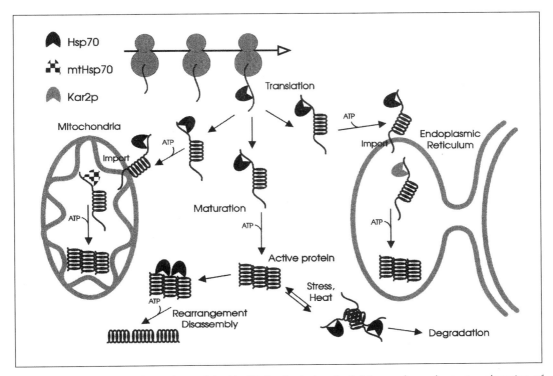

Fig. 3.3. Proposed roles of members of the Hsp70 family during the folding and membrane translocation of nascent polypeptides, molecular rearrangement or disassembly, stress protection and protein turnover. Adapted from ref. 79.

in binding to unfolded proteins, then catalyzing ATP-dependent folding of these proteins and assisting their assembly into higher-order protein structures (Fig. 3.4). The mitochondrial Hsp60p is a characteristic structure of 14 subunits of 60 kDa, each arranged in two heptameric rings stacked on top of each other (Fig. 3.4). In yeast it is encoded by the essential *MIF4* gene and is 54% homologous to the GroEL of *E. coli*.[84] There is also a large cytoplasmic complex that seems to fulfill the same role in the cytosol but, unlike mitochondrial Hsp60p of *S. cerevisiae*, it is not heat-inducible. This cytoplasmic complex in yeast is comprised of the essential Tcp1p and there is evidence for Tcp1p playing an important role in the biogenesis of tubulin and actin (see refs. 79-81 for literature).

The third class of yeast Hsps acting as chaperones are the Hsp90ps. Hsp90ps differ from both Hsp60ps and Hsp70ps in that they regulate the function of *specific*, *substantially folded* proteins. Also Hsp90ps act in an ATP-independent manner, earlier reports of an ATPase activity proving to be an artifact caused by the low protein kinase contamination of several Hsp90p preparations.[85] In vertebrates there is good evidence that Hsp90p binding regulates the activity of important regulatory proteins such as steroid hormone receptors and protein kinases.[1,79-81] In *S. pombe* Hsp90p regulates Wee1p tyrosine kinase activity and thus participates in cell cycle control.[86] However, the targets of its action in the *S. cerevisiae* cell still have yet to be identified. Yeast Hsp90p is an essential protein encoded by the two genes, *HSC82* and *HSP82*, whose products are 97% identical.[1] The elevation of Hsp90p levels by heat shock appears to reflects a need for higher levels of Hsp90p during high temperature growth. The three-dimensional structure of the 27 kDa N-terminal domain of the yeast protein containing three of the four

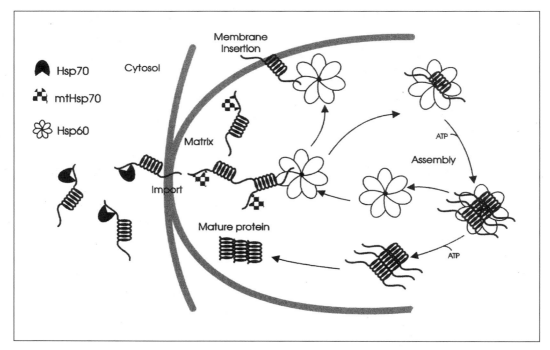

Fig. 3.4. Proposed roles of Hsp70 and Hsp60 during the import of protein precursors into mitochondria, their folding in the mitochondrial matrix and their re-export to the intermembrane space. Adapted from ref. 79.

highly-conserved regions of Hsp90p structure has recently been solved (Prodromou C, Piper PW, Pearl LH, manuscript in preparation). It may comprise the binding domain for other conserved proteins (e.g., immunophilins) needed for Hsp90p action.

4.2. HSP104P IS A CATALYST OF PROTEIN DISAGGREGATION OR REACTIVATION

Hsp104p is a member of a highly-conserved heat shock protein family (Hsp100ps[1]). Hsp100p proteins from bacteria, fungi, trypanosomes, plants and animals share approximately 60% amino acid similarity across their entire lengths and even higher similarity in regions of two ATP-binding site consensus elements. Biochemically, the best characterized member of this family is the *E. coli* ClpA protein which binds to and regulates the ClpP protease.[1] However, yeast Hsp104p is more closely related to *E. coli* ClpB, a protein which is not known to regulate protease action in *E. coli*.[1,86]

Yeast Hsp104p contributes to survival of many forms of stress, including severe heat stress and ethanol.[1,48,86-90] When wild-type and *hsp104* mutant strains are grown at 25°C and given a mild heat shock (30 minutes 37°C) before exposure to 50°C thermotolerance is induced in both of these strains. However, this tolerance is very transient in the *hsp104* mutant.[48] Hsp104p is also important for the elevated thermotolerance of *S. cerevisiae* cells growing by respiration as well as the high thermotolerance of stationary phase cells and spores.[48] Remarkably low levels of Hsp104p are sufficient to provide cells with high thermotolerance and the induction of high Hsp104p levels induces thermotolerance in the absence of all other inducible factors of the heat shock response (i.e., with no heat preconditioning).[90]

Parsell et al[87] used a temperature-sensitive *Vibrio harveyi* luciferase-fusion protein as a test substrate expressed in vivo in *hsp104* and wild-type strains. They found that the presence of Hsp104p did not protect this substrate against thermal denaturation or promote its proteolysis during heat shock.[87] Hsp104p instead assists the resolubilization of heat-inactivated luciferase from insoluble aggregates during a recovery period.[87] Remarkably, strains lacking either the constitutive or the heat-inducible forms of Hsp70p recover luciferase to almost wild-type levels, indicating this Hsp104p-mediated reactivation is independent of Hsp70p. This has led to the suggestion that the primary role of Hsp70p in thermotolerance (only clearly apparent in *hsp104* strains[73]) is to help prevent protein aggregates from forming whereas Hsp104p rescues proteins from these aggregates once they have formed.[87] Evidence that Hsp70p and Hsp104p may have complementary roles is suggested by the fact that *SSA* gene products assume an important role in tolerance to extreme temperatures in the absence of Hsp104p while—in cells with low levels of Hsp70p—Hsp104p assumes an important role in growth at normal temperatures.[73] Vegetative cells lacking Hsp104p show enhanced accumulation of heat-induced protein aggregates in electron micrographs when heat shocked at 44°C.[87] In normal wild-type cells these aggregates disappear during a 2 hour 25°C recovery period but in the *hsp104* mutant they persist.[87]

Hsp104p readily forms ring-shaped hexameric particles in the presence of adenine nucleotides while in the absence of ATP it is predominantly monomeric or dimeric.[88] Whether this oligomerization is important for its function is not yet clear. Both of the ATP binding sites on Hsp104p are essential for this protein to function in induced thermotolerance[86] and in luciferase reactivation in vivo.[87] The first of these sites appears to contribute the ATPase activity but to be nonessential for oligomerization whereas a single mutation in the second site totally eliminates Hsp104p oligomerization.[88]

An interesting role for Hsp104p has emerged from study of the hsr1-m3 mutant which has a nonsense mutation in the single gene for HSF (*HSF1*). Normally *HSP104* is not an essential gene, but in this mutant strain Hsp104p cannot be lost since it provides an essential function in propagation of the yeast [psi⁺] prion-like factor. [psi⁺] is needed in this strain since its presence causes nonsense suppression, thereby allowing HSF to be made from the mutant gene.[89,90] The yeast nonmendelian [psi⁺] factor has been suggested to be a self-modified protein analogous to mammalian prions. It becomes lost in yeast with overexpression or inactivation of *HSP104*.[90] This suggests a chaperone role for HSP-104p,[90] however, no such role could be demonstrated for luciferase-fusion protein reactivation in vitro.[88]

4.3. HSPs as Components of Systems for Protein Degradation

Heat shock increases the synthesis of certain components of the ubiquitination system for intracellular protein turnover (reviewed in ref. 91). This probably indicates a much greater requirement for turnover of abnormal proteins in cells recovering from heat stress. Production of ubiquitin itself is induced through heat-induced expression of the *UBI4* gene for polyubiquitin. Also, production of the Ubc4/5 ubiquitin conjugating enzymes is stimulated by heat. The proteasome (proteinase YscE) is almost certainly important in the stress response since the deletion of a gene that encodes a subunit of this protease causes sensitivity to stress conditions and accumulation of ubiquitin-protein conjugates.[92]

4.4. HSPs as Components of Antioxidant Defenses

There are several indications that the mitochondria of *S. cerevisiae* are especially sensitive to heat stress. High temperature growth causes a higher rate of appearance of spontaneous petites.[4] Growth on respiratory carbon sources is more adversely affected by temperatures above 34°C than is growth by the fermentation of glucose.[4] Also, diploid cells fail to sporulate at temperatures above about 34°C, possibly because sporulation requires respiratory competence. This high sensitivity of respiration to high temperatures is thought in part to reflect less efficient reduction of molecular oxygen by the mitochondrial respiratory chain, this in turn increasing endogenous production of superoxide free radicals.

The antioxidant defenses of yeast include free radical scavengers (notably glutathione); enzymes such as catalase and superoxide dismutase that directly catalyze the degradation of hydrogen peroxide (H_2O_2) and superoxide radicals, and diverse other activities that include systems of metal ion homeostasis (see chapter 6).[93] Heat shock does not alter glutathione levels[94] yet it increases the activities of at least two enzymes important for protection against oxidative damage. These are cytoplasmic catalase T encoded by the *CTT1* gene and the mitochondrial manganese form of superoxide dismutase (MnSod).[72,94] Yeast also has a cytoplasmic Cu,Zn-superoxide dismutase but the levels of this enzyme are not elevated by heat shock.

CTT1 is under STRE control (see above) with the result that catalase T is induced by heat shock except in cells with high protein kinase A activity.[72] Lack of catalase T causes a small thermotolerance reduction in both proliferating and stationary cells except when kinase A activity levels are high.[72] Loss of superoxide dismutase or cytochrome c peroxidase also renders cells more sensitive to the lethal effects of heat whereas overexpression of catalase or superoxide dismutase has the converse effect.[95] These influences of the levels of different antioxidant activities on thermotolerance are thought to reflect more severe oxidative damage to cellular proteins, nucleic acids and lipids caused by reactive oxygen species at higher temperatures, especially in cultures growing by respiration.[93-95] The increased thermotolerance of anaerobic yeast cultures[95] may reflect the absence of any adverse effects of superoxide radical formation by the mitochondrial respiratory chain.

4.5. HSPs IN SYSTEMS OF ENERGY GENERATION AND ENERGY CONSERVATION

A few of the mRNAs encoding yeast glycolytic enzymes show heat-inducibility.[1] Although this may be suggested to assist glycolytic flux in stressed cells, all of these enzymes (phosphoglycerate kinase, enolase, glyceraldehyde-3-phosphate dehydrogenase) are normally present in yeast in such large amounts that they are not limiting for glycolytic flux. Thus it is not clearly apparent as to why these mRNAs are heat-inducible. There can be little doubt, however, that heat stress imposes large demands for energy (ATP) generation by the cell. We have recently described evidence suggesting that yeast Hsp30p may have a role in energy (specifically ATP) conservation.[96] Hsp30p is the integral plasma membrane Hsp of *S. cerevisiae*. As described in section 5.2 below, Hsp30p acts to downregulate the stress-stimulation of plasma membrane H⁺-ATPase, an enzyme which affects thermotolerance[97] and also consumes much of the ATP generated by the cell.

5. PHYSIOLOGICAL EFFECTS OF HEAT STRESS AND THEIR IMPORTANCE FOR THE ADAPTIVE RESPONSE

5.1. CONSEQUENCES OF GROWING *S. CEREVISIAE* AT ITS MAXIMAL GROWTH TEMPERATURES

S. cerevisiae attains its maximal growth rates at around 34-37°C (Fig. 3.5). With sudden exposure to slightly higher temperatures that still permit growth there is a biphasic reduction in rate of exponential growth.[4] Biomass yields are also reduced, as the cells expend more energy counteracting the effects of the increased heat stress.[4] Imposition of an additional stress often shifts the temperature profiles for growth to lower temperatures, as shown for the effects of 6% ethanol in Figure 3.5A. The Van Uden laboratory has conducted detailed studies of the effects of temperature on growth, biomass yields and death of a number of different yeast species.[4] In addition to showing that these yeasts grow optimally over different temperature ranges they made the unexpected observation that most yeasts display significant rates of cell death at the maximum temperatures of growth. They categorized yeasts into some that show "associative" kinetics of growth and death and others ("nonassociative" yeasts) that do not display any appreciable loss of viability at their maximum growth temperature. Although the *S. cerevisiae* strains that they investigated displayed significant rates of cell death at the maximum temperatures of growth,[4] more recent work[98] has shown that certain *S. cerevisiae* strains display no appreciable death at their maximal temperatures of growth (Fig. 3.5B). Thus categorization of yeasts into *species* on the basis of whether they show "associative" or "nonassociative" temperature profiles of cell proliferation and death is possibly not valid. Whether *S. cerevisiae* shows appreciable death at its maximum temperatures of growth may be partly strain-dependent (Fig. 3.5).

5.2. THE EFFECTS OF HEAT STRESS ON THE REGULATION OF THE PLASMA MEMBRANE H⁺-ATPASE

As mentioned above, an important immediate effect of heat shock on the physiology of yeast cells is the stimulation of plasma membrane H^+-ATPase activity. This is partly a result of the increased permeability of the plasma membrane to protons acting to dissipate the electrochemical potential difference that the cell maintains at this membrane. This proton gradient—largely established and maintained through plasma membrane H^+-ATPase action—is essential for nutrient uptake, ion balance and intracellular pH regulation (see refs. 96 and 97 for literature).

The H^+-ATPase stimulation with heat stress is measurable both by direct enzyme assay of membrane preparations[96] and proton extrusion measurements on intact cells.[97] Through it the increased *passive* proton influx into the cell may be partly counteracted by increased *catalyzed* proton extrusion.

Another component in the stress-regulation of H^+-ATPase is Hsp30p, the plasma membrane protein induced by several stresses including heat shock, ethanol exposure, weak organic acids and glucose limitation.[10,31,96] Measurements of plasma membrane H^+-ATPase activity in heat shocked *hsp30* and wild-type cells have revealed that Hsp30p induction leads to downregulation of the stress-stimulation of this H^+-ATPase.[96] During prolonged heat stress cells reduce the levels of H^+-ATPase enzyme in the plasma membrane[31] and the presence of Hsp30p lowers the activity of that fraction of the enzyme that remains.[96] Since H^+-ATPase is a major consumer of total cellular ATP (especially in stressed cells) we have proposed that these two events may constitute a system of energy conservation under stress.[96] Consistent with this Hsp30p loss extends the time needed for adaptation to growth under several milder conditions of stress and also reduces the biomass yields and total ATP levels of yeast cultures.[96] Although plasma membrane H^+-ATPase activity is an important influence over thermotolerance levels[15,97] loss of Hsp30p does not appear to affect tolerances of severe stress.[96]

The plasma membrane H^+-ATPase has at its C-terminus a regulatory domain that includes a consensus site (RXXS/T) for phosphorylation by Ca^{2+}-calmodulin-dependent multiprotein kinase.[99] Phosphorylation/dephosphorylation at this domain governs the response of the enzyme to glucose metabolism and starvation. This domain of plasma membrane H^+-ATPase is also needed for Hsp30p action (R Braley, PW Piper, unpublished). While stress-induced depolarization of the plasma membrane is probably the major factor in the *immediate* increase in H^+-ATPase activity with heat shock, heat shock causes a Ca^{2+}-influx into the cell (discussed in section 5.3) that activates Ca^{2+}-calmodulin-dependent protein kinase. It can therefore be proposed that action of this kinase at the C-terminal regulatory domain of plasma membrane H^+-ATPase may influence H^+-ATPase activity during more prolonged heat stress.

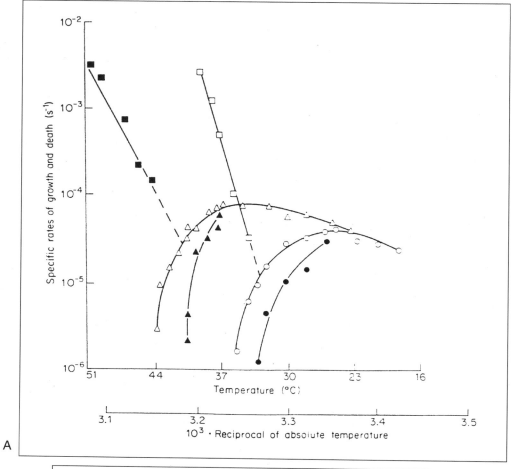

A

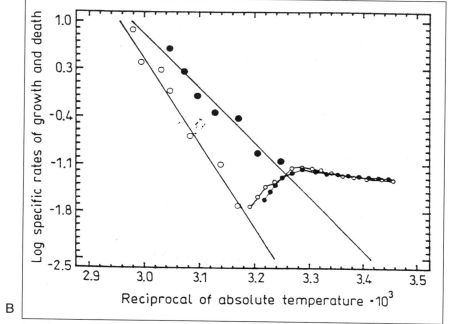

B

This may be in addition to the influences of Hsp30p induction and reduction of H+-ATPase protein levels mentioned above.

In addition to stimulating H+-ATPase the decline in intracellular pH with heat shock also stimulates the RAS-adenylate cyclase pathway.[100] Moderate increases in cAMP have been measured in cells subject to mild heat stress (see ref. 3 for literature) but whether these cAMP increases have any significant effect over thermotolerance levels has yet to be determined.

5.3. HEAT STRESS STIMULATES CALCIUM-ACTIVATED ENZYMES AND SIGNALING PATHWAYS

Evidence is now accumulating that Ca^{2+}-activated enzymes and signal transduction pathways are strongly stimulated by heat stress. It has been proposed that this might be a response to weakening of the cell wall during high temperature growth, since mutants in the PKC or cell integrity pathway (see chapter 4) lyse during growth at high temperatures.[101] Weakening of the cell wall at high temperatures might also be the secret behind the widespread use of a 42°C heat shock to greatly increase the efficiency of transformation of yeast by exogenously added DNA.[102]

Weakness of cell walls at high temperature might be detected by systems responding to stretch of the plasma membrane, such as mechanosensitive ion channels.[101] The opening of these ion channels may in turn lead to influxes of Ca^{2+} into the cytosol in response to mild heat shock (Fig. 3.6). This Ca^{2+} influx may then activate Ca^{2+}-regulated enzymes such as phospholipase C (PI-PLC), protein kinase C

(PKC) and Ca^{2+}/calmodulin-dependent protein kinase.

A 25-40°C heat shock may also stimulate endocytosis as shown by preliminary studies on the *end4* mutant (Zhao J-S and Piper PW, unpublished). It is possible that this heat stimulation of endocytosis may serve to counteract the effects of membrane stretch due to growth at stressful temperatures, thus reducing Ca^{2+}-influx and downregulating its associated effects in cells maintained at these temperatures. The plasma membrane protein encoded by *END4/MOP2/SLA2* appears to have several important roles including regulating the levels of plasma membrane H+-ATPase, maintaining the yeast cytoskeleton and in endocytosis.[103] This same protein may also participate in the regulation of the heat shock response since the *end4* mutation causes defects in both the trehalose induction and the downregulation of Hsp synthesis normally seen in cells maintained at mild heat shock temperatures (Zhao J-S and Piper PW, unpublished).

5.3.1. Heat stimulation of phospholipase C

The yeast phosphoinositidase C (phospholipase C or PI-PLC) is closely related to the PLCδ enzyme of higher organisms and strictly Ca^{2+}-dependent in its activity.[104-106] In response to an as yet unidentified signal (possibly intracellular Ca^{2+}) this enzyme catalyzes the splitting of phosphatidyl-inositol-4,5-bisphosphate (PtdIns(4,5)P$_2$) in the cell membrane into its hydrophilic and hydrophobic components, namely inositol-1,4,5-trisphosphate (InsP$_3$) and diacylglycerol (DAG),

Fig. 3.5 (opposite). Temperature profiles for specific growth and death rates. (A) Temperature profiles for Saccharomyces cerevisiae grown in liquid mineral medium. To the left are the profiles in the absence of added ethanol. Closed squares, specific death rates; open triangles, specific growth rates in the first exponential period; filled squares, net specific growth rates in the second exponential period. To the right are the profiles with the addition of ethanol to a final concentration of 6%. Open squares, specific death rates; open circles, specific growth rates in the first exponential period; filled circles, net specific growth rates in the second exponential period. (B) Temperature profile for specific growth and death rates. Filled large circles, death rate of strain 635; filled small circles, growth rate of strain 635; open large circles, death rate of strain BG7Fl; open small circles, growth rate of strain BG7Fl. Reproduced with permission from Van Uden N, Adv Microb Physiol 1984; 25:195-251 (A) and Sinigaglia M et al, Appl Microbiol Biotechnol 1993; 39:593-598 (B).

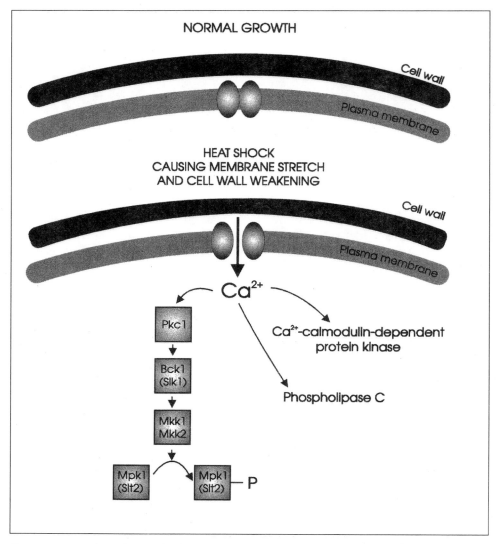

Fig. 3.6. Proposed mechanism for the stimulation of the PKC-pathway by heat stress. In this model heat stress causes a weakening of the cell wall and membrane stretching which in turn results in opening of mechanosensitive cation channels. This leads to an influx of calcium which stimulates the PKC-pathway as well as other calcium-sensitive enzymes involved in signal transduction. Adapted from ref. 101.

respectively. In higher organisms both of these compounds are well-established second messengers. $InsP_3$ mobilizes Ca^{2+} from intracellular stores, leading to stimulation of Ca^{2+}-activated enzymes. DAG activates protein kinase C (PKC) which in turn activates an important MAP kinase signaling pathway (see refs. 107, 108 for reviews).

S. cerevisiae strains that lack PI-PLC (*plc1* mutants) grow slowly at temperatures below about 35°C.[104] After a shift to 37°C they display a large increase in cell volume, become multibudded, appear unable to complete cytokinesis and/or cell separation, progressively lose viability and eventually lyse.[104-106] PI-PLC is therefore essential for growth of yeast under conditions of heat stress and may be stimulated upon heat shock by the rapid influx of Ca^{2+} ions into the cytoplasm. The *PLC1* gene also contains the HSE consensus in its promoter[105] and thus may show increased expression in stressed cells. The temperature-sensitive phenotype of one mutant that makes a defective PI-PLC enzyme (*plc1-1*) is suppressed by the addition of 100 mM calcium to the medium.[106] Stimulation of PI-PLC appears to be responsible for many of the longer term changes induced by ethanol in mammalian systems[109] but whether it also influences ethanol adaptation in yeast seems not to have been reported.

5.3.2. Heat stimulation of the PKC pathway

The PKC of *S. cerevisiae* is a homologue of the Ca^{2+}-stimulated PKC forms of mammalian cells and is essential under all conditions of growth (see also chapter 4).[110] One important PKC-regulated cascade involves four protein kinases: the MAPKKK homologue Bck1p, the redundant pair of MAPKK homologues Mkk1p and Mkk2p and the MAPK homologue Mpk1p. Loss of any of the genes for this pathway (*BCK1*, *MKK1/2* or *MPK1*) results in cells which lyse when cultured at high temperature but which are viable below 37°C. Because loss of *PKC1* results in a more severe phenotype (lysis at any temperature) than does loss of any of the genes of the Bck1p/(Mkk1p/Mkk2p)/Mpk1p cascade, it is thought that this cascade may be only one of those activated by PKC.[101,110] The Bck1p/(Mkk1p/Mkk2p)/Mpk1p cascade—in addition to being essential for high temperature growth—is also activated by high temperatures. Sublethal heat shock causes 170-fold activation of the Mpk1p MAPK after a short delay period.[101] This activation is maintained during growth at high temperatures and it may not be a consequence of increased PI-PLC activity (through direct DAG activation of PKC). It may result instead from the increased cytoplasmic levels of Ca^{2+} in cells suffering heat stress.[110] A number of *pkc1* mutant alleles show phenotypes of Ca^{2+}-dependency for growth or temperature-sensitivity that are suppressed by high exogenous calcium. They are also sensitive to hypertonic stress.[110]

Kamada et al[101] have provided three lines of evidence that strongly support the idea that the heat activation of Mpk1p (Fig. 3.6) is directly due to such membrane stretch: (i) while Mpk1p is normally activated by heat shock, this does not occur when cells are heat shocked in high osmolarity medium; (ii) a similar, though transient, Mpk1p activation can be induced by hypo-osmotic shock; and (iii) treatment of cells with chlorpromazine, an amphipathic drug that induces membrane stretch by inserting into the cytosolic face of the lipid bilayer, also strongly activates Mpk1p.

The Ca^{2+} influx with heat shock, acting to increase activities of both the PKC pathway and of several Ca^{2+}-activated enzymes (Fig. 3.6), is probably important for several of the events leading to the increased thermotolerance with heat shock. Thus it has been reported that mutants defective in the PKC pathway[101] and mutants lacking Ca^{2+}-calmodulin-dependent protein kinase[111] all show an impaired thermotolerance acquisition with heat shock. In addition, a constitutively-activated Bck1p kinase confers partial resistance to heat shock.[101] There is no evidence that the heat induction of Hsps is in any way impaired in these strains, indicating that signal transduction pathways activated by Ca^{2+} influx are not the key event in Hsp induction. Instead these pathways may contribute to Hsp-independent mechanisms of thermotolerance acquisition. Whether the PKC pathway and the Ca^{2+}-calmodulin-dependent protein kinase pathway are involved in the rapid induction of trehalose with heat shock has yet to be reported. We have found that the strains used for study of the effects of Ca^{2+}-calmodulin-dependent protein kinase[111] on thermotolerance (as with other strains of YPH499/500 genetic background) are defective in the trehalose induction with heat shock (Zhao J-S and Piper PW, unpublished). They may therefore be more heat-stress sensitive than normal.

6. FUTURE CHALLENGES

A major challenge for future research in this area is distinguishing the major heat stress-*sensing* mechanisms of yeast from what are effectively just temperature effects on other signal transduction pathways. Now that there is strong evidence that the cell membrane is involved in the heat induction of Hsps[17] it will be interesting to determine how the lipid composition of this membrane affects gene induction by heat shock. Also can lipid composition affect the several inducers of STRE elements? The temperature effects on pathways such as the PKC-pathway are probably important Hsp-independent systems for thermotolerance induction. Unfortunately there are often pleiotropic effects of altering the activity of such pathways, so that unraveling the key events in this induction will not be an easy task.

Other areas in which we can expect exciting developments are in our understandings of both functioning and regulation of the transcription factors binding to HSE and STRE sequences and in the mechanisms of action of different Hsps. Finally, at present there is comparatively little detailed three-dimensional structural information on Hsps (high resolution crystal structures of parts of mammalian Hsc70p and yeast Hsp90p; also a low resolution model of *E. coli* GroEL). However, this will change, and with it will come much more detailed understanding of how Hsp104p, Hsp90p, Hsp70p and Hsp60p operate as chaperones and in protein reactivation.

REFERENCES

1. Parsell DA, Lindquist S. The function of heat shock proteins in stress tolerance: degradation and reactivation of damaged proteins. Annu Rev Genet 1993; 27:437-496.
2. Mager WH, Moradas-Ferreira P. Stress response of yeast. Biochem J 1993; 290:1-13.
3. Piper PW. Molecular events associated with acquisition of heat tolerance by the yeast *Saccharomyces cerevisiae*. FEMS Microbiol Rev 1993; 11:1-11.
4. Van Uden N. Temperature profiles of yeasts. Adv Microb Physiol 1984; 25:195-251.
5. Kirk N, Piper PW. The determinants of heat shock element-directed *lacZ* expression in *Saccharomyces cerevisiae*. Yeast 1991; 7:539-546.
6. Coote PJ, Cole MB, Jones MV. Induction of increased thermotolerance in *Saccharomyces cerevisiae* may be triggered by a mechanism involving intracellular pH. J Gen Microbiol 1991; 137:1701-1708.
7. Grant CM, Firoozan M, Tuite MF. Mistranslation induces the heat shock response in yeast. Mol Microbiol 1989; 3:215-220.
8. Gropper T, Rensing L. Inhibitors of proteases and other stressors induce low molecular weight heat shock proteins in *Saccharomyces cerevisiae*. Exp Mycol 1993; 17:46-54.
9. Adams CC, Gross DS. The yeast heat shock response is induced by conversion of cells to sphaeroplasts and by potent transcriptional inhibitors. J Bacteriol 1991; 173:7429-7435.
10. Piper PW, Talreja K, Panaretou B, Moradas-Ferreira P, Byrne K, Praekelt UM, Meacock P, Regnacq M, Boucherie H. Induction of major heat shock proteins of *Saccharomyces cerevisiae*, including plasma membrane Hsp30, by ethanol levels above a critical threshold. Microbiology 1994; 140:3031-3038.
11. Piper PW. The heat shock and ethanol stress responses of yeast exhibit extensive similarity and functional overlap. FEMS Microbiol Lett 1995; 134:121-127.
12. Craig EA, Gross CA. Is hsp70 the cellular thermometer? Trends Biochem Sci 1991; 16:135-140.
13. Mager WH, De Kruijff AJJ. Stress-induced transcriptional activation. Microbiol Rev 1995; 59:506-531.
14. Maeda T, Takekawa M, Saito H. Activation of yeast PBS2 MAPKK by MAPKKKs or by binding of an SH3-containing osmosensor. Science 1995; 269:554-558.
15. Panaretou B, Piper PW. Plasma membrane ATPase action affects several stress tolerances of *Saccharomyces cerevisiae* and *Schizosaccharomyces pombe* as well as the extent and duration of the heat shock response. J Gen Microbiol 1990; 136:1763-1770.
16. Cheng L, Piper PW. Weak acid preservatives block the heat shock response and heat shock element directed *LacZ* expression of low pH *Saccharomyces cerevisiae* cultures, an inhibitory action partially relieved by respi-

ratory deficiency. Microbiology 1994; 140:1085-1096.

17. Carratu L, Franceschelli S, Pardini CL, Kobayashi GS, Horvath I, Vigh L, Maresca B. Membrane lipid perturbation modifies the set point of the temperature of heat shock response in yeast. Proc Natl Acad Sci USA 1996; 93:3870-3875.

18. Curran BPC, Khalawan SA. Alcohols lower the threshold temperature for maximal activation of a heat shock expression vector in the yeast *Saccharomyces cerevisiae*. Microbiology 1994; 140:2225-2228.

19. Colaco C, Sen S, Thangavelu M, Pinder S, Roser B. Extraordinary stability of enzymes dried in trehalose: simplified molecular biology. Bio Technology 1992; 10:1007-1011.

20. Hottiger T, De Virgilio C, Hall MN, Boller T, Wiemken A. The role of trehalose synthesis for the acquisition of thermotolerance in yeast. II. Physiological concentrations of trehalose increase the thermal stability of proteins in vitro. Eur J Biochem 1994; 219:187-193.

21. Wiemken A. Trehalose in yeast, stress protectant rather than reserve carbohydrate. Antonie van Leeuwenhoek 1990; 58:209-217.

22. Van Dijck P, Colavizza D, Smet P, Thevelein JM. Differential importance of trehalose is stress resistance in fermenting and nonfermenting *Saccharomyces cerevisiae* cells. Appl Env Microbiol 1995; 61:109-115.

23. DeVirgilio C, Hottiger T, Dominguez J, Boller T, Wiemken A. The role of trehalose synthesis for the acquisition of thermotolerance in yeast. I. Genetic evidence that trehalose is a thermoprotectant. Eur J Biochem 1994; 219:179-186.

24. Nwaka S, Kopp M, Burgert M, Deuchler I, Kienle I, Holzer H. Is thermotolerance of yeast dependent on trehalose accumulation? FEBS Lett 1994; 344:225-228.

25. Winkler K, Kienle I, Burgert M, Wagner J-C, Holzer H. Metabolic regulation of the trehalose content of vegetative yeast. FEBS Lett 1991; 291:269-272.

26. Neves M-J, Francois J. On the mechanism by which a heat shock induces trehalose accumulation in *Saccharomyces cerevisiae*. Biochem J 1992; 288:859-864.

27. Londesborough J, Vuorio OE. Purification of trehalose synthase from baker's yeast. Its temperature-dependent activation by fructose-6-phosphate and inhibition by phosphate. Eur J Biochem 1993; 216:841-848.

28. Winderickx J, de Winde JH, Crauwels M, Hino A, Hohmann S, Van Dijck P, Thevelein JM. Expression regulation of genes encoding subunits of the trehalose synthase complex in *Saccharomyces cerevisiae*. Mol Gen Genet 1996; (in press).

29. Hottiger T, De Virgilio C, Bell W, Boller T, Wiemken A. The 70-kDa heat-shock proteins of the *SSA* subfamily negatively regulate heat shock-induced accumulation of trehalose and promote recovery from heat stress in the yeast *Saccharomyces cerevisiae*. Eur J Biochem 1992; 210:125-132.

30. Hazell BW, Nevalainen H, Attfield PV. Evidence that the *Saccharomyces cerevisiae* CIF1 (GGS1/TPS1) gene modulates the heat shock response positively. FEBS Lett 1995; 377:457-460.

31. Panaretou B, Piper PW. The plasma membrane of yeast acquires a novel heat shock protein (Hsp30) and displays a decline in proton-pumping ATPase levels in response to heat shock and the entry to stationary phase. Eur J Biochem 1992; 206:635-640.

32. Watson K. Membrane lipid adaptation in yeast. In: Kates M, Manson LA, eds. Membrane Fluidity. Plenum Press, 1984:517-542

33. Johnston GC, Singer RA. Ribosomal precursor RNA metabolism and cell division in the yeast *Saccharomyces cerevisiae*. Mol Gen Genet 1980; 178:357-360.

34. Reed SI. G1-specific cyclins: in search of an S phase-promoting factor. Trends Genet 1991; 7:95-99.

35. Wittenberg C, Sugimoto K, Reed SI. G1-specific cyclins of *S. cerevisiae*: cell cycle periodicity, regulation by mating pheromone, and association with the p34[CDC28] protein kinase. Cell 1990; 62:225-237.

36. Rowley A, Johnston GC, Butler B, Werner-Washburne M, Singer RA. Heat shock-mediated cell cycle blockage and G1 cyclin expression in the yeast *Saccharomyces cerevisiae*. Mol Cell Biol 1993; 13:1034-1041.

37. Miller MJ, Xuong NH, Geiduschek EP. Quantitative analysis of the heat shock re-

sponse of *Saccharomyces cerevisiae*. J Bacteriol 1982; 151:311-327.

38. Craig EA, Jacobsen K. Mutations in the heat-inducible 70-kilodalton genes of yeast confer temperature-sensitive growth. Cell 1984; 38:841-849.

39. Ruis H, Schüller C. Stress signalling in yeast. BioEssays 1995; 17:959-965.

40. Belazzi T, Wagner A, Wieser R, Schanz M, Adam G, Hartig A, Ruis H. Negative regulation of transcription of the *Saccharomyces cerevisiae* catalase T (CTT1) gene by cAMP is mediated by a positive control element. EMBO J 1991; 10:585-592.

41. Marchler G, Schuller C, Adam G, Ruis H. A *Saccharomyces cerevisiae* UAS element controlled by protein kinase A activates transcription in response to a variety of stress conditions. EMBO J 1993; 12:1997-2003.

42. Schuller C, Brewster JL, Alexander MR, Gustin MC, Ruis H. The HOG pathway controls osmotic regulation of transcription via the stress response element (STRE) of the *Saccharomyces cerevisiae CTT1* gene. EMBO J 1994; 13:4382-4389.

43. Boorstein WR, Craig EA. Structure and regulation of the SSA4 HSP70 gene of *Saccharomyces cerevisiae*. J Biol Chem 1990; 265:18912-18921.

44. Varela JCS, Praekelt UM, Meacock PA, Planta RJ, Mager WH. The *Saccharomyces cerevisiae HSP12* gene is activated by the high osmolarity glycerol pathway and negatively regulated by protein kinase A. Mol Cell Biol 1995; 5:6232-6245.

45. Boorstein WR, Craig EA. Transcriptional regulation of SSA3, an HSP70 gene from *Saccharomyces cerevisiae*. Mol Cell Biol 1990; 10:3262-3269.

46. Watt R, Piper PW. *UBI4*, the polyubiquitin gene of *Saccharomyces cerevisiae*, is a heat shock gene that is also subject to catabolite derepression control. Mol Gen Genet 1996; (in press).

47. Pillar TM, Bradshaw RE. Heat shock and stationary phase induce transcription of the *Saccharomyces cerevisiae* iso-2-cytochrome c gene. Curr Genet 1991; 20:185-188.

48. Sanchez Y, Taulein J, Borkovich KA, Lindquist S. Hsp104 is required for tolerance to many forms of stress. EMBO J 1992;

11:2357-2364.

49. Sorger PK. Heat shock factor and the heat shock response. Cell 1991; 65:363-366.

50. Yang W, Gahl W, Hamer D. Role of heat shock transcription factor in yeast metallothionein gene expression. Mol Cell Biol 1991; 11:3676-3681.

51. Wisniewski J, Orosz A, Allada R, Wu C. The C-terminal region of *Drosophila* heat shock factor contains a constitutively functional trans-activation domain. Nucl Acids Res 1996; 24:367-374.

52. Jakobsen BK, Pelham HR. Constitutive binding of yeast heat shock factor to DNA in vivo. Mol Cell Biol 1988; 8:5040-5042.

53. Gross DS, English KE, Collins KW, Lee SW. Genomic footprinting of the yeast *HSP82* promoter reveals marked distortion of the DNA helix and constitutive occupancy of the heat shock and TATA elements. 1990; J Mol Biol 1990; 216:611-631.

54. Sorger PK, Lewis MJ, Pelham HR. Heat shock factor is regulated differently in yeast and HeLa cells. Nature 1987; 329:81-84.

55. Sorger PK, Pelham HR. Purification and characterisation of a heat shock element binding protein from yeast. Cell 1988; 54:855-864.

56. Flick KE, Gonzalez L, Harrison CJ, Nelson HCM. Yeast heat shock transcription factor contains a flexible linker between the DNA-binding and trimerisation domains. J Biol Chem 1994; 269:12475-12481.

57. Harrison CJ, Bohm AA, Nelson HCM. Crystal structure of the DNA binding domain of heat shock transcription factor. Science 1994; 263:224-227.

58. Nieto-Sotelo J, Wiederrecht G, Okuda A, Parker CS. The yeast heat shock transcription factor contains a transcriptional activation domain whose activity is repressed under nonshock conditions. Cell 1990; 62:807-817.

59. Bonner JJ, Heyward S, Fackenthal DL. Temperature-dependent regulation of a heterologous transcriptional activation domain fused to yeast heat shock transcription factor. Mol Cell Biol 1992; 12:1021-1030.

60. Sorger PK. Yeast heat shock factor contains separable transient and sustained response

transcriptional activators. Cell 1990; 62:793-805.

61. Chen Y, Barlev NA, Westergaard O, Jakobsen BK. Identification of the C-terminal activator domain in yeast heat shock transcription factor: independent control of transient and sustained transcriptional activity. EMBO J 1993; 12:5007-5018.

62. Hoj A, Jakobsen BK. A short element required for turning off heat shock transcription factor: evidence that phosphorylation enhances deactivation. EMBO J 1994; 13:2617-2624.

63. Jakobsen BK, Pelham HR. A conserved heptapeptide restrains the activity of the yeast heat shock transcription factor. EMBO J 1991; 10:369-375.

64. Werner-Washburne M, Stone DE, Craig EA. Complex interactions among members of an essential subfamily of Hsp70 genes in *Saccharomyces cerevisiae*. Mol Cell Biol 1987; 7:2568-2577.

65. Sorger PK, Pelham HRB. Yeast heat shock factor is an essential DNA binding protein that exhibits temperature-dependent phosphorylation. Cell 1988; 54:855-864.

66. McDaniel D, Caplan AJ, Lee M-S, Adams CC, Fishel BR, Gross DS, Garrard WT. Basal-level expression of the yeast *HSP82* gene requires a heat shock regulatory element. Mol Cell Biol 1989; 9:4789-4798.

67. Smith BJ, Yaffe MP. A mutation in the yeast heat shock factor gene causes temperature-sensitive defects in both mitochondrial protein import and the cell cycle. Mol Cell Biol 1991; 11:2647-2655.

68. Smith BJ, Yaffe MP. Uncoupling thermotolerance from the induction of heat shock proteins. Proc Natl Acad Sci USA 1991; 88:11091-11094.

69. Kobayashi N, McEntee K. Identification of cis and trans-components of a novel stress regulatory pathway in *Saccharomyces cerevisiae*. Mol Cell Biol 1993; 13:248-256.

70. Martinez-Pastor MT, Marchler G, Schuller C, Marchler-Bauer A, Ruis H, Estruch F. The *Saccharomyces cerevisiae* zinc finger proteins Msn2p and Msn4p are required for transcriptional induction through the stress-response element (STRE). EMBO J 1996; 15:2227-2235.

71. Sanchez Y, Lindquist SL. HSP104 is required for induced thermotolerance. Science 1990; 248:1112-1115.

72. Wieser R, Adam G, Wagner A, Schuller C, Marchler G, Ruis H, Krawiec Z, Bilinski T. Heat shock factor-independent heat control of transcription of the *CTT1* gene encoding the cytosolic catalase T of *Saccharomyces cerevisiae*. J Biol Chem 1991; 266:12406-12411.

73. Sanchez Y, Parsell DA, Taulein J, Vogel JC, Craig EA, Lindquist S. Genetic evidence for a functional relationship between Hsp104 and Hsp70. J Bacteriol 1993; 175:6484-6489.

74. Gounalaki N, Thireos G. Yap1p, a yeast transcriptional activator that mediates multidrug resistance, regulates the metabolic stress response. EMBO J 1994; 13:4036-4041.

75. Bentley NJ, Fitch IT, Tuite MF. The small heat shock protein Hsp26 of *Saccharomyces cerevisiae* assembles into a high molecular weight aggregate. Yeast 1992; 8:95-106.

76. Praekelt UM, Meacock PA. *HSP12*, a new small heat shock gene of *S. cerevisiae*: analysis of structure, regulation and function. Mol Gen Genet 1990; 223:97-106.

77. Jang Y-J, Park S-K, Yoo H-S. Isolation of an *HSP12*-homologous gene of *Schizosaccharomyces pombe* suppressing a temperature-sensitive mutant allele of *cdc4*. Gene 1996; 172:125-129.

78. Russo P, Simonen M, Uimari A, Teesalu T, Makarow M. Dual reguation by heat and nutrient stress of the yeast *HSP150* gene encoding a secretory glycoprotein. Mol Gen Genet 1993; 239:273-280.

79. Gething M-J, Sambrook J. Protein folding in the cell. Nature 1992; 355:33-45.

80. Becker J, Craig EA. Heat shock proteins as molecular chaperones. Eur J Biochem 1994; 219:11-23.

81. Hartl FU. Molecular chaperones in cellular protein folding. Nature 1996; 381:571-580.

82. Ziegelhoffer T, Lopez-Buesa P, Craig EA. The dissociation of ATP from Hsp70 of *Saccharomyces cerevisiae* is stimulated by both Ydj1p and peptide substrates. J Biol Chem 1995; 270:10412-10419

83. Flaherty KM, DeLuca-Flaherty C, McKay DB. Three dimensional structure of the

ATPase fragment of a 70L heat-shock cognate protein. Nature 1990; 346:623-628.

84. Reading DS. Hallberg RL, Myers AM. Characterisation of the yeast *HSP60* gene coding for a mitochondrial assenbly factor. Nature 1989; 337:655-659.

85. Shi Y, Brown ED, Walsh CT. Expression of recombinant human casein kinase II and recombinant heat shock protein 90 in *Escherichia coli* and characterisation of their interactions. Proc Natl Acad Sci USA 1994; 91:2767-2771.

86. Aligue R, Akhavan-Niak H, Russell P. A role for Hsp90 in cell cycle control: Wee1 tyrosine kinase activity requires interaction with Hsp90. EMBO J 1994; 13:6099-6106.

86a. Parsell DA, Sanchez Y, Stitzel JD, Lindquist S. Hsp104 is a highly conserved protein with two essential nucleotide binding sites. Nature 1991; 353:270-273.

87. Parsell DA, Kowal AS, Singer MA, Lindquist S. Protein disaggregation mediated by heat-shock protein Hsp104. Nature 1994; 372:475-478.

88. Parsell DA, Kowal AS, Lindquist S. *Saccharomyces cerevisiae* Hsp104 protein; purification and characterisation of ATP-induced structural changes. J Biol Chem 1994; 269:4480-4487.

89. Lindquist S, Kim G. Heat-shock protein 104 expression is sufficient for thermotolerance in yeast. Proc Natl Acad Sci USA 1996; 93:5301-5306.

90. Chernoff YO, Lindquist SL, Ono B, Inge-Vechtemov SG, Liebman SW. Role of chaperone protein Hsp104 in propagation of the yeast prion-like factor [psi+]. Science 1995; 268:880-883.

91. Finley D, Chau V. Ubiquitination. Annu Rev Cell Biol 1991; 7:25-69.

92. Heinemeyer W, Kleinschmidt JA, Saidowsky J, Escher CH, Wolf D. Proteinase yscE, the yeast proteasome/multicatalytic-multifunctional proteinase. EMBO J 1991; 10:555-562.

93. Moradas-Ferriera P, Costa V, Piper P, Mager W. The molecular defences against reactive oxygen species in yeast. Mol Microbiol 1996; 19:651-658.

94. Costa V, Reis E, Quintanilha A, Moradas-Ferreira P. Acquisition of ethanol tolerance

in *Saccharomyces cerevisiae:* The key role of the mitochondrial superoxide dismutase. Arch Biochem Biophys 1993; 300:608-614.

95. Davidson JF, Whyte B, Bissinger PH, Schiestl RH. Oxidative stress is involved in heat-induced cell death in *Saccharomyces cerevisiae.* Proc Natl Acad Sci USA 1996; 93:5116-5121.

96. Piper PW, Ortiz-Calderon C, Holyoak C, Coote P, Cole M. Hsp30, the integral plasma membrane heat shock protein of *Saccharomyces cerevisiae*, is a stress-inducible regulator of plasma membrane H^+-ATPase. Cell Stress and Chaperones 1996; (in press).

97. Coote PJ, Jones MV, Seymour IJ, Rowe DL, Ferdinando DP, McArthur AJ, Cole MB. Activity of the plasma membrane H^+-ATPase is a key physiological determinant of thermotolerance in *Saccharomyces cerevisiae.* Microbiology 1994; 140:1881-1890.

98. Sinigaglia M, Gardini F, Guerzoni ME. Relationship between thermal behaviour, fermentation performance and fatty acid composition in two strains of *Saccharomyces cerevisiae.* Appl Microbiol Biotechnol 1993; 39:593-598.

99. Portillo F, Eraso P, Serrano R. Analysis of the regulatory domain of yeast plasma membrane H+-ATPase by directed mutagenesis and intragenic suppression. FEBS Lett 1991; 287:71-74.

100. Thevelein JM. Signal transduction in yeast. Yeast 1994; 10:1753-1790.

101. Kamada Y, Jung US, Piotrowski J, Levin DE. The protein kinase C-activated MAP kinase pathway of *Saccharomyces cerevisiae* mediates a novel aspect of the heat shock response. Genes Dev 1995; 9:1559-1571.

102. Gietz D, St Jean A, Woods RA, Schiestl RH. Improved method for high efficiency transformation of intact yeast cells. Nucl Acids Res 1992; 20:1425-1426.

103. Na S, Hincapie M, McCusker JH, Haber JE. *MOP2 (SLA2)* affects the abundance of the plasma membrane H+-ATPase of *Saccharomyces cerevisiae.* J Biol Chem 1995; 270:6815-6823.

104. Payne WE, Fitzgerald-Hayes M. A mutation in *PLC1*, a candidate phosphoinositide-specific phospholipase C gene from *Saccharomyces cerevisiae*, causes aberrant mitotic

chromosome segregation. Mol Cell Biol 1993; 13:4351-4364.

105. Flick JS, Thorner J. Genetic and Biochemical characterisation of a phosphatidylinositol-specific phospholipase C in *Saccharomyces cerevisiae*. 1993; Mol Cell Biol 13:5861-5876.

106. Yoko-o T, Matsui Y, Yagisawa H, Nojima H, Uno I, Toh-e A. The putative phospho-inositide-specific phospholipase C gene, *PLC1*, of the yeast *Saccharomyces cerevisiae* is important for cell growth. Proc Natl Acad Sci USA 1993; 90:1804-1808.

107. Berridge MJ. Inositol triphosphate and calcium signalling. Nature 1993; 361:315-325.

108. Lee SB, Rhee S-G. Significance of PIP2 hydrolysis and regulation of phospholipase C isozymes. Curr Opin Cell Biol 1995; 7:183-189.

109. Hoek JB, Taraschi TF. Cellular adaptation to ethanol. Trends Biochem Sci 1988; 13:269-274.

110. Levin DE, Fields FO, Kunisawa R, Bishop JM, Thorner J. A candidate protein kinase C gene, *PKC1*, is required for the *S. cerevisiae* cell cycle. Cell 1990; 62:213-224.

111. Iida H, Ohya Y, Anraku Y. Calmodulin-dependent protein kinase II and calmodulin are required for induced thermotolerance in *Saccharomyces cerevisiae*. Curr Genet 1995; 27:190-193.

CHAPTER 4

SHAPING UP: THE RESPONSE OF YEAST TO OSMOTIC STRESS

Stefan Hohmann

1. INTRODUCTION

1.1. THE NEED TO ADAPT TO CHANGES IN WATER AVAILABILITY

Water plays a central role in life as a solvent for most of the components of cells, thus determining the structure of proteins, nucleic acids and membranes. Water also plays a pivotal role in biochemical reactions, either by providing the appropriate environment as a solvent and/or by being directly involved as a reactive agent. Finally, the water content of the cell determines turgor and hence, size and shape of the cell and of the entire organism.[1]

Thus it appears to be essential for all types of cells as well as for entire multicellular organisms to maintain and to control their water content within certain limits. Indeed, all types of cells that have been investigated have developed mechanisms to adapt to changes in the relative water content (osmolarity) of their environment. This is obvious for unicellular organisms which expose their entire surface to an environment which, in most cases, can change rapidly. But also cells that are part of multicellular organisms have maintained such response mechanisms even though one would expect that the body of such organisms should provide a more or less constant environment. The osmolarity of body fluids and hence, osmoregulation of the entire organism, is under hormonal control in both animals[2] and plants.[3] However, even within this fairly stable environment changes in the osmolarity of the body fluids occur. The individual cell needs to adapt to such changes in order to

Yeast Stress Responses, edited by Stefan Hohmann and Willem H. Mager.
© 1997 R.G. Landes Company.

maintain cellular turgor and shape as well as water transport through the organism. In addition, specialized cells involved directly in osmoregulation of the entire organism, like kidney cells in animals, are in fact exposed to dramatic changes in surrounding osmolarity.[4]

The natural environment of yeasts are plants and their fruits and sugar-containing juices. Yeast cells probably have to survive long periods in the absence of an aqueous environment and they do so as meiotic spores. The yeast *Saccharomyces cerevisiae* multiplies within aqueous solutions containing sugars, organic acids, amino acids, ions and other nutrients and, of course, the ethanol produced during fermentation. When the cell is transferred by wind, insects or rain droplets onto a grape in a sunny climate it will experience a solution with a sugar content as high as 30% and more. On the other hand, after a rain shower the cell may find itself in virtually distilled water. These drastic changes in water availability document that the cell needs mechanisms that respond very quickly in order to stay alive and to return to productivity as soon as possible.

During the fermentation of a grape the cell is exposed to more than 20% of sugar at the beginning of the process and to a sugar-free solution containing 10% ethanol at the end of the fermentation. During this entire process the cell maintains its shape and its metabolic activity. Thus the cell must not only have mechanisms that sense and respond to rapid alterations in external osmolarity but also must be able to keep up with the continuous changes caused by its own metabolic activity.

In conclusion, it can be expected that due to its natural environment, yeast should possess all the mechanisms that a eukaryotic cell requires to respond and to adapt to changes in the osmolarity of the environment. The powerful tools of yeast genetics and molecular biology have been applied over the last couple of years in order to unravel the molecular mechanisms involved in such adaptive processes. As

with many other key processes in cell biology, it is expected that the principle molecular mechanisms operating in yeast osmoregulation may serve as a model for the situation in plants and animals. The first glimpse that we have obtained up to now suggest that this is indeed at least partially the case.

In this chapter I will focus on the purely osmotic effects imposed on the cell, even if salt (sodium chloride, NaCl) is commonly used as an experimental osmolyte. However, salt causes both hyperosmotic stress as well as effects due to salt toxicity. The basis of salt toxicity, the consequences of salt stress and the response mechanisms employed by yeasts are discussed in chapter 5.

1.2. OSMOTIC SHOCK, HYPEROSMOTIC STRESS, HYPO-OSMOTIC STRESS

In order to maintain turgor and shape the yeast cell requires the internal osmolarity to be higher than that of the surrounding medium. In standard laboratory medium which contains about 50 g/l of sugar, yeast extract and peptone, the internal solute concentration is about 0.5 M.[5]

S. cerevisiae is known to grow only at moderately high external osmolarity, and this ability is highly variable among different laboratory yeast strains. Some strains do not even grow in the presence of 2% NaCl. The strains most commonly used by laboratories studying osmotic stress responses—like W303-1A or the YPH strains—still grow at 10% NaCl. In such strains usually about 7% NaCl (1.2 M) giving a water activity of 0.960 are applied to cause a hyperosmotic shock. In terms of water activity this amount of NaCl is equivalent to 65% sucrose (1.8 M). The need for these enormous amounts of sugars necessary to impose hyperosmotic stress explains why NaCl is being used preferentially despite the additional toxic effects caused by sodium and chloride (see chapter 5). While usually such high levels of osmolytes are required to monitor growth defects of hyper-osmosensitive mutants, much smaller amounts are sufficient to trigger the osmoresponsive high osmolarity glycerol (HOG)-signaling pathway or the closing of the Fps1p channel (see further).

Screening for mutations that affect growth at high external osmolarity is fairly straightforward. Several such screens for mutants defective for growth in the presence of about 1 M NaCl have been performed, and the analysis of the mutants obtained form the basis of our present knowledge. Such mutants are per definition sensitive to hyperosmotic stress. However, no distinction has been made in such screens for a defect in the survival of a hyperosmotic shock or for an inability to adapt to high osmolarity and to resume growth. These may be two independent processes (see further).

Tests for the sensitivity to a hypo-osmotic shock, i.e., to a sudden drop in external osmolarity are less simple since the standard growth medium is already hypotonic. Thus the cells have to be pregrown at high osmolarity and then transferred to a medium with low osmolarity and survival or the survival rate needs to be scored. No such screen has yet been reported. There are, however, a number of mutants known that are unable to grow on standard medium unless the osmolarity is raised by the addition of, e.g., 0.5 M sorbitol to isotonic conditions. Such mutants are for various reasons defective in cell wall assembly and not truly hypo-osmosensitive.

The response of yeast cells to (mainly) hyperosmotic stress has been reviewed from various viewpoints. Brown summarized the pioneering stages of the physiological analysis of fungal osmoregulation.[6-8] Blomberg and Adler provided an excellent theoretical and physiological background before the impact of molecular genetic analysis revolutionized the field.[5] Varela and Mager focused on signaling and gene expression in two reviews where the second one updates the earlier and provides a connection to the general stress response.[9,10] Prior and Hohmann discussed osmoregulation within the context

of a review in glycerol metabolism.[11] Overviews on salt tolerance mechanisms also cover aspects of osmoregulation.[12,13] Numerous reviews on signal transduction discuss, among other aspects, osmostress-induced signaling in yeast (for example, see refs. 14-19).

2. EFFECTS CAUSED BY OSMOTIC SHOCK

Intuitively one can expect that the adaptation of a cell to changes in external osmolarity involves many cellular processes of which certainly only a small number are being recognized at the moment. Furthermore, it is likely that the cell has developed mechanisms that allow survival upon an osmotic shock, i.e., immediate rescue mechanisms that prevent cell death after a sudden change in osmolarity. After the cell has managed to actually survive it should then initiate long-term adaptation processes that eventually allow the cell to resume growth. Very little is known at present with respect to the timing of responses, although a few indications indeed suggest that the response mechanisms can be divided into early SOS-responses and sustained adaptation responses. This appears to be a general theme in most responses to environmental changes.

2.1. CELL VOLUME CHANGES

Cell volume regulation is studied intensively in animal cells at the physiological level, but more and more also at the molecular level.[20] Animal cells do not have a cell wall that could counteract drastic volume alterations and hence their volume changes quickly even upon minor changes in external osmolarity. Nevertheless, despite their rigid cell wall[21] yeast cells show volume changes upon osmotic shock.

A hyperosmotic shock, i.e., the addition of salt or sugar to the growth medium, causes an instant and rapid loss of water and cell shrinkage (reviewed in ref. 5; Fig. 4.1). The process of shrinking takes about 1 min[22] and the cell volume attained inversely correlates with solute concentrations.[23,24] Measurements using a Coulter

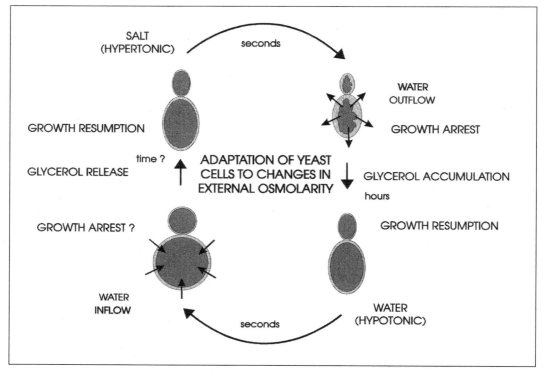

SALT
(HYPERTONIC)

seconds

WATER
OUTFLOW

GROWTH RESUMPTION

GROWTH ARREST

GLYCEROL RELEASE

time ?

**ADAPTATION OF YEAST
CELLS TO CHANGES IN
EXTERNAL OSMOLARITY**

GLYCEROL ACCUMULATION

hours

GROWTH ARREST ?

GROWTH RESUMPTION

WATER
INFLOW

WATER
(HYPOTONIC)

seconds

Fig. 4.1. The impact of hyper- and hypo-osmotic stress on yeast cells and the basic response mechanisms.

counter indicate that after a shock with 7% NaCl (water activity 0.960) the cell shrinks to about 30-35% of its original volume.[24] Clear indications of plasmolysis as is observed in plant cells have not been obtained in yeast.[5] During the subsequent recovery period the cells increase their volume again but interestingly, they do not reach their initial volume. Cells growing in 7% NaCl appear to do so at a cell volume that is roughly 50% of that of cells growing in the absence of salt. Again, the higher the solute concentration in the medium the smaller the cell size.[24]

When cells incubated in the presence of high solute concentrations are placed into water, i.e., when a hypo-osmotic shock is imposed, they rapidly increase their volume (Fig. 4.1).[5,22] Little is known about the extent of the volume increase. The time span of the recovery period after a hypoosmotic shock also has not been investigated.

2.2. IMMEDIATE RESCUE MECHANISMS

As mentioned above yeast cells are well capable of growing at salt concentrations up to 10%. However, a large portion of the cells does not survive even a relatively mild shock (reviewed in ref. 5). A survey of different yeast strains has shown that only 10% of the cells survive a shock with about 6% salt. The basis for this "osmotic hypersensitivity" is not understood. The phenomenon appears to be growth phase-dependent and cells are particularly sensitive during active growth, a common observation in yeast stress responses (see chapter 7). There does not seem to be a correlation between the ability of a given strain to grow at a high salt level and the survival rate upon a hyperosmotic shock.[25] Thus the survival of an osmotic shock and the ability to adapt to and to grow at high osmolarity may have different genetic bases. In extension, one might interpret the mechanism forming the basis for osmotic hypersensitivity as a rapidly operating rescue system. This system should be expressed in the cell constitutively, i.e., before osmotic shock, and should be (largely) independent of a prerequisite for the adaptation mechanisms which prepare the cell for growth resumption. It will be interesting to study mutants known to be affected for growth under high-osmolarity for their ability to survive a hyperosmotic shock. This may allow a further genetic dissection of mechanisms involved in the immediate rescue mechanism and in long-term adaptation.

One rapid rescue mechanism appears to require a functional vacuole. Mutants affected in vacuolar integrity are viable but die within 10 seconds after hyperosmotic shock.[26] The yeast vacuole is fairly large and may take up to 50% of the cellular volume.[27] It is supposed to actually resemble lysosomes of animal cells with its very low internal pH and the presence of hydrolytic enzymes. It also serves as a storage compartment for amino acids, polyphosphate and ions. Another possible role of the vacuole could be that of a water reservoir which might be used when water is lost from the cytosol upon the addition of an osmolyte to the growth medium. Such a role has been proposed for plant vacuoles which can take up more than 90% of the cellular volume. The vacuolar membrane (the tonoplast) exhibits abundant water channels (see further) which are likely to serve the role of equilibrating the water content of the cytosol during water stress.[28,29]

Approximately 20% of the cells from the wild type W303-1A are able to form colonies after a hypotonic osmotic shock from 10% NaCl into distilled water.[30] Thus these osmotic shock conditions are accompanied by substantial cell death. Whether this effect is strain- or growth-phase-specific has not been investigated. Further work is required to study the basic physiology of yeast cells after a hypoosmotic shock.

2.3. PLASMA MEMBRANE AND CYTOSKELETON

The drastic changes in cell volume caused by osmotic shock can be expected

to have an impact on several cellular components like the plasma membrane and the cytoskeleton. This is indeed the case.

The existence of ion channels in the yeast plasma membrane sensitive to mechanical triggers has been demonstrated in patch-clamp studies.[31] These channels transport both cations and anions and appear to be the only calcium channels in the yeast plasma membrane. It has also been shown that the function of these mechanosensitive channels is voltage-dependent. The molecular nature of the channels is not known.

The impact on the yeast cytoskeleton has been studied in more detail. It has been known for a long time that diploid cells carrying only one copy of the single actin gene *ACT1* are hyperosmosensitive.[32] Temperature-sensitive *act1* mutants are also osmosensitive.[32] Staining of actin with rhodamine-phalloidin showed that the actin cytoskeleton becomes disassembled after a hyperosmotic shock within the first minute.[33] Restoration of the actin cytoskeleton took about 2 hours under the rather mild shock conditions tested, and coincided with growth resumption. This suggests that the re-establishment of the actin cytoskeleton may be the factor that determines the time span until growth can resume.

One consequence of the disassembly of the actin cytoskeleton appears to be a temporary loss of cell polarity (Fig. 4.2). Polarity of a yeast cell is most obvious during the process of budding at one pole of the cell and is determined in part through cortical patches of actin filaments in the growing bud.[34-36] Haploid cells perform an axial budding pattern, i.e., the new bud always appears in the vicinity of the previous bud at the same pole.[37] While wild type cells are able to resume growth of this bud after recovery from a hyperosmotic shock, mutants affected in the HOG-pathway (see further) abandon their bud and form a new one. This remarkable behavior also suggests a connection to cell cycle control since apparently the cell cycle machinery is being reset after an osmotic shock (M.C. Gustin, personal communication). The

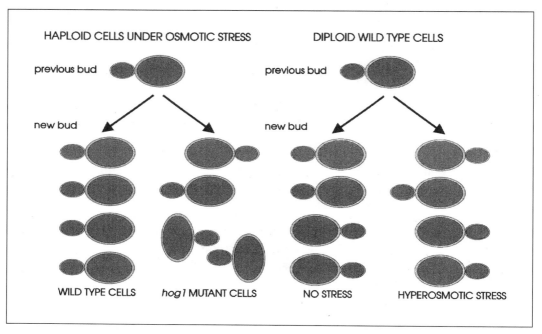

HAPLOID CELLS UNDER OSMOTIC STRESS   DIPLOID WILD TYPE CELLS

previous bud   previous bud

new bud   new bud

WILD TYPE CELLS   *hog1* MUTANT CELLS   NO STRESS   HYPEROSMOTIC STRESS

Fig. 4.2. Cell polarity as manifested in yeast budding pattern and the consequences of hyperosmotic stress on the position of bud appearance. Haploid cells show a unipolar budding pattern where the new bud usually appears close to the scar of the previous bud. Diploid cells exhibit a bipolar pattern where the new bud usually develops from the pole opposite to the scar of the previous bud.

cells also show a more random budding pattern when forming a new bud.[38] Diploid cells bud in an axial fashion, i.e., new buds always appear on the opposite pole with respect to the previous bud (Fig. 4.2.).[37] Even wild type diploid cells appear to become disoriented after a hyperosmotic shock and resume growth with a more random budding pattern.[38] These observations demonstrate very nicely the impact of an osmotic shock on the cytoskeleton. They also suggest that the actin cytoskeleton "remembers" at least for some time the set-up and the polarity that it had before it became disassembled. It is unclear why there is a difference in this respect between haploid and diploid cells.

A mutation suppressing the temperature and osmotic sensitivity of the *act1-1* mutant has been studied in more detail. The role of the corresponding gene *SAC3* remains unclear, however. Deletion of *SAC3* does not cause the same suppressive effect as the *sac3* point mutation.[39] Interestingly, suppressor mutations of *act1* mutants that specifically suppressed either the tem-

perature or the osmosensitivity could be isolated, suggesting that actin interacts with different proteins and/or in a different way with the same protein(s) after a heat or an osmotic shock.[32,33] These mutations may lead to the identification of novel actin-binding proteins or to as yet unknown mechanisms controlling the interaction of actin with already known actin-binding proteins.

Recently a strain defective in the *KAR3* gene was isolated from a collection of osmosensitive mutants.[40] *KAR3* has been initially identified through mutants that are impaired in karyogamy during mating.[41] Kar3p is a minus-end-directed motor kinesin that is also involved in mitosis, specifically by functioning together with centromere binding proteins in the kinetochor.[42] Expression of the *KAR3* gene is induced by mating pheromone[41] and apparently also by hyperosmotic stress.[40] The mutant cells arrest growth after a hyperosmotic shock with an aberrant morphology. The molecular details of the role of Kar3p in the resumption of growth after an osmotic shock are unknown. Perhaps the spindle apparatus can function fairly well even without Kar3p under optimal growth conditions but not when osmotic stress is imposed. *Kar3* mutants are not sensitive to other stress conditions.[40]

2.4. ACCUMULATION OF COMPATIBLE SOLUTES

The response to osmotic stress that has attracted most attention—probably due to its rather simple biochemical detection—is the accumulation of compatible solutes.[43] The strategy to accumulate such solutes is shared by probably all types of cells. The purpose of accumulating compatible solutes is likely to increase internal osmolarity and thereby enable the cell to retrieve water more efficiently from the environment.[43] Some compatible solutes may also be able to replace water or to increase its surface tension and thereby stabilize the water around macromolecules.[1,43,44] The term "compatible" refers to the most important property of those osmolytes: they do not affect the physical and biochemical processes within the cell. Compatible solutes accumulated by different organisms include certain ions like potassium, sugars, sugar alcohols as well as amino acids and their derivatives.[43]

The compatible solute is either produced by the cell or it is taken up actively from the environment or both.[5,43,45,46] The specific solute employed is species-, cell type- and stage-specific. Certain organisms use only one, others use several compatible osmolytes and the complexity of the response in this respect is unrelated to the complexity of the organism: *Escherichia coli*, apparently in a well defined order, uses potassium ions, trehalose and glycine betaine as compatible solutes during the response to hyperosmotic stress.[45,47] This also shows that the different classes of osmolytes are not restricted to certain groups of organisms. Yeasts appear to employ preferentially sugar alcohols and *S. cerevisiae* uses glycerol as its sole compatible solute.[5-8]

The more tolerant yeasts like *Zygosaccharomyces rouxii* use in addition to glycerol polyols such as mannitol[5,8] as a compatible solute. They also differ in their strategy to accumulate glycerol. *S. cerevisiae*, however, appears to control glycerol accumulation mainly at the level of production[48,49] and membrane permeability[50] (e.g., *Z. rouxii* or *Debaryomyces hansenii* employ active uptake systems for glycerol that allow accumulation of glycerol against a 1000-fold concentration gradient.[51,52]) It is, however, unclear whether the ability to actively accumulate glycerol is the basis for the much higher osmotolerance of these yeasts. It appears that the intracellular glycerol concentrations attained under hyperosmotic stress (about 1-1.5 M) do not differ substantially between yeasts that possess an active glycerol uptake system and those that do not.[5]

While the observation that yeast cells produce glycerol under hyperosmotic stress was many years ago, only recently has genetic analysis confirmed that glycerol production actually constitutes an important step in the adaptation to high external osmolarity. Mutants unable to produce high levels of glycerol, like *gpd1Δ* mutants, (see further) are hyperosmosensitive[53,54] and mutants unable to accumulate any glycerol at all, like *gpd1Δ gpd2Δ, ανδ* double mutants are even more sensitive.[55]

3. GLYCEROL METABOLISM

Glycerol is produced by yeast cells and can be utilized as a carbon source. The pathways for glycerol production and degradation employ different enzymes (Fig. 4.3).[56,57] Both pathways share glycerol-3-phosphate as an intermediate and it is not known how the glycerol-3-phosphate pools for utilization and production are separated. The enzymes converting glycerol-3-phosphate and dihydroxyacetonephosphate are located in different compartments and employ different cofactors. Moreover, the equilibrium of both dehydrogenase reactions very strongly favors the reaction required within the corresponding pathway. The pathways for glycerol production and utilization both use the glycolytic intermediate dihydroxyacetonephosphate as link to glycolysis and gluconeogenesis.[5,56,57]

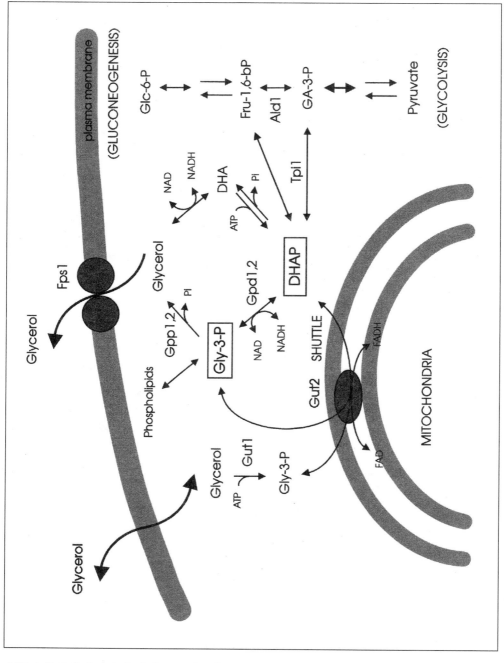

Fig. 4.3. Yeast glycerol metabolism. Gut1p: Glycerol kinase; Gut2p: FAD-dependent glycerol-3-phosphate dehydrogenase. Gpd1p,2p: NAD-dependent glycerol-3-phosphate dehydrogenase. Gpp1p,2p: Glycerol-3-phosphatase. Fps1p: Glycerol facilitator. Tpi1p: Triosephosphate isomerase. Ald1p: Fructose-1,6-bisphosphate aldolase. DAH: Dihydroxyacetone. DHAP: Dihydroxyacetonephosphate. The relevance of the pathways in glycerol production and utilization via DHA in S. cerevisiae is unknown but genes encoding at least the enzymes for glycerol breakdown appear to exist.

Other yeast such as *Schizosaccharomyces pombe* may use alternative pathways in which glycerol and dihydroxyacetone are interconverted by glycerol dehydrogenase. In glycerol utilization dihydroxyacetone is phosphorylated by a specific kinase to dihydroxyacetonephosphate. A specific dihydroxyacetonephosphatase is involved in glycerol production.[11,56,57]

3.1. GLYCEROL UTILIZATION

No transport system has been identified for glycerol utilization and it is assumed that glycerol is taken up by passive diffusion.[56] Mutants lacking the yeast glycerol facilitator Fps1p are not impaired in the utilization of different glycerol concentrations or under different growth temperatures.[50] This is in contrast to the situation in *Escherichia coli* where a homologue of Fps1p—the glycerol facilitator *glpF*—is part of the *glp* operon encoding components of the glycerol utilization pathway. Lack of GlpF abolishes growth on glycerol.[58-60]

Glycerol is phosphorylated by glycerol kinase encoded by the *GUT1* gene[61,62] and oxidized to dihydroxyacetonephosphate by a mitochondrial, FAD[+]-dependent glycerol-3-phosphate dehydrogenase encoded by *GUT2*.[61,63] Mutants defective in either of the two genes *GUT1* or *GUT2* are unable to utilize glycerol as a carbon source, suggesting that this pathway is the only route for glycerol utilization in *S. cerevisiae*.[61-63]

Surprisingly, however, recent evidence in the genome of *S. cerevisiae* implies the existence of genes encoding glycerol dehydrogenase and dihydoxyacetonekinase, enzymes of the alternative route of glycerol utilization employed, e.g., by *S. pombe*.[64] The role of those gene products is presently being investigated specifically with the possibility in mind that this pathway may be used in the degradation of glycerol produced under osmotic stress conditions.

3.2. GLYCEROL PRODUCTION

The first step in glycerol production is the reduction of dihydroxyacetonephosphate to glycerol-3-phosphate catalyzed by a cytosolic NAD[+]-dependent glycerol-3-phosphate dehydrogenase.[56,65,66] Two genes have been identified which encode the two isoenzymes *GPD1*[53,54] and *GPD2*.[55,67] There has been some confusion about the map location of *GPD2*, which has lead to the proposal of the existence of a third gene designated *GPD3*.[67,68] However, the recent completion of the sequence of the yeast genome revealed only two genes: *GPD1* on chromosome IV and *GPD2* on chromosome XV with the latter being referred to as *GPD3* in ref. 68. In addition, mutants lacking both *GPD1* and *GPD2* do not produce any detectable glycerol[55]; this confirms that the products of these two genes are the only glycerol-3-phosphate dehydrogenases and that this pathway is most likely the only pathway for glycerol production in *S. cerevisiae*.

Glycerol-3-phosphate is dephosphorylated by an apparently specific glycerol-3-phosphatase.[69] This enzyme also appears to be encoded by two highly homologous genes: *GPP1* and *GPP2*.[70] A mutant lacking these two proteins has not yet been reported. It is therefore unclear at present whether unspecific phosphatases may contribute to glycerol production from glycerol-3-phosphate.

Glycerol is exported from the cell probably in two different ways. Passive diffusion has been thought for many years to be the only mechanism by which glycerol penetrates the plasma membrane.[56] However, the larger part of the glycerol produced leaves the cell by facilitated diffusion through the Fps1p channel protein.[50] Lack of Fps1p causes poor growth in strains that overproduce glycerol or under anaerobic conditions when high levels of glycerol are produced.[71]

3.3. DUAL ROLE OF GLYCEROL PRODUCTION IN YEAST PHYSIOLOGY: OSMOREGULATION AND REDOX-BALANCING

Biochemical and physiological analysis led to the conclusion that in yeast glycerol metabolism has two very different roles in osmoregulation and in redox-balancing.

The role of glycerol as a compatible solute has been described above.

Yeast sugar catabolism is very much restricted to ethanol fermentation even under aerobic conditions.[57] Alcoholic fermentation is formally a redox-balanced pathway, i.e., all the NADH + H[+] produced in glycolysis by the activity of glyceraldehyde-3-phosphate dehydrogenase is re-oxidized by alcohol dehydrogenase. However, intermediates of the lower part of glycolysis are also being used for biosynthetic purposes like pyruvate for amino acid biosynthesis. Hence there is always some excess of NADH + H[+]. In the presence of oxygen the reduced nucleotide is re-oxidized by respiration. However, under anaerobic conditions other mechanisms have to take over. It has been proposed that glycerol production recycles NAD[+] (reviewed in ref. 72; Fig. 4.4) since under

such conditions higher glycerol levels are produced.

Analysis of the phenotypes of mutants affected in the genes *GPD1* and/or *GPD2* has confirmed the role of glycerol production in both osmoregulation and redox-balancing and has also revealed that the two genes play distinct roles in these two physiological processes. *Gpd1Δ* mutants are specifically affected for growth at high osmolarity, but the phenotype is not very strong and needs to be monitored on plates with about 8% NaCl in a strain background like W303-1A.[53-55,73] Growth of *gpd1Δ* mutants in the absence of oxygen is not impaired.[55] On the other hand *gpd2Δ* mutants show poor growth under anaerobic conditions but normal growth under hyperosmotic stress.[55] The double mutant *gpd1Δ gpd2Δ* is highly sensitive to osmotic stress, i.e., it does not grow on plates containing 4%

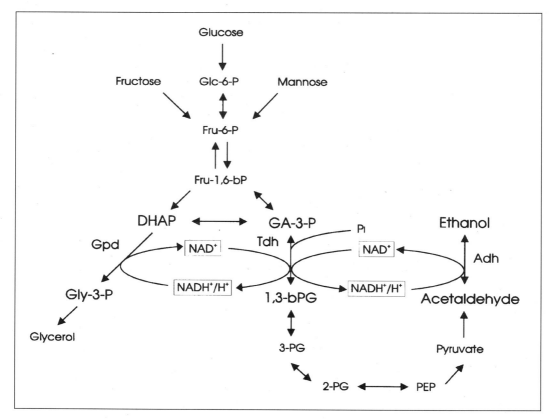

Fig. 4.4. The role of glycerol production in redox-regulation. Gpd, Glycerol-3-phosphate dehydrogenase; Tdh, Glyceraldehyde-3-phosphate dehydrogenase; Adh, Alcohol dehydrogenase.

NaCl and does not grow at all under anaerobic conditions.[55] These results suggest that Gpd1p and Gpd2p have overlapping but distinct functions (Fig. 4.5). Overexpression of either *GPD1* or *GPD2* from a multi-copy plasmid can rescue both the osmosensitivity and the inability to grow under anaerobic conditions of a *gpd1Δ gpd2Δ* double mutant, demonstrating that there is no specific functional difference between the two isoenzymes.

The basis for the distinct roles of the two proteins is differential expression of the genes *GPD1* and *GPD2*.[55] Expression of *GPD1* is stimulated by hyperosmotic stress[54,55,73,74] but not during anaerobic growth.[55] Expression of *GPD2*, however, is stimulated during growth in the absence of oxygen but not under hyperosmotic stress. The level of *GPD2* mRNA appears to be rather diminished during osmostress-induction of *GPD1* expression.[55] Thus the two very different physiological stimuli appear to trigger different signaling path-

ways controlling the expression of the two genes *GPD1* and *GPD2*. The two isoenzymes, however, appear to contribute to the same pool of glycerol-3-phosphate dehydrogenase activity (Fig. 4.5). Remarkably, however, *gpd1Δ* mutants can still increase glycerol production under osmotic stress to some extent[54] and *gpd2Δ* mutants show a stimulation of glycerol production under anaerobic conditions[55], although the enzyme activity does not increase in either situation. Thus in addition to enhanced enzyme production other mechanisms such as control of glycolytic flux must exist that regulate glycerol production under osmotic stress and anaerobic conditions, respectively.

It will be interesting to see whether this dual pathway of glycerol production is also apparent at the level of glycerol-3-phosphatase. Indeed, expression of *GPP2* but not of *GPP1* is stimulated by high external osmlarity[70,74] (note that the genes *HOR2* and *RHR2*, referred to as of un-

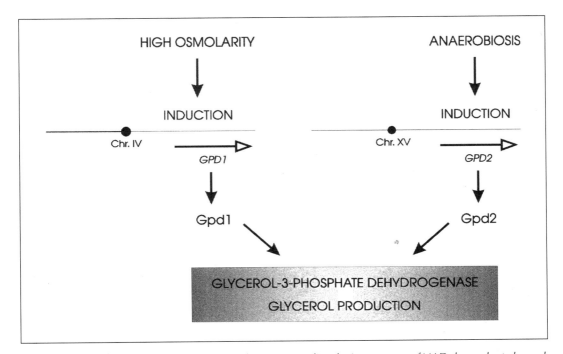

Fig. 4.5. *The expression of the genes* GPD1 *and* GPD2 *encoding the isoenzymes of NAD-dependent glycerol-3-phosphate dehydrogenase is controlled by different stimuli, i.e., by osmotic stress and by anaerobic growth, respectively. However, it appears that both isoenzymes contribute to the same pool of enzyme and are functionally interchangeable for glycerol production.*

known function in ref. 74, are the same as *GPP2* and *GPP1* respectively[70]). Whether expression of *GPP1* responds to the oxygen level has not been studied. It should, however, be noted that the glycerol-3-phosphatase activity is not really essential for redox regulation and therefore it is well possible that *GPP1* is expressed at a constitutive level.

The signaling pathway(s) involved in the control of expression of *GPD2* is entirely unknown and expression of this gene is not affected by known regulators putatively involved in redox-regulation like Rox1p.[55,75] Normal induction of expression by hyperosmotic stress of the genes *GPD1* and *GPP2* depends on an active HOG-pathway (see further and refs. 54, 70, 74).

4. OSMOTIC AND OXIDATIVE STRESS

Redox-regulation and response to oxidative stress are probably largely distinct mechanisms. The former is understood as those control circuits that ensure a balance between oxidized and reduced NAD and NADP nucleotides, while the latter includes the mechanism acting upon damage and stress induced by oxygen radicals (see chapter 6).

However, an overlap has been observed for the responses to hyperosmotic and oxidative stress (see also chapter 6). Out of 16 complementation groups identified in a screen for mutants sensitive to hydrogen peroxide, 9 groups were not only sensitive to oxidative stress but also to high salt concentrations. Of those 9 groups, 6 were also heat sensitive which may point to some defect in a general stress response mechanism (see chapter 7). The phenotype of the other 3 groups—sensitivity to both oxidative and hyperosmotic but not to heat stress—may be indicative of an overlap in the osmotic and oxidative stress response mechanisms.[76]

Such an overlap between water stress and oxidative damage is well established in plants. The photosynthetic reactions are the major source of reactive oxygen species in plants. Since water stress affects cellular metabolism and chloroplast function in particular, in several ways it appears that water stress also increases the production of reactive oxygen species. Hence water stress in plants also requires the establishment of mechanisms that protect the cell against oxidative damage.[46,77] Whether water stress in yeast can also lead to oxidative stress via disturbance of mitochondrial metabolism has not been investigated. However, it has been demonstrated that mutants in catalase T, which plays an important role in the disposal of reactive oxygen species, are also sensitive to very severe forms of salt stress. This might be taken as a (preliminary) hint towards an overlap between both forms of stress.[78]

5. SIGNAL TRANSDUCTION UNDER OSMOTIC STRESS

Signal transduction pathways stimulated by osmotic stress are among the best understood stress-induced signaling pathways, at least in yeast. Remarkably, both yeast signaling pathways that respond to changes in osmolarity—the HOG (high osmolarity glycerol)-pathway and the cell integrity or PKC (protein kinase C) pathway—use as their central signal transduction sequence a cascade of protein kinase which is generally referred to as MAP (mitogen activated protein)-kinase pathways.

5.1. YEAST MAP-KINASE PATHWAYS

The HOG-pathway is one of probably five MAP-kinase pathways in *S. cerevisiae* (Fig. 4.6). These pathways are the mating pheromone response pathway, the pseudohyphal development pathway, the cell integrity or PKC-pathway, the sporulation pathway and the HOG-pathway (reviewed in refs. 15-19). All these pathways have in common a cascade of three protein kinases which are referred to as the MAP-kinase (MAPK), the MAP kinase kinase (MAPKK) and the MAP kinase kinase kinase (MAPKKK). The yeast MAP-kinase signaling cascades share this feature with similar pathways from animals[18] and plants.[79] All three protein kinases have been identified by genetic analysis for three of the five yeast MAP kinase pathways. No MAPK has been described up to now for the pseudohyphal development pathway, and the sporulation pathway is still lacking the MAPKK and the MAPKKK (Fig. 4.6).[18] However, since such kinases can be identified by specific sequence homologies the knowledge of the entire yeast genome sequence should reveal those missing links in the near future.

Another upstream protein kinase which controls the MAP kinase cascade has been identified for all yeast MAP kinase pathways. It appears, however, that the basic mechanisms for the input upstream of this controlling kinase is different for each signaling pathway. In three cases—the mating response pathway, the pseudophyhal development pathway and the cell integrity (PKC) pathway—one or several GTP-binding proteins appear to be involved in signaling between the sensor and the protein kinases. The actual receptor proteins are only known for the mating pheromone response pathway and for the HOG-pathway.[15-18]

Remarkably, two pathways—the mating pheromone response pathway and the pseudohyphal development pathway—share the MAPKK and the MAPKKK as well as the upstream kinase (see also chapter 1). This surprising finding, as well as the homology of the kinases to each other in general, imply the important question of how specificity of these pathways is controlled. Keeping the activity and the level of all components of the pathway well controlled appears to be one strategy to support specificity. Gain of function mutations of Ste7p—the MAPKK of the pheromone response pathway—can stimulate the PKC cell integrity pathway when the mutant protein is overexpressed.[80] An amazing alternative mechanism to achieve pathway specificity has been discovered in the mating pheromone response pathway which possesses the specific scaffold protein Ste5p, which appears to bind the entire MAP kinase cascade as well as upstream components (reviewed in refs. 81, 82). It has been demonstrated that Ste5p indeed contributes

to the specificity of the pathway since the same gain of function mutation of Ste7p mentioned above can activate the PKC cell integrity pathway in the absence of Ste5p even when expressed to normal levels.[80] However, such a scaffold protein has not yet been identified for the other MAP kinase pathways.

5.2. THE HIGH OSMOLARITY GLYCEROL RESPONSE (HOG) PATHWAY

The known components of the HOG-pathway (Fig. 4.7) have been identified by different approaches and in different genetic screens. *HOG1*—the MAP kinase homologue—was first identified as a mutation conferring sensitivity to high external osmolarity, and the gene was cloned by complementation of this phenotype.[83] *PBS2*, which encodes the MAPKK, had been identified well before in a screen for mutants sensitive to the antibiotic polymyxin B which damages membranes and causes cell lysis.[84,85] Subsequently this gene has been found in at least three independent collections of osmosensitive mutants.[83,86,87] In addition, both *HOG1* and *PBS2* were identified as mutations suppressing the lethality of the *sln1* mutation.[88] Sln1p is a protein which bears homologies to both the receiver and transmitter domains of bacterial two component systems.[89-91] *SLN1* was first identified in a screen for mutations that require for viability the *UBR1* gene which encodes a protein involved in the first step in protein degradation.[92] Ubr1p recognizes the N-terminus of proteins according to the N-end rule for sequences mediating protein degradation and initiates the process of ubiquitination and ubiquitin-dependent protein degradation.[93] Subsequently *SLN1* was also identified by screening for mutations that require the protein phosphatase Ptp2p for survival.[94] Although the *sln1* mutation turned out to be lethal by itself, a relation to Ptp2p could be established since overexpression of *PTP2* suppressed the lethality of the *sln1* mutation.[88] In addition, the screen for synthetic lethality

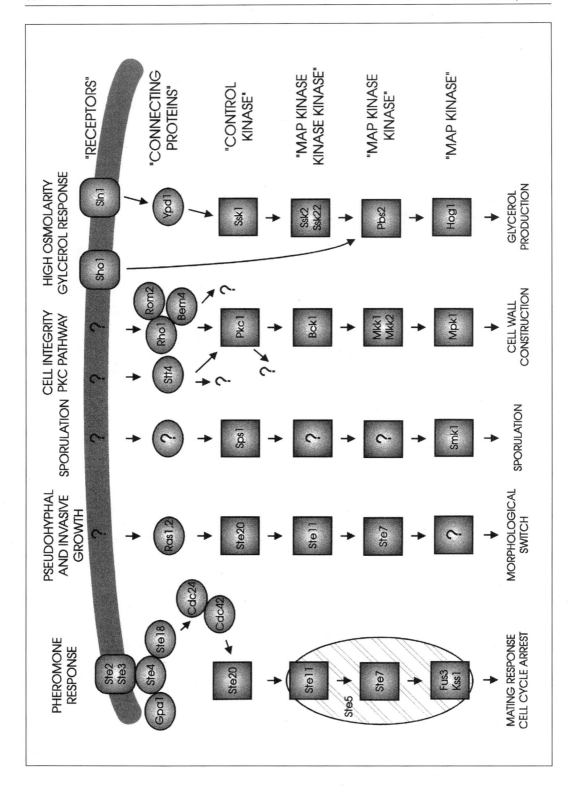

with *ptp2Δ* lead to the identification if *YPD1* encoding a novel protein involved in signal transduction from Sln1p (see further).[95]

Other mutations suppressing the lethal phenotype of *sln1* mutations in addition to *hog1* and *pbs2*, are *ssk1* and *ssk2*. Ssk1p appears to be the third protein in the Sln1p-Ypd1p-Ssk1p osmosensor-signaling complex since it contains a receiver domain as well as a putative output domain.[88,95] Ssk2p has homology to MAPKKK and appears to function in the HOG-pathway redundantly together with Ssk22p, which is 69% identical to Ssk2p within the protein kinase domain, and also shares homology to Ssk2p over the entire protein.[96] Surprisingly, and in contrast to *hog1Δ* and *pbs2Δ* mutants, the double mutant *ssk2Δ ssk22Δ* is not osmosensitive. Hence mutations were selected that caused osmosensitivity when in combination with mutations in *SSK2* and *SSK22*. This approach identified a novel mutation in *PBS2* as well as a new gene: *SHO1*. Sho1p is predicted to be a membrane protein and is thought on this basis to be a second osmosensor in addition to Sln1p/Ssk1p.[96]

In addition to the tyrosine phosphatase Ptp2p two further genes were shown to suppress the lethality of *sln1* mutants when overexpressed: *PTC1* and *PTC3* encoding serine/threonine phosphatases of the PP2C type.[88,94]

All these components of the HOG-pathway were identified within the last three to four years. This demonstrates how powerful genetic approaches in yeast can be for the identification of components of signaling pathways.

5.3. OPERATION OF THE HOG-PATHWAY

Although many details about its control and the interplay with other pathways remains to be explored, an impressive genetic analysis combined with biochemical approaches has shed substantial insight into how the HOG-pathway operates.

The sensor Sln1p appears to control the pathway negatively and lack of Sln1p causes lethality due to an overactive pathway. This is inferred from the observations that mutations blocking the pathway downstream of the sensor as well as overexpression of three different protein phosphatases suppress the lethality caused by lack of Sln1p.[88,96] This proposal suggests that under normal growth conditions the Sln1p protein kinase is active and it turns the output domain of Ssk1p silent. It has indeed been demonstrated that a Ssk1p variant which cannot be phosphorylated due to mutation of Asp554 to Asn is functional since a double mutant *sln1Δ ssk1(Asp554Asn)* is not viable while a *sln1Δ ssk1Δ* strain is.

The histidine kinase domain of Sln1p autophosphorylates His576 of the protein and from there the phosphate group is transferred to the Sln1p receiver domain on Asp1144. The phosphate is not directly transferred to Ssk1p from this asparate

Fig. 4.6 (opposite). *MAP kinase signaling pathways in yeast. Five pathways have been identified. For several of the pathways protein phosphatases that may control the activity of the protein kinases negatively have been identified but are not depicted in this scheme (see Figs. 4.7 and 4.9). The general topology of the pathways predicts the existence of a module of three kinases controlled by an upstream kinase. This upstream kinase appears to be linked to the receptors in different, pathway-specific ways. Often G-proteins (and proteins regulating their activity via GTP/GDP binding and GTP hydrolysis) are involved in upstream signaling such as in the pheromone response pathway, the pseudohyphal development pathway or in the PKC-pathway. Ste5p is a scaffold protein that binds several components of the pheromone response pathway. Stt4p is not a G-protein but a phosphatidylinositol-4-kinase, suggesting that more proteins involved in phosphatidylinositol metabolism may play a role at this step. The phenotype of mutants defective in different components of the PKC-pathway suggest that there are several pathways branching off from the upstream components (see also Fig. 4.9). Ypd1p is a protein that transfers a phosphate group from Sln1p to Ssk1p (Fig. 4.7).*

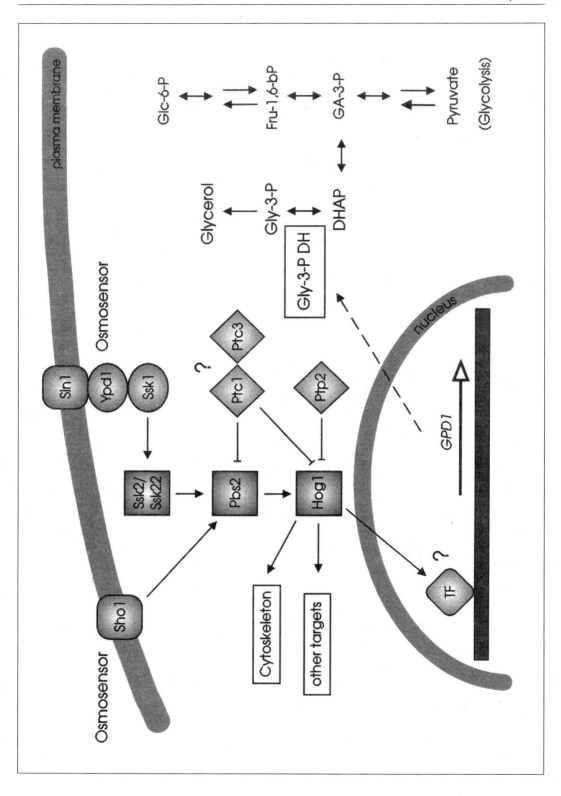

group. In contrast to classical two-component systems, the yeast osmosensor complex appears to consist of (at least) three proteins. The phosphate group from Sln1p Asp1144 is first passed further to His64 on Ypd1p and then reaches Asp554 on Ssk1p. This complex phosphorelay mechanism has been studied by elegant genetic analysis of the three components involved.[95] Why this mechanism is so elaborate is unclear but it may provide additional levels of regulation and fine tuning. The mechanism may also apply to other nonconventional "two-component systems" in particular in eukaryotes.[95]

Sln1p is predicted to have two transmembrane domains in an arrangement similar to the bacterial osmosensor EnvZ which is also part of a two-component system. However, the proteins are not homologous except for the histidine kinases domain and for the phosphorylated residues.[89,90] Both EnvZ and Sln1p have a domain between the two transmembrane segments which is predicted to face the periplasmic space and hence could be involved in the sensing process. However, these domains differ very much in size (EnvZ: about 125 amino acids; Sln1p: about 290 amino acids) and they do not show any homology. It is not understood in either system[45,97] how hyperosmolarity induces transmembrane signaling and modulation of the histidine kinase domain. In addition, Sln1p differs from EnvZ in that the former has its own response regulator domain while in the bacterial EnvZ-OmpR system only OmpR has such a domain. In addition, there is no evidence for the involvement of a third protein as in the yeast Sln1p-Ypd1p-Ssk1p complex.[88,90,95]

As outlined above Ssk1p appears to be inactivated by Sln1p-Ypd1p under normal growth conditions. Whether this is the only mechanism inhibiting the activity of Ssk1p, i.e., if the Ssk1p(Asp544Asn) nonphosphorylatable mutation has a lethal effect even in the presence of a wild type Sln1p, has not been reported. Upon osmotic stimulation of Sln1p (a portion of) Ssk1p becomes dephosphorylated and active. Ssk1p directly interacts with the MAPKKK Ssk2p and Ssk22p in the two hybrid system, suggesting that the putative output domain of Ssk1p provides the link between the sensor system and the MAP-kinase cascade. In the case of Ssk2p it appears that the N-terminal region of the protein is sufficient for this interaction.[96] An effect similar to mutation of *sln1* can also be caused by deletion of the N-terminal inhibitory domain of either Ssk2p or Ssk22p which causes the MAPKKK to be constitutively active. The lethality of this manipulation can also be suppressed by deletion of *HOG1* or *PBS2*.[96] This confirms that overstimulation of the HOG-pathway is causing the lethality of mutations in *SLN1*. This finding also confirms the proposed order of the kinases in the pathway in that Ssk2p/Ssk22p must be upstream of Hog1p and Pbs2p. Thus according to genetic and two-hybrid analysis the following order of events can be summarized: The default state of Sln1p is active and it inactivates via Ypd1p Ssk1p. Upon stimulus Sln1p is (partially) inhibited, leading to (partial) activation of Ssk1p which then (partially) relieves the inhibition of the MAPKKKs.[96]

Osmostress-induced phosphorylation of Hog1p depends on Pbs2p[83] but not on Ssk2p/Ssk22p.[96] Furthermore, *hog1* and *pbs2* mutants are osmosensitive[83] while neither *ssk1* nor *ssk2* and *ssk22* single or double mutants are sensitive to hyper-osmolarity.[96] These observations have led to the discovery of Sho1p which is proposed to be an alternative osmosensor based on the presence of four predicted transmembrane

Fig. 4.7 (opposite). Model of the HOG-pathway in S. cerevisiae. *Two osmosensors stimulate at different steps and by different mechanisms a MAP-kinase cascade which in turn is controlled by at least three protein phosphatases. No target has been identified at the molecular level but it is expected that at least one transcription factor (TF) should be controlled by Hog1p. See text for details.*

domains. How Sho1p might operate in osmosensing is entirely unknown. It is also not known whether Sho1p is actually localized in the plasma membrane or in an intracellular membrane. Sho1p does not show strong homologies to any known proteins except for a SH3 domain. The SH3 domain interacts directly with a putative proline-rich SH3-binding motif in Pbs2p, as has been demonstrated by two hybrid analysis. A mutation in the SH3-binding motif of Pbs2p which has been isolated in the same screen for osmosensitive mutants in a ssk2Δ ssk22Δ background that also identified *SHO1* abolishes this interaction. How interaction between Sho1p and Pbs2p activates Pbs2p is not known and the possibility of involvement of other proteins is not excluded.[96] Thus it has been proposed that two signaling pathways converge at the level of the MAPKK Pbs2p: one triggered by the Sln1p/Ssk1p sensor transmitting the signal via a MAPKKK (Ssk2p/Ssk22p) and the other triggered by Sho1p (Fig. 4.7).

The signal transmitted either by the Sho1p branch or by Ssk2p has been followed by measuring phosphorylation of Hog1p in ssk2Δ ssk22Δ and sho1Δ ssk22Δ mutants, respectively. These experiments suggest that the Sln1p/Ypd1p/Ssk1p/Ssk2p/Pbs2p pathway responds to lower osmolarity (saturated at 100 mM NaCl) and faster (within 1 minute) than the Sho1p/Pbs2p pathway (saturated at 300 mM NaCl; 2 minutes). Thus it has been suggested that the role of the two sensors might be in modulating the response over a broader range of osmolarity changes and over different time spans.[96]

As mentioned above, tyrosine-phosphorylation of Hog1p upon a hyperosmotic shock is very fast and can be monitored within 1 minute after addition of the osmolyte.[83,88] The same type of data also suggest that activation of the pathway is transient, i.e., phosphorylation of Hog1p is reduced to a lower level[83] or entirely cancelled[88] within about 20 minutes after the shock. Finally, it has been shown that Hog1p phosphorylation appears to be

maximal at about 300-400 mM salt and higher concentrations did not further stimulate the pathway, although stimulating mutation such as sln1Δ and N-terminal deletions of Ssk2p or Ssk22p cause a much higher tyrosine-phosphorylation.[83,88,96] These observations have profound consequences for the physiological role of the pathway, especially in the induction of gene expression as discussed in section 6.2.

No molecular target of the Hog1p protein kinase has been identified up to now. However, since a variety of genes are stimulated by hyperosmotic stress in a Hog1p-dependent fashion (see further) and since many MAP-kinases have been shown to probably directly control transcription factors[18], it is generally anticipated that one important target is a (or several) transcription factor(s).

Interestingly, the sensor Sln1p, but not the protein kinase cascade of the HOG-pathway, has recently been related to the transcription factor Mcm1p which controls a variety of genes, e.g., in mating response. *SLN1* has been identified as a mutation which stimulates the activity of Mcm1p but Hog1p does not transmit the signal.[98] This suggests that Sln1p controls other pathways in addition to the HOG-pathway kinase cascade.

MAPKK are dual-specific kinases that phosphorylate the MAPK on both threonine and tyrosine. MAPKKK, however, are specific for serine and threonine. Thus the Ptc1p and Ptc3p serine/threonine, specific phosphatases are expected to control both Hog1p and Pbs2p while the tyrosine-specific phosphatase Ptp2p may be responsible for inactivation of Hog1p. There is biochemical evidence that Ptp2p affects the phosphorylation state of Hog1p (M.C. Gustin, personal communication).

As mentioned above, *SLN1* and *PBS2* were originally identified in genetic screens entirely unrelated to the osmotic stress response. The reason for this is still unclear in both cases. A mutant in *sln1* was found to cause synthetic lethality to the protein degradation mutant *ubr1*.[89,92] Since mutation of *SLN1* appears to cause a hyperactive HOG-pathway, it is possible that ubiquitin-dependent protein degradation is involved in a process downregulating the pathway or inactivating a target of the HOG-pathway. There is no report available addressing this issue.

PBS2 has been identified as a mutant that confers hypersensitivity to polymyxin B which causes membrane destruction and cell lysis.[84,85] It is possible that this effect is enhanced in a mutant with defective osmoregulation, however, this has not yet been studied.

The HOG-pathway can apparently also be stimulated by low pH-stress since tyrosine-phosphorylation of Hog1p also increases under such conditions, albeit only weakly.[78] It has not been tested whether the same upstream components of the pathway that are involved in osmosensing also trigger this effect.

5.4. THE "HOG-PATHWAY" IN *SCHIZOSACCHAROMYCES POMBE*

MAP-kinase pathways appear to occur in all eukaryotic cell types and they transmit signals generated by a variety of stimuli and by different mechanisms.[18] A highly important question is whether pathways not only structurally but also functionally related to the *S. cerevisiae* HOG-pathway exist in other eukaryotic organisms. This appears to be the case.

In the fission yeast *S. pombe,* which appears to be evolutionary quite distant from baker's yeast, components of an osmosensing MAP-kinase pathway have been discovered in studies on cell cycle control (Fig. 4.8). *Wis1*+ encodes a Pbs2p homolog[99-101] and *spc1*+[102]—also called *sty1*+[101] and *phh1*+[103]—encodes a Hog1p homologue. *Wis1*+ was first identified as a regulator of mitosis.[99] Subsequently the same gene was found in screens for suppressors of the lethality of mutants lacking the tyrosine phosphatases Pyp1p and Pyp2p[101,102] as well as the CaCl$_2$-sensitivity of mutants lacking the serine/threonine phosphatases Ptc1p and Ptc3p.[100] Both those screens also yielded *spc1*+.[101,102] Genetic as well as biochemical evidence showed that Spc1p is

activated by the Wis1p MAPKK[101,102] and inhibited by the two tyrosine phosphatases Pyp1p and Pyp2p.[101] Pyp2p has been shown in the two hybrid system to interact physically with Spc1p.[104] In addition, the serine/threonine phosphatases of the 2C type Ptc1p, Ptc2p and Ptc3p appear to downregulate the pathway as well.[100,102] Thus the situation is similar to that in *S. cerevisiae* where the two PP2C-type phosphatases Ptc1p and Ptc3p as well as the tyrosine phosphatase Ptp2p control the HOG-pathway.[88,94] More such phosphatases may exist in *S. cerevisiae* as well. Several protein phosphatases are implicated in the control of other yeast MAP kinase pathways like the mating pheromone response pathway[105] and perhaps the cell integrity PKC-pathway[106-109] as well as of MAP-kinase pathways in higher eukaryotes.[110] It can be speculated that the phosphorylation/dephosphorylation of targets on the MAPK will be controlled in a similar fashion by kinase-phosphatase interplay. It should be noted that while the mechanisms for activation of the protein kinases of MAP-kinase pathways are, at least in terms of the pathway involved, fairly well understood, little is known how the phosphatases are controlled. It is of course possible that such phosphatases are constitutively active and as soon as the stimulation of the pathway is diminished the kinases (and their substrates?) become inactivated. Indeed, there is evidence from the phenotype of mutants lacking certain phosphatases that the HOG-pathway (in *S. pombe*) is operative even under normal osmolarity but is downregulated by the phosphatases (see further). However, closer investigation usually shows that regulatory proteins do not work uncontrolled in living cells. Accordingly, expression of the gene encoding the Pyp2p phosphatase in *S. pombe* is strongly induced by hyperosmotic stress and this expression is dependent on the Wis1p/Spc1p pathway.[101,104] This suggests that stimulation of the protein kinase cascade also leads at least to an increased potential for its inactivation. It is well known that the mating pheromone response becomes

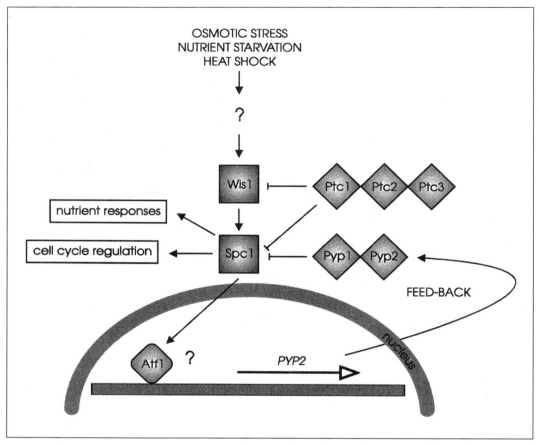

Fig. 4.8. Model of the HOG-pathway in S. pombe. In contrast to the situation in budding yeast this pathway is stimulated by different stress conditions in a, however, unknown fashion. In addition to expression of gpd1⁺ (like in S. cerevisiae) it has been shown in S. pombe that the expression of at least one of the genes encoding a phosphatase is controlled by the pathway, suggesting a feedback loop as depicted in the illustration.

attenuated after some time even if exposure to the pheromone persists. Thus downregulation of MAP-kinase pathways may include active and well regulated mechanisms.[105] Further work is required to achieve a full understanding of how the kinases and phosphatases cooperate in the fine-tuning of MAP-kinase pathways in general and in osmoregulation in particular.

No elements of the Wis1p/Spc1p pathway upstream of the MAPKK Wis1p have been reported. However, and in contrast to the situation with the HOG-pathway in *S. cerevisiae,* a candidate for a transcription factor regulated by the pathway in *S. pombe* has recently been identified. The factor Atf1p belongs to the group of basic leucine zipper (bZIP) proteins and has been

shown to be required for sexual development and for the acquisition of stress tolerance in stationary phase. Those phenotypes have also been described for mutants affected in the kinases of the Wis1p/Spc1p-pathway. There appears to be also evidence that mutants lacking Atf1p are defective in osmostress-induced transcription. However, this evidence is still unpublished.[18,111] There is no close homologue to Atf1p in the *S. cerevisiae* genome. Another transcription factor possibly controlled by the Wis1p/Spc1p pathway may be Pcr1p.[112]

Remarkably, the phenotype of mutants defective in the *S. pombe* HOG-pathway appears to be more complex than that in *S. cerevisiae.* Mutants lacking Wis1p or Spc1p are not only sensitive to high external osmolarity. In addition, they are heat-sensitive, show an elongated cell morphology at normal osmolarity, appear to be defective in nutrient control of the cell cycle, undergo mitosis at a larger cell size and are partially sterile.[99-103] Overexpression of *pyp1*[+] or *pyp2*[+] causes a similar effect at least with respect to cell size.[101,113] Mutation of either *pyp1*[+] or *pyp2*[+] or in *ptc1*[+] and *ptc3*[+] cause the opposite phenotype, i.e., cell division at a smaller cell size. Such mutants also show aberrant cell morphology.[100-102,113] The morphology and cell size phenotype of *ptc1 ptc3* mutants can be rescued by osmotic stabilizers such as 1 M sorbitol.[100] This suggests that such mutants are adapted to higher external osmolarity and hence have a growth defect at normal osmolarity, i.e., that they are osmophilic mutants. This also suggests that the *S. pombe* HOG-pathway is operating even on medium with normal external osmolarity, but is inhibited by the phosphatases. The lethality of the *pyp1Δ pyp2Δ* double mutant cannot be rescued by addition of 1 M sorbitol either because the consequences of high HOG-pathway activity are too drastic or because of additional targets of these two phosphatases.

The more severe and pleiotropic phenotype of mutants affected in the *S. pombe* pathway may be due to a somewhat different role of the pathway as compared to the *S. cerevisiae* counterpart. The Wis1p/Spc1p-pathway in *S. pombe* is not only stimulated by high osmolarity but also by heat and oxidative stress.[104] In fission yeast the target genes of the pathway such as *GPD1* and *TPS1* (for trehalose-6-phosphate synthase) are consequently also stimulated by all these stress conditions.[104] In budding yeast *GPD1* expression is specifically induced by osmotic stress via the HOG-pathway and other unknown pathways, while *TPS1* transcription is controlled by multiple stresses via at least the HOG- and the Ras-cAMP pathway (see section 6).

Recently it has been shown that the Wis1p/Spc1p pathway is involved in carbon source-dependent control of gene expression of genes encoding gluconeogenic

enzymes such as *fdp1*[+] and in nutrient control of sporulation. In this regulatory context there appears to be a very close interplay between the *S. pombe* HOG-pathway and protein kinase A. It is unclear at which level this interplay occurs.[114,115] While a close interplay between the HOG-pathway and the Ras-cAMP protein kinase A pathway in the control of general stress responsive genes (see chapter 7) appears to occur in *S. cerevisiae* as well, no evidence has yet been provided for a role of baker's yeast HOG-pathway in nutrient control of gene expression or growth.

Other effects observed in mutants in fission yeast HOG-pathway may be due to *S. pombe* being more sensitive to high external osmolarity than budding yeast (A. Blomberg, personal communication). Cell size could certainly be affected by turgor control of the cell and even wild type cells of *S. cerevisiae* are known to grow at a smaller cell size in high osmolarity medium.[24] Cell size differences are also more readily monitored in *S. pombe* and investigators that have discovered the Wis1p/Spc1p pathway in fission yeast have actually searched for mutants undergoing mitosis at altered cell size. An interesting questiontherefore may be if the HOG-pathway monitors cell size or directly osmolarity, however, this problem might be difficult to address.

Finally, the partial sterility phenotype of *S. pombe* HOG-pathway mutants may have its basis in defective osmoregulation. In *S. cerevisiae* mutants with diminished glycerol export have been found to be affected in mating (J. Philips, I. Herskowitz, personal communication). Thus future analysis will show how similar are the architecture, the control and the physiological roles of the HOG-pathways in these two yeasts. This will have important consequences for the understanding of the role of similar pathways in higher eukaryotes.

5.5. MAMMALIAN SYSTEMS

Two mammalian MAP-kinases—JNK1 and p38—have been shown to be tyrosine-phosphorylated upon hyperosmotic shock.

Moreover, both kinases have been expressed in budding yeast and can partially substitute for Hog1p in a *hog1Δ* mutant.[116,117] The pathways which control these two kinases appear to be stimulated by different types of stress, growth factors, cytokines and mitogens (reviewed in refs. 18, 19). Thus it might be possible that these mammalian pathways are more similar to the HOG-pathway from *S. pombe* than from *S. cerevisiae*. At this stage, however, we canonly conclude that apparently also animal cells transmit stress signals via MAP-kinase pathways.[18,19]

5.6. PLANTS

Plants have been proposed to possess several signaling pathways operating under different aspects of water stress such as drought, high salt or cold. At least one of those pathways leads to the production of the stress hormone abscissic acid (ABA) while two pathways may be stimulated by the hormone.[79] In addition, three pathways may lead to effects like altered gene expression directly without employing ABA. Those pathways might be candidates for HOG-pathway-like signaling cascades in plants.

Plants appear to be a very rich source for MAP-kinase pathways.[79,118] They also have been shown to possess two component sensor-signaling devices.[79,119] Thus it is tempting to speculate that pathways similar to the HOG-pathway should be involved in the sensing and signaling of cellular stress imposed by drought, salt and cold. However, a large number of attempts to clone plant components of a putative HOG-pathway or any other MAPK by complementation of the *S. cerevisiae hog1Δ* mutant have been unsuccessful. This is remarkable since a MAPK from alfalfa has been shown to complement a yeast mutant lacking Mpk1p—the MAP kinase from the cell integrity PKC-pathway[120]—and an *Arabidopsis* MAPKKK could replace a *S. cerevisiae ste11* mutant lacking the corresponding kinase from the mating pheromone response pathway.[121] Considering the large number of MAP-kinases already

known from plants, the fact that such proteins can function even in an inappropriate pathway when overproduced and the large success of cloning plant genes for very different proteins by complementation in yeast[122], it is a real disappointment that this approach has failed up to now in osmoregulatory signal transduction. One possible explanation might be that plant kinases overstimulate the HOG-pathway in yeast which is, as discussed above, lethal. Thus future screens should address this problem. One could attempt to clone upstream kinases that confer when expressed in yeast *HOG1* dependent lethality.

5.7. THE CELL INTEGRITY PKC-PATHWAY

In some sense loss of the PKC-pathway in yeast has phenotypic consequences just opposing those caused by mutations in the HOG-pathway. While mutants of the HOG-pathway are sensitive to high osmolarity, mutants affected in the PKC-pathway require osmotic stabilizers. However, it appears that these effects are less related than it might look at first glance and mutants in the PKC-pathway are probably not truly hypo-osmosensitive or osmophilic. Instead it appears that they have a defect in cell wall assembly and hence require isotonic conditions for survival. This is in contrast to the phenotype of mutants lacking certain phosphatases in the Wis1p/Spc1p-pathway of *S. pombe* which also require osmotic stabilizers but which do not have a cell wall defect.[100] In any case the requirement of mutants in the PKC-pathway for increased osmolarity has prompted Gustin and co-workers to study the response of the PKC-pathway to changed osmolarity. They found that the MAPK Mpk1p is rapidly and strongly tyrosine-phosphorylated in response to a hypo-osmotic shock.[123] Thus we consider the PKC-pathway in this review as an osmosensing signaling pathway. However, while the HOG-pathway appears to be specifically stimulated by a drop in external osmolarity, alternative stimuli for the PKC-pathway have been reported: heat shock,[124] polarized cell growth[125,126] and cell cycle progression.[127] The pathway has also been implicated in nutrient sensing.[128]

The first protein of the PKC-pathway (Fig. 4.9) identified was Pkc1p itself, which is a homologue of mammalian α, β and γ isoforms of protein kinase C.[129-131] As mentioned above, mutants lacking *PKC1* are only viable when osmotic stabilizers are added to the medium (0.5 M sorbitol).[132] This effect is most probably due to a substantial weakening of the cell wall.[133] Isolation of suppressor mutations and genes suppressing in multi-copy the cell lysis phenotype of *pkc1* mutants revealed the existence of a MAP-kinase pathway downstream of Pkc1p, consisting of the MAPKKK homologue Bck1p,[134] a pair of probably redundant MAPKK encoded by *MKK1* and *MKK2*[134] and the Mpk1p MAPK.[135] Bck1p was initially described in a genetic screen for mutants affected in polarized cell growth and named Slk1p.[136] Mpk1p was first isolated as a suppressor of a mutation showing a lysis defect and was called Slt2p.[137,138] Epistasis[134,135,139,140] as well as biochemical[123,133,141] and two-hybrid analysis[141] have confirmed the order of kinases in this pathway as predicted according to other MAP-kinase pathways.

The cell lysis defect of *pkc1Δ* mutants is much more pronounced than that of any of the mutants lacking a downstream kinase of the pathway. Thus it has been proposed that Pkc1p has additional targets probably involved in cell wall assembly as well. Candidates for proteins functioning in this postulated branch are Bck2p, a protein of unknown function,[107] as well as the two phosphatases Ppz1p and Ppz2p.[106-108] Mutation lacking both Ppz1p and Ppz2p show a phenotype similar to mutants defective in the Pkc1p-mediated MAP kinase cascade. However, a triple mutant *ppz1Δ ppz2Δ mpk1Δ* has a more severe phenotype resembling that of the *pkc1Δ* mutant. This suggests that Ppz1p and Ppz2p on the one hand and Mpk1p on the other are not acting within the same signaling route. The *bck2Δ* mutation did not confer a phenotype on its own while it aggravated that of an *mpk1Δ* but not of the *ppz1Δ ppz2Δ* mutant. Therefore it has been proposed that Bck2p functions in the same branch as Ppz1p and Ppz2p.[107] Mutations in the *BRO1* gene cause synthetic phenotypes when combined with mutations affecting protein kinases of the PKC-pathway, including *PKC1* itself. This suggests a functional or regulatory relation of Bro1p to the PKC-pathway but the role of Bro1p is not understood at this stage.[142]

Pkc1p has been demonstrated to physically interact with Rho1p, a small GTP-binding protein implicated in the control of polarized growth.[143,144] *RHO1* is an essential gene which can be replaced partially by mammalian RhoA and a strain producing Rho1A instead of Rho1p is temperature-sensitive. This phenotype was used to select suppressor mutations which identified a dominant allele of *PKC1*. Interaction between Pkc1p and Rho1p was confirmed in the two hybrid system. However, the dominant *PKC1* allele could not suppress the *rho1Δ* mutant, suggesting that Rho1p controls other processes in addition to the protein kinase C signaling.[143,144] For instance it appears that Rho1p, being a regulatory subunit of this enzyme, is controlling glucan synthase activity involved in cell wall biosynthesis.[145]

Several additional proteins that interact with Rho1p have been identified. Rom1p and Rom2p are GTP/GDP exchange factors[146] for Rho1p and Rdi1p, a GDP dissociation inhibitor.[147] Rom7p/Bem4p is a protein that has been implicated in the budding process and it has recently been shown to interact with Rho1p as well.[148]

Staurosporine is a specific inhibitor of protein kinase C and mutations or conditions lowering protein kinase C activity confer sensitivity to the drug, while overexpression of protein kinase C confers resistance. Staurosporine sensitive mutants (*STT* mutants) have identified the *PKC1* gene itself (named *STT1*[149]) and putative upstream activators. *STT4* encodes a phosphaditylinositol-4-kinase. Deletion of *STT4* confers a phenotype similar to that of

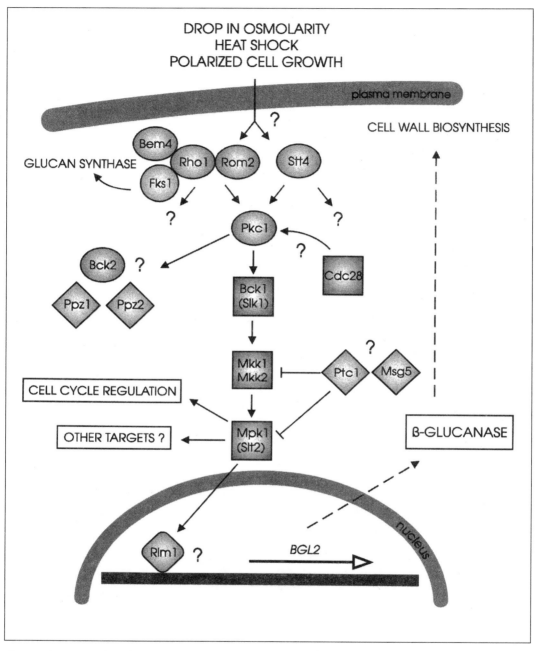

Fig. 4.9. The PKC-pathway in S. cerevisiae. The pathway is stimulated by various conditions that seem to affect cell shape or size. The role of the Stt4p phosphatidylinositol-4-kinase and of the Rho1p G-protein in stimulation of protein kinase C is not known but Rho1p directly interacts with Pkc1p. In addition, Rho1p interacts with a number of other proteins. Pkc1p appears to be stimulated also by the Cdc28p protein kinase which plays a central role in cell cycle control. Pkc1p appears to stimulate the MAP-kinase cascade which may be controlled by two or more phosphatases. Several genes have been found whose mutation causes phenotypes similar to mutation in the MAP-kinase cascade but genetic analysis places those genes in one of (possibly) several parallel pathways which are perhaps controlled by the same upstream components. See text for further details.

pkc1Δ.[150] Genetic analysis suggests that Pkc1p may be one but not the only downstream target of Stt4p.[150,151] It appears that Stt4p also affects other signaling pathways as might be expected from its enzymatic activity.[151] The roles of other *STT* genes like of *STT3*—an essential gene[152]—and of *STT10*—whose gene product affects vacuolar pH homeostasis[153]—are less well understood at present.

Finally, the PKC-pathway appears to be under control of the cell cycle machinery. The Cdc28p cyclin-dependent protein kinase (the Cdc2p homologue of baker's yeast) appears to stimulate the pathway in an unknown fashion.[127] Thus there is a wide range of mechanisms that appear to stimulate the PKC-pathway. A possible unifying theme in all of these stimuli might be morphogenesis and growth which leads or may lead to a remodeling of the cell wall.

It is not surprising that the PKC-pathway appears to interact with several different signaling routes since growth, morphogenesis and cell wall remodeling are certainly highly complex processes. It has been shown that the Ca^{2+}/calmodulin-dependent protein phosphatase calcineurin becomes essential in mutants affected in the PKC-pathway.[154,155] The Ptc1p protein phosphatase—which according to genetic analysis functions in the HOG-pathway[88]—was also identified as a suppressor mutation for a *pkc1* allele and deletion of *PTC1* is lethal in combination with *mpk1Δ*.[109] While the suppressive effect on the *pkc1* mutation is difficult to explain, one might imagine that deletion of *PTC1* activates the HOG-pathway, causing overproduction of glycerol which in turn could be deleterious to a cell with a weaker cell wall such as *mpk1Δ* mutant cells. A significant interplay between the HOG- and the PKC-pathway has been also observed by Gustin and co-workers.[123]

In contrast to the situation in the HOG-pathway, putative targets for the PKC-pathway have been reported. The just mentioned screen for suppressors of a *pkc1* allele also identified a gene called *KCS2*

which encodes a putative basic leucine zipper (bZIP) transcription factor.[109] Multicopy suppression of the lethality of the double mutation *slt2/mpk1Δ spa2Δ* (Spa2p is required for polarized growth) revealed the gene *NHP6A* and its homologue *NHP6B* which encode chromatin associated high mobility group (HMG) proteins. Genetic analysis suggests that they function downstream of Slt2p/Mpk1p.[156] Finally, expression of the gene *BGL2* appears to be negatively controlled by the PKC-pathway.[157] This gene encodes a β-glucanase probably involved in cell wall degradation and remodeling during growth. The transcription factor controlling *BGL2* expression is not known. A putative candidate might be Rlm1p whose gene has been identified via a mutation that suppresses the toxicity of overexpression of an activated allele of *MKK1*. Rlm1p is homologous to serum response factor.[140]

The effect of the PKC-pathway on *BGL2* expression is in line with the statement that the PKC-pathway is involved in processes related to the remodeling of the cell wall and hence in morphogenesis. How could this be explained in light of the observation that phosphorylation of Mpk1p is stimulated by a drop in osmolarity? A drop in external osmolarity will lead to water uptake and hence to cell expansion. Cell expansion is also the result of cellular growth and the cell might monitor during growth the ratio of the volume of the cytoplasm and the surrounding cell wall, i.e., whether the cell still fits into its suit. Thus cellular expansion could give in some way—perhaps by sensors in the plasma membrane that are stimulated by touching the cell wall—a signal to partially hydrolyze and remodel the cell wall to a bigger size or towards the development of a bud. Such processes might link growth and morphogenesis to osmoregulation.

Relatively little is known about a PKC-pathway in *S. pombe*. Two genes encoding putative protein kinase C homologues have been identified from fission yeast. It has been demonstrated that like in *S. cerevisiae* these two proteins appear to affect cell morphology and cell wall formation.[158,159]

6. GENE EXPRESSION UNDER OSMOTIC STRESS

Based on experiments employing two-dimensional (2-D) gel electrophoresis it has been estimated that the levels of about 200 proteins respond significantly to salt or osmotic stress.[64,160] Interestingly, most of the proteins that were produced at a rate differing more than 4-fold after a salt shock exhibited increased levels. On the other hand if proteins whose production rate was altered 4-fold or less were considered, then about an equal number showed induced or diminished production. This suggests that the adaptation is largely due to the establishment of new systems or the stimulated production of systems that play a minor role under normal growth conditions. Probably there is no need to shut off genes that are highly expressed under normal growth conditions. Interestingly, however, several enzymes from the lower part of glycolysis appear to be produced at a reduced rate under hyperosmotic stress. This may indicate a long-term adaptation of glycolytic flux for stimulated glycerol production.[64]

Two-dimensional gel electrophoresis as well as Northern analysis has shown that probably each gene displays its own well fine-tuned pattern of response with respect to the nature of the osmolyte (salt or sorbitol) or to its level, the timing of the response (rapid-late and transient-sustained) and to the stress-specificity.[64,74,160,161] This of course suggests the involvement of several signaling pathways and a complex interplay between general and specific transcription factors at the level of the promoters of stress controlled genes. In addition, posttranscriptional processes may well be involved in establishing different mRNA or protein levels under osmotic stress but this has not been studied yet.

It appears—not unexpectedly—that genes whose expression is induced under osmotic control can be classified according to two main parameters which in turn might be relevant for the mechanisms of this effect: genes which are controlled under a general stress response and those specifically controlled by osmotic stress. Details of the former class will be discussed in chapter 7 and only very briefly here. Genes whose expression responds to high salt can be split further. First there are genes whose expression is salt-responsive but do not respond to other osmolytes; these genes are discussed in chapter 5. Finally, this chapter will focus on the genes whose expression is specifically stimulated by osmolarity. The expression of not a single gene that is negatively controlled by hyperosmotic stress has been studied at the level of its mRNA, not to speak of the mechanisms involved in the regulation.

6.1. OSMOTIC STRESS AND THE GENERAL STRESS RESPONSE

The general stress response is characterized by genes whose expression is induced by hyperosmotic stress but also by other stress conditions such as heat stress, oxidative stress, nutrient starvation, weak acids and probably any condition that cause a slow-down or arrest of cell proliferation. To those genes belong *CTT1* encoding catalase T,[78,162] *HSP12* encoding a small heat shock protein,[163,164] *HSP104* encoding a larger heat shock protein,[78,165] *GAC1* encoding a regulatory protein in glycogen metabolism,[78,166] *UBI4* encoding polyubiquitin,[78,167,168] *DDR2* involved in repair of DNA-damage,[78,169] *CYC7* encoding an isoform of iso-1-cytochrome C,[170] genes encoding proteins involved in trehalose production,[171-173] and probably more (for review see chapter 7 and also refs. 10, 174, 175). All those genes appear to have in common a sequence element called STRE for stress response element.[78,162,169,175-178] This element with the sequence CCCCT (in either orientation) is necessary and sufficient for stress-induced gene expression of a reporter gene. The factors binding to this element may be Msn2p and Msn4p.[175,179,180] Another candidate for a factor that mediates the general stress response is Rox3p which has been initially identified as controlling gene expression under oxygen limitation.[170,181] STRE-mediated transcription strongly responds to the level of activity of cAMP-dependent protein kinase (protein kinase A), suggesting that the transcription factors binding to this element may be targets of this kinase.[78,162,164,172,177,178] High protein kinase A activity has a repressive effect while low protein kinase A activity stimulates expression from STRE. Recently it has been shown that the induction by hyperosmotic stress of *CTT1* as well as that of a reporter gene driven by STRE depends on a functional HOG-pathway. It has been proposed that these two pathways converge at the level of STRE since the effects conferred by protein kinase A and those triggered by the HOG-pathway were found to be independent of each other.[78] However, the available data are also consistent with the HOG-pathway controlling protein kinase A activity in a manner independent of the regulatory subunit of protein kinase A (Bcy1p) which is controlled by cAMP. There is accumulating evidence that control mechanisms exist which do not involve Bcy1p exist (chapter 1 and ref. 182). Any other scenario, including common factor(s) downstream of protein kinase A and Hog1p and upstream of STRE, is of course also possible.

Remarkably, there appear to be variations of the general stress response allowing the modulation of the response to a specific type of stress, even for genes controlled by STRE. The expression of the *SSA3* gene encoding a 70 kDa heat shock protein (chapter 3) responds perfectly well to a variety of stress conditions as well as to protein kinase A activity changes but is not inducible by hyperosmotic stress.[173] Thus even if there is a common theme in general stress response, specific mechanisms must exist that modulate or even prevent the response by certain genes to some types of stress.

6.2. SPECIFIC INDUCTION OF GENE EXPRESSION UNDER OSMOTIC STRESS

The expression of two genes involved in the production of the compatible solute glycerol—*GPD1* and *GPP2*—encoding

isoenzymes for glycerol-3-phosphate dehydrogenase and glycerol-3-phosphatase, respectively is induced specifically by hyperosmotic stress but not by other types of stress[70,74,183] (note that *HOR2* in ref. 74 is *GPP2*[70]). There is also evidence that expression of genes encoding enzymes of the upper part of glycolysis, namely sugar transporters and sugar kinases, is stimulated by osmotic stress which may point to an adaptation of glycolytic flux to enhanced glycerol production.[74,184] Remarkably, however, while expression of *GPD1* and *GPP2* appears to be controlled in a very similar fashion, the two genes *HXT1* (encoding a glucose transporter) and *GLK1* (encoding glucokinase) are controlled differently. *GLK1* (as well as *HXK1* encoding hexokinase I) expression appears to be controlled by different stresses and by protein kinase A and thus might belong to the family of genes controlled by STRE. *HXT1* on the other hand is specifically controlled by osmotic stress in a HOG-pathway dependent fashion but it does not respond to protein kinase A.[74,184,185]

Recently a locus on chromosome XIII has been identified that contains four genes whose expression is induced by high osmolarity. These genes—*ALD2* encoding a aldehyde dehydrogenase probably involved in redox-regulation during high level glycerol production under osmotic stress, *PAI3* encoding a protease inhibitor, *SIP18* encoding an unknown protein and *DDR48* encoding a protein involved in DNA-damage repair—are arranged in two divergent pairs. However, the expression of each gene within the pairs appears to be controlled in different ways. The osmotic induction of three of the four genes—*ALD2*, *PAI3* and *SIP18*—depends on the HOG-pathway. Remarkably, the osmotic induction of *DDR48* does not require the HOG-pathway.[161]

The expression of *GPD1* as well as the promoter of the gene have been studied intensively. The upstream region of *GPD1* appears to contain two physically separated elements which confer osmotic induction. One of these two elements mediates osmotic induction in dependence of the HOG-pathway. This region contains two matches to the CCCCT motif of STRE. However, it has not been studied whether these putative STREs are functional. The second region is located further upstream, it does not contain STRE-like sequences and it mediates osmotic induction independent from the HOG-pathway.[186] This situation is somewhat similar to that of the *ALD2* promoter where the upstream sequences required for regulated expression of the gene also do not contain consensus STREs.[161] Further, in contrast to genes controlled by STRE, expression of the *GPD1* gene appears to involve the general transcription factor Rap1p[187] which is also required for the expression of genes encoding glycolytic enzymes and ribosomal proteins as well as in gene silencing and other regulatory processes.[188,189] However, Rap1p does not mediate osmotic induction.[186,187] Finally, stress-induced expression via STRE appears to be an induction process, i.e., STRE confers a positive control.[78,178] However, the stimulation of expression of *ALD2* by osmotic stress appears to be due to derepression since a negative promoter element has been identified in the *ALD2* promoter which might confer this effect.[161] Also, osmostress-induction of *GPD1* may be regulated at least in part via a negative control system as can be inferred from the phenotype of *rgs1* mutants which show enhanced expression of *GPD1* (see further). In conclusion, the mechanisms regulating *GPD1* and *ALD2* expression may differ fundamentally from those of genes controlled by the general stress response as well as from each other.

Additionally, the expression pattern of *GPD1* does not suggest control by a general stress-responsive mechanism. While it has been reported that the specific activity of the gene product glycerol-3-phosphate dehydrogenase is slightly increased during growth on nonfermentable carbon sources[24] and under heat stress[183], no evidence for carbon source[183] or temperature-dependent increase in the mRNA level could be found.[74,183] In addition, the mRNA level also does not increase during nitrogen star-

vation or entry into stationary phase.[183] Somewhat conflicting results have been obtained with respect to effects conferred by altered protein kinase A activity. While one report did not find any effect of high levels of protein kinase A on the *GPD1* mRNA level[74], we[183] and others (T. Miyakawa, personal communication) have observed that the basal level of the mRNA is increased in strains with low protein kinase A and decreased in strains with high protein kinase A. In addition, a strain with low activity of adenylate cyclase also showed high activity of glycerol-3-phosphate dehydrogenase.[24] However, the effects of protein kinase A activity at the mRNA level were far less pronounced than for the classical STRE-controlled gene *CTT1* and most importantly did not affect induction by hyperosmolarity. In contrast, we found that expression of *CTT1* could not be induced any more by high osmolarity in a strain having very high protein kinase A activity. Thus we concluded that expression of *GPD1* is controlled specifically by osmotic stress and its basal level may be affected by protein kinase A activity.[183] Again the situation is slightly different for the *ALD2* as well as for *PAI3* and *SIP18* genes where high protein kinase A activity prevented osmotic induction and low activity lead to derepression which could not be further stimulated by hyperosmotic stress.[161] In this respect as well as in their dependence on the HOG-pathway for induction by high osmolarity, these three genes resemble genes controlled by STRE.

Deletion of the two genes encoding the catalytic subunit of the phosphatase calcineurin which is involved in the adaptation to salt tolerance (see chapter 5) causes a reduction in the basal level of *GPD1* mRNA. Overexpression of one of the two genes stimulates expression of *GPD1*. As has been observed for the influence by protein kinase A, induction by osmotic stress was unaffected by altered activity of calcineurin (T. Miyakawa, personal communication).

The induction of *GPD1* expression by hypertonic conditions has been reported to depend on a functional HOG-pathway.[54]

However, time course experiments, the use of mutants and studies using different salt concentrations suggest that either the HOG-pathway is not the only pathway controlling *GPD1* expression under osmotic stress or may not be directly involved in this process at all.

The degree of induction of the *GPD1* mRNA level depends on osmolyte concentrations gradually increasing until at least 1.7 M NaCl and 1.5 M sorbitol.[55,67,74] This pattern does not correlate with tyrosine-phosphorylation of Hog1p which appears to be maximal already at about 0.3 M NaCl.[83,96] In the case of *ENA1* it has been shown that the HOG-pathway is responsible for induction by low salt concentrations but at higher concentrations another pathway which includes the calcineurin phosphatase takes over[190] (see chapter 5). The situation appears to be different for expression of *GPD1*: at low salt concentrations induction of the mRNA level is more transient in a *hog1Δ* mutant than in the wild type. At higher concentrations the lag-phase between addition of the osmolyte and the rise in the mRNA level is prolonged but induction eventually occurs even in a *hog1Δ* mutant. Thus the HOG-pathway is not absolutely required for *GPD1* induction.[183] A similar observation in the *hog1Δ* mutant has been reported for *CYC7* expression under osmotic stress.[170]

There are more indications in line of this assumption. In addition to being very rapid, induction of tyrosine-phosphorylation of Hog1p is only transient and drops after about 20 minutes to a value comparable to or only slightly higher than that before stimulation.[83,96] However, induction of the *GPD1* mRNA level usually requires a lag-phase which is longer at higher osmolyte concentrations.[183] Thus the timing of stimulation of the Hog1p protein kinase and the apparent increase in the *GPD1* mRNA level do not fit and Hog1p phosphorylation precedes *GPD1* induction—depending on the salt concentration—by some minutes or even up to several hours.

Another strong argument for a HOG-pathway independent pathway stimulating expression of genes like *GPD1* comes from the identification of a *GPD1* promoter element that confers induction by osmotic stress even in a *hog1Δ* mutant (see above).[186]

We (and others; M.C. Gustin, personal communication) have recently isolated mutations that partially suppress the osmosensitivity of the *hog1Δ* mutant. One such recessive mutation is *rgs1* (for regulator of glycerol synthesis). The mutation causes constitutive but still further inducible glycerol production and the mutant is actually osmophilic. Neither the *rgs1* single nor the *hog1Δ rgs1* double mutant show elevated *GPD1* mRNA levels under normal growth conditions but a very rapid and much stronger response when salt is added to the cell.[183] The nature of the *rgs1* mutation is unknown and hence it is also unclear whether the *RGS1* gene product is directly involved in the control of *GPD1* expression. However, as outlined above, the *hog1Δ* mutation appears to affect the timing of the response and the timing can be restored by the *rgs1* mutation. This can be interpreted such that the HOG-pathway is involved in rapid recovery processes after the osmotic shock such as cytoskeleton re-establishment[38] and that the initiation or completion of those processes is a prerequisite for *GPD1* induction to occur. This assumption does not contradict an also direct involvement of the HOG-pathway in *GPD1* expression. Such a direct role is actually supported by the finding that a temperature-sensitive *sln1* mutant shows strong elevation of *GPD1* mRNA levels when shifted to the restrictive temperature.[74] On the other hand, an osmostress-induced increase in *GPD1* mRNA levels entirely independent of the HOG-pathway is also observed during the recovery from a hypo-osmotic shock. Under such conditions the *GPD1* mRNA level first drops almost below detection and then rises again.[183] Apparently the cell needs to "test" a new steady state level and this occurs to the same degree and in the same time scale in wild type and *hog1Δ* mutant. Thus despite the general anticipation that the HOG-pathway is controlling *GPD1* expression there is accumulating evidence that this assumption is at least incomplete. Undoubtedly alternative pathways must exist that control osmostress-induced expression of genes whose osmotic induction is affected by mutation in the HOG-pathway.

Finally, at least two genes have been identified whose expression is induced by high osmolarity entirely independent of the HOG-pathway: *HOR7*, a gene of unknown function[74] and *DDR48* (see above).[161] Nothing is known about the mechanisms that control the expression of those genes. It is remarkable that the induction of the gene *DDR48,* which requires 1 M NaCl, is still operating in a *hog1Δ* mutant which does not grow and exhibits a strongly aberrant morphology under such high osmolarity.[38,161]

An obvious candidate pathway for being involved in regulated gene expression under osmotic stress is the cell integrity PKC-pathway which has been shown to be stimulated by a hypo-osmotic shock.[123] However, this expectation has also been disappointed and the results were much clearer in this case. Mutations in *PKC1* or in *MPK1* do not affect induction by increased osmolarity of *GPD1* nor of any other of the seven genes tested (*GPP2, HXT1, GLK1, ENA1, HSP12, HOR7*).[74,183] This has been investigated for *GPD1* at different levels of salt, with sorbitol as an osmolyte and in time scale experiments.[183] The *mpk1Δ* mutation also does not affect the transient drop of the *GPD1* mRNA level after a hypo-osmotic shock. Such conditions might have been even more likely to affect gene expression in an Mpk1p-dependent way than hyperosmotic stress since a hypo-osmotic shock has been shown to stimulate Mpk1p phosphorylation.[183]

In conclusion, amazingly little knowledge with respect to the control of gene expression after an osmotic shock is well established and assumptions that seemed appealing might be misleading or superficial. The detailed characterization of osmostress-controlled promoters such as those of *GPD1, GPP2* and *ALD2* (and maybe others) may provide novel information. Promoter elements involved in osmostress-controlled expression should be used

in genetic and biochemical approaches to search for genes/proteins involved in their regulation. This might then lead to the identification of the actual signaling pathway(s) involved in osmoregulated gene expression.

7. CONTROL OF GLYCEROL AND WATER TRANSPORT

Water and glycerol pass biological membranes well by passive diffusion, at least according to textbooks of biochemistry. However, the transport of these molecules which are of central importance in osmoregulation is controlled by specific channel proteins which remarkably belong to the same protein family. While very little is known at present about the role of water channels in microbial physiology beyond their mere existence the central role of the yeast glycerol facilitator in osmoregulation has recently been established.

7.1. THE ROLE OF THE YEAST GLYCEROL FACILITATOR FPS1P

Glycerol has long been thought to pass the yeast plasma membrane solely by passive diffusion.[56] However, the fact that osmotolerant yeasts can accumulate glycerol against high concentration gradients by means of active transport suggests that there should also be mechanisms for glycerol retainment.[5,51,191] In addition, studies on *S. cerevisiae* have also indicated that under hyperosmotic stress cells can retain glycerol better than nonstressed cells.[48] The discovery of the Fps1p channel protein has now led to an explanation for this phenomenon.

The *FPS1* gene was isolated by coincidence as a suppressor of the growth defect on glucose of the *fdp1* mutant (hence the *fdp1* suppressor).[192] This mutant is affected in the gene *TPS1* encoding trehalose-6-phosphate synthase. It suffers from a defect in the regulation of glycolysis resulting in the accumulation of metabolites from the upper part of glycolysis.[193,194] This in turn stimulates glycerol production.[50] Thus the suppressive effect of overproduction of Fps1p might be due to stimulated glycerol export. However, suppression of *tps1* mutants turned out to be strain background specific and it may be that the original *fdp1* mutant used was actually also defective in the *fps1Δ* gene. *Tps1Δ fps1Δ* double mutants show an even stronger growth defect on glucose than *tps1Δ* single mutants.[50]

Studies on uptake of radioactive glycerol have shown that glycerol influx can be separated into probably two components. The first component has all characteristics of passive diffusion and is probably not mediated by any proteins. The second component is entirely dependent on Fps1p and appears to be due to facilitated diffusion. Despite this clear effect, there has been no evidence that Fps1p plays a role in glycerol utilization.[50]

Measurements of the ratio between intra- and extracellular glycerol in cells that overproduce glycerol have demonstrated that about 80% of the glycerol produced leave the cell via Fps1p.[50] This conclusion is supported by observations that mutants lacking Fps1p have a growth problem under conditions when glycerol is overproduced, e.g., in *rgs1* mutants[183] and under anaerobic conditions.[71] Excessive glycerol accumulation in *fps1Δ* mutants may affect proper osmoregulation.

These observations suggest that Fps1p may have a role in controlling glycerol accumulation under osmotic stress conditions. This is indeed the case. Upon a hyperosmotic stress Fps1p closes within 1-2 minutes apparently in order to allow glycerol accumulation.[30,50] A mutant of Fps1p which is unable to close accumulates only about 25% of the amount of glycerol compared to the wild type. A strain carrying this mutant allele of *FPS1* is about as sensitive to high osmolarity as a *gpd1Δ* mutant.[30] On the contrary, Fps1p opens upon a hypo-osmotic shock within 1-2 minutes in order to allow release of glycerol.[30] About 50% of the accumulated glycerol exits in the cell within the first five minutes after the hypo-osmotic shock.[50] Mutants lacking Fps1p survive a shift from 10% NaCl to distilled water at about a 50-fold lower ratio than the wild type.[30] If *FPS1* is deleted in a strain carrying a

deletion in *mpk1Δ*, which causes a weakening of the cell wall (see above), even a mild hypo-osmotic shock from 1 M sorbitol to normal growth medium is already lethal.[30] Thus *FPS1* is the only known yeast gene, mutations of which, can cause either sensitivity to an increase or to a decrease in external osmolarity. This underscores the importance of Fps1p in yeast osmoregulation.

Transporters or facilitators that control the exit of a compatible solute upon a hypo-osmotic shock appear to exist in bacterial and animal cells as well. An efflux system for the compatible solute glycine betaine has been described in bacteria.[195] In addition, it has been proposed that stretch-activated channels mediate the loss of metabolites from bacteria. However, those channels appear to be rather unselective.[196] In mammalian cells which lack a cell wall and hence exhibit significant swelling upon a hypo-osmotic shock, the process of regulated volume decrease is thought to be accompanied by the release of compatible solutes.[197] It has been proposed that stretch-activated anion channels also transport organic osmolytes upon hypo-osmotic shock.[198-200] Such a channel has been named VSOAC for volume-sensitive organic osmolyte anion channel. A candidate gene encoding this protein—I_{cln}—has been isolated but due to the difficulties in genetic approaches in mammalian cells the identity to VSOAC has not been confirmed yet.[200,201]

Recently mutants defective in Fps1p have been isolated in search for functions required for cell fusion during mating (J. Philips and I. Herskowitz, personal communication). Yeast strains of opposite mating type attract each other by means of a mating pheromone and then grow towards each other (for a review see ref. 202). During fusion the cells obviously need to degrade their cell wall at the position of fusion. It is thought that Fps1p is required at this step to adjust intracellular osmolarity and prevent bursting. The mating defect can be suppressed by addition of 1 M sorbitol to the growth mediu, further m supporting this view. In addition,

reduction of glycerol production by deletion of *GPD1* also suppresses this defect
(J. Philips and I. Herskowitz, personal
communication). However, mating is usually tested on normal growth medium
where little glycerol accumulates. Thus either
glycerol is produced during polarized cell growth
and mating—maybe in order to allow the
cell to increase its volume—or Fps1p is
also involved in the export of other metabolites than glycerol. The observation
that *S. pombe* mutants with an increased or
reduced activity of the HOG-pathway have
altered cell size and morphology could also
be interpreted as a role of osmoregulation
in shaping of the cell. Future work
will hopefully address this interesting
perspective.

7.2. THE MIP-FAMILY

Fps1p is a member of the MIP family
of channel proteins. This family has recently received much attention and is probably the most rapidly growing family of
proteins.[203-205] More than 70 members have
been described up to now. All these proteins
share a common topology consisting of six
transmembrane domains (Fig. 4.10). It appears that this family derives from an an-
cestor with three transmembrane domains
which has duplicated and the first and the
last three transmembrane domains became
arranged as direct repeats.[203] However, the
homology between these two repeats is
rather weak as is the homology between
different proteins of the family. All these
proteins, however, share the short consensus sequence NPA located between the second and the third and between the fifth
and the sixth transmembrane domain.
These two motifs are predicted to be on
opposite sides of the membrane and their
surrounding is fairly hydrophobic. It is
thought that these two domains dip from
opposite sides into the membrane and
thereby form the pore.[205,206]

The members of the MIP-family have
recently been classified on the basis of
phylogenic analysis of the sequences as well
as on functional analysis[204] into:

- glycerol- and metabolite-facilitators of
  bacterial origin which are involved in

the utilization of glycerol and maybe
of other polyols (at least 9 proteins);
- the nodulins which are supposed to
  transport metabolites between the nitrogen fixating bacteriods and the plant
  host (4);
- the plasma membrane intrinsic proteins
  from plants which are water channels,
  controlled by water stress at the transcriptional level and expressed in a tissue-specific fashion (18);
- the tonoplast intrinsic proteins from
  plants which are water channels as
  well, controlled by water stress at the
  transcriptional and probably also at the
  protein level; they appear to function
  in cytoplasmic-vacuolar water transport
  (15); and
- the animal water channels (17).

Thus most members of this family are
actually water channels. This function has
been confirmed for many of those proteins
by expression in *Xenopus* oocytes and
measurement of osmotic swelling.[204,205]

Interestingly,Fps1p does not fit well
into any of those groups listed above although it appears closest to the bacterial
glycerol facilitators. The most striking difference that distinguishes Fps1p from most
members of the MIP-family is size. While
the bacterial glycerol facilitators and the
water channels are about 230 amino acids
long consisting mainly of the region of six
transmembrane domains, Fps1p has long
N- and C-terminal extensions of 250 and
150 amino acids respectively. Only two
other proteins of the MIP-family have been
reported to have such long extensions: the
Drosophila big brain protein BIB[207] and an
open reading frame of unknown function
identified in the course of the yeast genome
sequencing.[208] The extensions of Fps1p are
not homologous to those of the other two
proteins nor to any other sequence in the
data bases.

7.3. REGULATION OF FPS1P

We have recently started to analyze the
molecular mechanism of the closing and
opening of Fps1p in response to changes
in external osmolarity. The two osmosensing signaling pathways in yeast—the

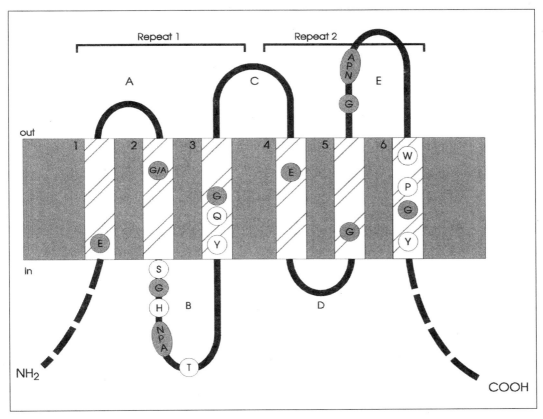

Fig. 4.10. Topology of proteins belonging to the MIP-family. Six transmembrane domains are arranged as a duplication of a three transmembrane domain module (repeats 1 and 2). Amino acids conserved throughout the family are highlighted in white circles while dark circle point to residues also conserved between the two repeats. The N- and C-terminal extensions are only a few amino acids long for most members like the aquaporins but Fps1p has extensions of 250 and 150 amino acids, respectively.

HOG- and the PKC-pathway, respectively—are not involved in the process of closing and opening.[30,50] Of course it is well possible that Fps1p responds to mechanical forces like membrane stretching and then it might not need the assistance of any other protein. If this turns out to be the case Fps1p could have a sensor function and might even control other proteins.

In any case the extensions of the protein are of course suspicious and to this end we have deleted these parts of the protein.[30] Deletion of the N-terminal extensions renders the protein insensitive to hyperosmotic stress, i.e., the channel cannot be closed. As mentioned above, this results in an inability to accumulate glycerol to high levels and in hyperosmosensitivity. Deletion of the C-terminus abolishes function of the channel but presently it is not clear whether this is due to actual loss of function or to an inability to open the channel. Future work will define the roles of the extension in the regulation of Fps1p in more detail.

Interestingly, the ClC-2 chloride channel from mammalian cells operates in a somewhat similar fashion as Fps1p, although its sequence and topology is quite different from that of Fps1p. Analysis of the N-terminal extension of this protein has revealed that certain small sequences are required for the closing of the channel. It has been proposed that the extension functions as an arm with a ball at the end that can physically close the channel.[200,209]

7.4. WATER CHANNELS

Passage of water through cellular membranes appears to be a well controlled process. It has been known for a long time that different membranes exhibit very different capacity for water passage. The basis for this phenomenon was obscure until the discovery of CHIP28, a water channel of the MIP family in the cytoplasma membrane of erythrocytes.[210] Since then it became clear that water channels are important factors of osmoregulation at the cellular as well as at the level of the entire organism.[204,205,211-214] The hormone regulated water channels of kidney involved in urine concentration have been studied especially intensively.[205,211-213] In plants water channels appear to be controlled both at the transcriptional as well as at the protein level.[28,29,204] They control water transport through the plasma membrane of cells in a variety of tissues. In addition, they also regulate water transport between the cytosol and the vacuole, the latter being probably an important water reservoir in plant cells.[28,29]

Until recently, water channels have not been reported for microorganisms. This situation has also changed. The discovery of a water channel from *Escherichia coli* has been reported. Its function as a water channel has been confirmed in the *Xenopus* oocyte system but its role in cellular physiology has not yet been determined.[215] The yeast sequencing project has revealed two sequences which are likely to encode water channels.[208] One of the open reading frames appears to be incomplete, probably due to a sequencing error. Considering this possibility the two gene products are about 80% identical and show highest homology to plant plasma membrane water channels. It will be interesting to study the role of these proteins in yeast osmoadaptation.

8. CONCLUSIONS AND PERSPECTIVES

The studies on osmoregulation in yeast have focused up to now on the response of cells to hyperosmotic stress and—to a much lesser extent—on those to a hypo-

osmotic shock. Very little is known about how cells manage to survive the first minutes after a hyperosmotic shock. Future work should address this problem by testing osmosensitive mutants for their ability to either survive an osmotic shock or to adapt to and to grow under increased osmolarity. It is clear by now that the production and the retention of the compatible solute glycerol are of central importance for the adaptation to and growth under high external osmolarity. The mechanisms that control glycerol production under stress are, however, only very poorly understood. The same holds true for the role of the HOG-pathway. Although the name of this pathway implies that it controls glycerol production under osmotic stress, it is clear by now that the HOG-pathway and stimulated gene expression under hyperosmotic stress occur at different phases. In addition, osmotic induction of many genes is only delayed in the absence of the HOG-pathway while other genes appear to be induced by hyperosmolarity in a fashion entirely independent of the HOG-pathway. Thus alternative and/or additional mechanisms must exist that stimulate gene expression during the adaptation to high external osmolarity.

An amazing body of information has been accumulated over the last three years on the structure and function of the HOG-pathway. However, at present we still do not know at the molecular level how this pathway is stimulated by the two proposed osmosensors. We know little about the control of this pathway and its targets by phosphorylation and dephosphorylation and we even do not know any of its targets. The tools, i.e., the genes for probably most of the components of the pathway, are at hand and many strong laboratories are studying this pathway. Hence rapid further progress can be expected. The next few years should also show whether the general expectation concerning the existence of similar osmosensing signaling pathways in mammalian and plant systems will hold true.

Neither the HOG- nor the PKC-pathway is involved in the control of membrane permeability for glycerol. However, the control of glycerol efflux plays a key role in osmoadaptation. Closing of the Fps1p channel is required for adaptation and for growth at high osmolarity while opening of Fps1p is needed for survival after a hypo-osmotic shock. The next few years will certainly yield interesting information concerning the control mechanisms of glycerol efflux and the mechanisms that regulate the activity of the Fps1p glycerol channel. These mechanisms may serve as a model for swelling-, stretch- or volume-regulated channels in higher organisms where genetic approaches cannot rigorously be applied. Finally, novel members of the MIP-family of channel proteins involved in water and metabolite transport have been identified by systematic sequencing of the yeast genome and the studies on their role will certainly unravel novel pathways involved in osmoregulation.

Once the mechanisms involved in the adaptation to osmotic stress are better understood we might also learn more about the role of osmoregulation in the control of cellular growth, morphology and cell division. A couple of seemingly unrelated observations discussed in this review based on work on both fission and budding yeast point to such roles of osmoregulatory mechanisms.

ACKNOWLEDGMENTS

I thank Johan Thevelein, Bernard Prior and Lennart Adler in whose departments my own work has been conducted and for their support and many stimulating discussions. In addition, I am indebted to the numerous colleagues who have communicated results prior to publication and who have contributed to this review through many discussions. I also acknowledge support from the Commission of the European Union via contract BIO4-CL95-0161.

REFERENCES

1. Wiggins PM. Role of water in some biological processes. Microbial Rev 1990; 54:432.
2. Bourque CW, Oliet SHR, Richard D. Osmoreceptors, osmoreception, and osmoregulation. Front Neuroendoc 1994; 15:231-274.
3. Chandler PM, Robertson M. Gene expression regulated by abscisic acid and its relation to stress tolerance. Ann Rev Plant Physiol Mol Biol 1994; 45:113-141.
4. Garcia-Perez A, Burg MB. Physiol Rev 1991; 71:1081-1115.
5. Blomberg A, Adler L. Physiology of osmotolerance in fungi. Adv Microbial Phys 1992; 33:145-212.
6. Brown AD. Microbial water stress. Bact Rev 1976; 40:803-846.
7. Brown AD. Compatible solute and extreme water stress in eukaryotic microorganisms. Adv Microbial Physiol 1978; 17:181-242.
8. Brown AD, Edgley M. Osmoregulation in yeast. In: Rains DW, Valentine RC, Hollander A, eds. Symposium on Genetic Engineering of Osmoregulation. Vol. 14. New York: Plenum Press, 1980:75-90.
9. Mager WH, Varela JCS. Osmostress response of the yeast *Saccharomyces*. Mol Microbiol 1993; 10:253-258.
10. Varela JCS, Mager WH. Response of *Saccharomyces cerevisiae* to changes in external osmolarity. Mol Microbiol 1996; 142: 721-731.
11. Prior BA, Hohmann S. Glycerol production and osmoregulation. In: Zimmermann FK, ed. Yeast Sugar Metabolism. Lancaster: Technomic Publishing Co. Inc., 1997: (in press).
12. Adler L. Fungi and salt stress. In: Frankland MG, Magan N, Gadd GM, eds. Fungi and Environmental Change. Cambridge: Cambridge University Press. 1996:217-234.
13. Serrano R. Salt tolerance in plant and microorganisms: toxicity targets and defense responses. Int Rev Cytol 1996; 165:1-52.
14. Cooper JA. MAP kinase pathways: Straight and narrow or tortuous and intersecting? Curr Biol 1994; 4:1118-1121.
15. Herskowitz I. MAP kinase pathways in yeast: for mating and more. Cell 1995; 80:187-197.
16. Levin DE, Errede B. The proliferation of MAP kinase signaling pathways in yeast. Curr Opin Cell Biol 1995; 7:197-202.

17. Schultz J, Ferguson B, Sprague GFJ. Signal transduction and growth control in yeast. Curr Opin Genet Develop 1995; 5:31-37.

18. Treisman R. Regulation of transcription by MAP kinase cascades. Curr Opin Cell Biol 1996; 8:205-215.

19. Kyriakis JM, Avruch J. Protein kinase cascades activated by stress and inflammatory cytokines. BioEssays 1996; 18:567-577.

20. Hallows KR, Knauf PA. Principles of cell volume regulation. In: Strange K, ed. Cellular and Molecular Physiology of Cell Volume Regulation. Boca Raton, FL: CRC Press, 1994:3-29.

21. Klis FM. Review: cell wall assembly in yeast. Yeast 1994; 10:851-869.

22. Morris GJ, Winters L, Coulson GE, Clarke KJ. Effect of osmotic stress on the ultrastructure and viability of the yeast *Saccharomyces cerevisiae*. J Gen Microbiol 1986; 132:2023-2034.

23. Meikle AJ, Reed RH, Gadd GM. Osmotic adjustment and the accumulation of organic solutes in whole cells and protoplasts of *Saccharomyces cerevisiae*. J Gen Microbiol 1988; 134:3049-3060.

24. Albertyn J, Hohmann S, Prior BA. Characterization of the osmotic stress response in *Saccharomyces cerevisiae*: osmotic stress and glucose repression regulate glycerol-3-phosphate dehydrogenase independently. Current Genetics 1994; 25:12-18.

25. Blomberg A. The osmotic hypersensitivity of the yeast *Saccharomyces cerevisiae* is strain and growth media dependent: quantitative aspects of the phenomenon. Yeast 1996; (in press).

26. Latterich M, Watson DM. Evidence for a dual osmoregulatory mechanism in the yeast *Saccharomyces cerevisiae*. Biochem Biophys Res Commun 1993; 191:1111-1117.

27. Klionsky DJ, Herman PK, Emr SD. The fungal vacuole: composition, function and biogenesis. Microbiol Rev 1990; 54: 266-292.

28. Chrispeels MJ, Maurel C. Aquaporins: the molecular basis of facilitated water movement through living plant cells? Plant Physiol 1994; 105:9-13.

29. Chrispeels MJ, Agre P. Aquaporins: water channel proteins in plant and animal cells. TIBS 1994; 19:421-425.

30. Luyten K, Tamas M, Sutherland FCW, Albertyn J, Prior BA, Kilian SG, Ramos J, Thevelein JM, Hohmann S. Yeast Fps1p, a member of the MIP-family of channel proteins, determines the accumulation of the compatible solute glycerol under osmotic stress. In preparation.

31. Gustin MC, Zhou X-L, Martinac B, Kung C. A mechanosensitive ion channel in the yeast plasma membrane. Science 1988; 242:762-765.

32. Novick P, Botstein D. Phenotypic analysis of temperature-sensitive yeast actin mutants. Cell 1985; 40:415-426.

33. Chowdhury S, Smith KW, Gustin MC. Osmotic stress and the yeast cytoskeleton: phenotype-specific suppression of an actin mutation. J Cell Biol 1992; 118:561-571.

34. Drubin DG. Development of cell polarity in budding yeast. Cell 1991; 65:1093-1096.

35. Welch MD, Holtzman DA, Drubin DG. The yeast actin cytoskeleton. Curr Opin Cell Biol 1994; 6:110-119.

36. Lew DJ, Reed SI. Cell cycle control of morphogenesis in budding yeast. Curr Opin Genet Develop 1995; 5:17-23.

37. Chant J, Pringle JR. Budding and cell polarity in *Saccharomyces cerevisiae*. Curr Opin Genet Develop 1991; 1:342-350.

38. Brewster JL, Gustin MC. Positioning of cell growth and division after osmotic stress requires a MAP kinase pathway. Yeast 1994; 10:425-439.

39. Bauer A, Kölling R. Characterization of the *SAC3* gene of *Saccharomyces cerevisiae*. Yeast 1996; 12:965-975.

40. Schoch CL, Brüning ARNE, Pretorius GHJ, Prior BA. Mutants defective in the kinesin Kar3p are sensitive to high osmolarity. Current Genetics 1997; (in press).

41. Meluh PB, Rose MD. *KAR3*, a kinesin-related gene required for yeast nuclear fusion. Cell 1990; 60:1029-1041.

42. Middleton K, Carbon J. *KAR3*-encoded kinesin is a minus-end-directed motor that functions with centromere binding proteins (CBF3) on an in vitro yeast kinetochore. Proc Natl Acad Sci USA 1994; 91:7212-7216.

43. Yancey PH, Clark ME, Hand SC, Bowlus RD, Somero GN. Living with water stress:

evolution of osmolyte systems. Science 1982; 217:1214-1222.

44. Galinski EA. Compatible solutes of halophilic eubacteria: molecular principles, water-solute interaction, stress protection. Experientia 1993; 49:487-495.

45. Csonka LN, Hansen AD. Prokaryotic osmoregulation. Ann Rev Microbiol 1991; 45:569-606.

46. Bohnert HJ, Nelson DE, Jensen RG. Adaptation to environmental stress. Plant Cell 1995; 7:1099-1111.

47. Csonka LN. Physiological and genetic responses of bacteria to osmotic stress. Microbiol Rev 1989; 53:121-147.

48. Blomberg A, Adler L. Roles of glycerol and glycerol-3-phosphate dehydrogenase (NAD⁺) in acquired osmotolerance of *Saccharomyces cerevisiae*. J Bacteriol 1989; 171:1087-1092.

49. André L, Hemming A, Adler L. Osmoregulation in *Saccharomyces cerevisiae*. Studies on the osmotic induction of glycerol production and glycerol 3-phosphate dehydrogenase (NAD⁺). FEBS Letts 1991; 286:13-17.

50. Luyten K, Albertyn J, Skibbe WF, Prior BA, Ramos J, Thevelein J, Hohmann S. Fps1, a yeast member of the MIP family of channel proteins, is a facilitator for glycerol uptake and efflux and is inactive under osmotic stress. EMBO J 1995; 14:1360-1371.

51. Van Zyl PJ, Kilian SG, Prior BA. The role of an active transport mechanism in glycerol accumulation during osmoregulation by *Zygosaccharomyces rouxii*. Appl Microbiol Biotechnol 1990; 34:231-235.

52. Lucas C, da Costa M, van Uden N. Osmoregulatory active sodium-glycerol cotransport in the halotolerant yeast *Debaryomyces hansenii*. Yeast 1990; 6:187-191.

53. Larsson K, Eriksson P, Ansell R, Adler L. A gene encoding *sn*-glycerol 3-phosphate dehydrogenase (NAD⁺) complements an osmosensitive mutant of *Saccharomyces cerevisiae*. Mol Microbiol 1993; 10:1101-1111.

54. Albertyn J, Hohmann S, Thevelein JM, Prior BA. *GPD1*, which encodes glycerol-3-phosphate dehydrogenase is essential for growth under osmotic stress in *Saccharomyces cerevisiae* and its expression is regulated by the high-osmolarity glycerol response pathway. Mol Cell Biol 1994; 14:4135-4144.

55. Ansell R, Granath K, Hohmann S, Thevelein J, Adler L. The two isoenzymes for yeast NAD-dependent glycerol 3-phosphate dehydrogenase encoded by *GPD1* and *GPD2*, have distinct roles in osmoadaption and redox regulation. Submitted for publication.

56. Gancedo C, Gancedo JM, Sols A. Glycerol metabolism in yeasts. Pathways of utilization and production. Eur J Biochem 1968; 5:165-172.

57. Gancedo C, Serrano R. Energy-yielding metabolism. In: Rose AH, Harrison JS, ed. The Yeasts. 2nd ed. vol. 3. New York: Academic Press, 1989:205-257.

58. Lin ECC. Glycerol dissimilation and its regulation in bacteria. Ann Rev Microbiol 1976; 30:535-578.

59. Heller KB, Lin ECC, Wilson TH. Substrate specificity and transport properties of the glycerol facilitator of *Escherichia coli*. J Bacteriol 1980; 144:274-278.

60. Muramatsu S, Mizuno T. Nucleotide sequence of the region encompassing the *glpKF* operon and its upstream region contain a bent sequence of *Escherichia coli*. Nucl Acids Res 1989; 17:4378.

61. Sprague GF, Cronan Jr. JE. Isolation and characterization of *Saccharomyces cerevisiae* mutants defective in glycerol catabolism. J Bacteriol 1977; 129:1335-1342.

62. Pavlik P, Simon M, Schuster T, Ruis H. The glycerol kinase (*GUT1*) gene of *Saccharomyces cerevisiae*: cloning and characterization. Curr Genet 1993; 24:21-25.

63. Rønnow B, Kielland-Brandt M. *GUT2*, a gene for mitochondrial glycerol 3-phosphate dehydrogenase of *Saccharomyces cerevisiae*. Yeast 1993; 9:1121-1130.

64. Norbeck J, Blomberg A. Metabolic and regulatory changes associated with growth of *Saccharomyces cerevisiae* in 1.4 M NaCl: Evidence for osmotic induction of glycerol dissimilation via the dihydroxyacetone pathway and identification of genes encoding dihydroxyacetone kinase and glycerol dehydrogenase (NADP+). J Biol Chem, in press.

65. Chen S-M, Trumbore MW, Osinchak JE, Merkel JR. Improved purification and some molecular and kinetic properties of *sn*-glycerol 3-phosphate dehydrogenase from *Sac-*

charomyces cerevisiae. Prep Biochem 1987; 17:435-436.

66. Albertyn J, van Tonder A, Prior BA. Purification of glycerol-3-phosphate dehydrogenase of *Saccharomyces cerevisiae.* FEBS Lett 1992; 308:130-132.

67. Eriksson P, Andre L, Ansell R, Blomberg A, Adler L. Molecular cloning of *GPD2*, a second gene encoding *sn*-glycerol 3-phosphate dehydrogenase (NAD⁺) in *Saccharomyces cerevisiae*, and its comparison to *GPD1.* Mol Microbiol 1995; 17:95-107.

68. Mannhaupt G, Vetter I, Schwarzlose C, Mitzel S, Feldmann H. Analysis of a 26 kb region on the left arm of yeast chromosome XV. Yeast 1996; 12:67-76.

69. Tsuboi KK, Hudson PB. Acid phosphatases. VI. Kinetic properties of purified yeast and erythrocyte phosphomonoesterase. Arch Biochem Biophys 1956; 61:197-210.

70. Norbeck J, Påhlmann AK, Akhtar N, Blomberg A, Adler L. Purification and characterization of two isoenzymes of DL-glycerol 3-phosphatase from *Saccharomyces cerevisiae.* Identification of the corresponding *GPP1* and *GPP2* genes and evidence for osmotic regulation of Gpp2p expression by the osmosensing MAP kinase signal transduction pathway. J Biol Chem 1996; 271:13875-13881.

71. Larsson C, Valadi H, Hohmann S, Gustafsson L. Unpublished results.

72. Van Dijken JP, Scheffers WA. Redox balances in the metabolism of sugars by yeasts. FEMS Microbiol Rev 1986; 32:199-225.

73. Wang H-T, Rahaim P, Robbins P, Yocum RR. Cloning, sequence and disruption of the *Saccharomyces diastaticus DAR1* gene encoding a glycerol-3-phosphate dehydrogenase. J Bacteriol 1994; 176:7091-7095.

74. Hirayama T, Maeda T, Saito H, Shinozaki K. Cloning and characterization of seven cDNAs for hyperosmolarity-responsive (*HOR*) genes of *Saccharomyces cerevisiae.* Mol Gen Genet 1995; 249:127-138.

75. Deckert J, Rodriguez-Torres AM, Simon JT, Zitomer RS. Mutational analysis of Rox1, a DNA-bending repressor of hypoxic genes in *Saccharomyces cerevisiae.* Mol Cell Biol 1995; 15:6109-6117.

76. Krems B, Charizanis C, Entian K-D. Mutants of *Saccharomyces cerevisiae* sensitive to oxidative and osmotic stress. Curr Genet 1995; 27:427-434.

77. Bohnert HJ, Jensen RG. Strategies for engineering water-stress tolerance in plants. Trends in Biotechnology 1996; 14:89-97.

78. Schüller G, Brewster JL, Alexander MR, Gustin MC, Ruis H. The HOG pathway controls osmotic regulation of transcription via the stress response element (STRE) of the *Saccharomyces cerevisiae CTT1* gene. EMBO J 1994; 13:4382-4389.

79. Shinozaki K, Yamaguchi-Shinozaki K. Molecular responses to drought and cold stress. Curr Opin Biotechnol 1996; 7:161-167.

80. Yashar B, Irie K, Printen JA, Stevenson BJ, Sprague GF, Matsumoto K, Errede B. Yeast MEK-dependent signal transduction: Response thresholds and parameters affecting fidelity. Mol Cell Biol 1995; 15:6545-6553.

81. Elion EA. Ste5: a meeting place for MAP kinases and their associates. Trends Cell Biol 1995; 5:322-327.

82. Faux MC, Scott JD. Molecular glue: kinase anchoring and scaffold proteins. Cell 1996; 85:9-12.

83. Brewster JL, de Valoir T, Dwyer ND, Winter E, Gustin MC. An osmosensing signal transduction pathway in yeast. Science 1993; 259:1760-1763.

84. Boguslawski G, Polazzi JO. Complete nucleotide sequence of a gene conferring polymyxin B resistance on yeast: Similarity of the predicted polypeptide to protein kinases. Proc Natl Acad Sci USA 1987; 84:5848-5852.

85. Boguslawski G. *PBS2*, a yeast gene encoding a putative protein kinase, interacts with the *RAS2* pathway and affects osmotic sensitivity of *Saccharomyces cerevisiae.* J Gen Microbiol 1992; 138:2425-2432.

86. Akhtar N, Blomberg A, Adler L. Cloning and deletion of *OSG2/PBS2*, encoding a MAP kinase activator homologue involved in osmosensing in *S. cerevisiae.* Analysis of defects in osmoregulation and osmotic induction of proteins in *pbs2Δ.* In preparation

87. Brüning ARNE, Entian KD, Prior BA. Unpublished results.

88. Maeda T, Wurgler-Murphy SM, Saito H. A two-component system that regulates an

osmosensing MAP kinase cascade in yeast. Nature 1994; 369:242-245.

89. Ota IM, Varshavsky A. A yeast protein similar to bacterial two-component regulators. Science 1993; 262:566-569.

90. Alex LA, Simon MI. Protein histidine kinases and signal transduction in prokaryotes and eukaryotes. Trends Genet 1994; 10:133-138.

91. Morgan BA, Bouquin N, Johnston LH. Two-component signal-transduction systems in budding yeast MAP a different pathway? Trends in Cell Biology 1995; 5:453-457.

92. Ota IM, Varshavsky A. A gene encoding a putative tyrosine phosphatase suppresses lethality of an N-end rule-dependent mutant. Proc Natl Acad Sci USA 1992; 89:2355-2359.

93. Varshavsky A. The N-end rule. Cell 1992; 69:725-735.

94. Maeda T, Tsai AYM, Saito H. Mutations in a protein tyrosin phosphatase gene (*PTP2*) and a protein serine/threonine phosphatase gene (*PTC1*) cause a synthetic growth defect in *Saccharomyces cerevisiae*. Mol Cell Biol 1993; 13:5408-5417.

95. Posas F, Wurgler-Murphy SM, Maeda T, Witten EA, Thai TC, Saito H. Yeast HOG1 MAP kinase cascade is regulated by a multistep phosphorelay mechanism in the LSN1-YPD1-SSK1 "two-component" osmosensor. Cell 1996; 86:865-875.

96. Maeda T, Takekawa M, Saito H. Activation of yeast PBS2 MAPKK by MAPKKKs or by binding of and SH3-containing osmosensor. Science 1995; 269:554-558.

97. Tokishita S, Mizuno T. Transmembrane signal transduction by the *Escherichia coli* osmotic sensor, *EnvZ*: intermolecular complementation of transmembrane signalling. Mol Microbiol 1994; 13:435-444.

98. Yu G, Deschenes RJ, Fassler JS. The essential transcriptiopn factor, Mcm1, is a downstream target of Sln1, a yeast "two-component" regulator. J Biol Chem 1995; 270:8739-8743.

99. Warbrick E, Fantes P. The Wis1 protein kinase is a dose-dependent regulator of mitosis in *Schizosaccharomyces pombe*. EMBO J 1991; 10:4291-4299.

100. Shiozaki K, Russell P. Counteractible roles of protein phosphatase 2C (PP2C) and a MAP kinase kinase homolog in the osmoregulation of fission yeast. EMBO J 1995; 14:492-502.

101. Millar JBA, Buck V, Wilkinson MG. Pyp1 and Pyp2 PTPases dephosphorylate an osmosensing MAP kinase controlling cell size at division in fission yeast. Genes Develop 1995; 9:2117-2130.

102. Shiozaki K, Russell P. Cell-cycle control linked to extracellular environment by MAP kinase pathway in fission yeast. Nature 1995; 378:739-743.

103. Kato TJ, Okazaki K, Murakami H, Stettler S, Fantes P, Okayama H. Stress signal, mediated by a Hog1-like MAP kinase, controls sexual development in fission yeast. FEBS Lett 1996; 378:207-212.

104. Degols G, Shiozaki K, Russell P. Activation and regulation of the Spc1 stress-activated protein kinase in *Schizosaccharomyces pombe*. Mol Cell Biol 1996; 16:2870-2877.

105. Doi K, Gartner A, Ammerer G, Errede B, Shinkawa H, Sugimoto K, Matsumoto K. *MSG5*, a novel protein phosphatase promotes adaptation to pheromone response in *S.cerevisiae*. EMBO J 1994; 13:61-70.

106. Posas F, Casamayor A, Arino J. The PPZ phosphatses are involved in the maintenance of osmotic stability of yeast cells. FEBS Lett 1993; 318:282-286.

107. Lee KS, Hines LK, Levin DE. A pair of functionally redundant yeast genes (*PPZ1* and *PPZ2*) encoding type 1-related protein phophatases function within the *PKC1*-mediated pathway. Mol Cell Biol 1993; 13:5843-5853.

108. Posas F, Camps M, Arino J. The PPZ protein phosphatases are important determinants of salt tolerance in yeast cells. J Biol Chem 1995; 270:13036-13041.

109. Huang KN, Symington LS. Suppressors of a Saccharomyces cerevisiae pkc1 mutation identify alleles of the phosphatase gene *PTC1* and of a novel gene encoding a putative basic leucine zipper protein. Genetics 1995; 141:1275-1285.

110. Alessi DR, Gomez N, Moorhead G, Lewis T, Keyse SM, Cohen P. Inactivation of p42 MAP kinase by protein phosphatase 2A and

a protein tyrosine phosphatase, but not CL100, in various cell lines. Curr Biol 1995; 5:283.295.

111. Takeda T, Toda T, Kominami K-i, Kohnosu A, Yanagida M, Jones N. *Schizosaccharomyces pombe atf1+* encodes a transcription factor required for sexual development and entry into stationary phase. EMBO J 1995; 14:6193-6208.

112. Watanabe Y, Yamamoto M. *Schizosaccharomyces pombe pcr1+* encodes a CREB/ATF protein involved in regulation of gene expression for sexual development. Mol Cell Biol 1996; 16:704-711.

113. Millar JBA, Russel P, Dixon JE, Guan K-L. Negative regulation of mitosis by two functionally overlapping PTPases in fission yeast. EMBO J 1992; 11:4943-4952.

114. Dal Santo P, Blanchard B, Hoffman CS. The *Schizosaccharomyces pombe* pyp1 protein tyrosine phosphatase negatively regulates nutrient monitoring pathways. J Cell Science 1996; 109:1919-1925.

115. Stettler S, Warbrick E, Prochnik S, Mackie S, Fantes P. The wis1 signal transduction pathway is required for expression of cAMP-repressed genes in fission yeast. J Cell Science 1996; 109:1927-1935.

116. Galcheva-Gargova Z, Derijard B, Wu I-H, Davies RJ. An osmosensing signal transduction pathway in mammalian cells. Science 1994; 265:806-808.

117. Han J, Lee J-D, Bibbs L, Ullevitch RJ. A MAP kinase targeted by endotoxin and hyperosmolarity in mammalian cells. Science 1994; 265:808-811.

118. Jonak C, Heberle-Bors E, Hirt H. MAP kinases: universal multi-purpose signaling tools. Plant Mol Biol 1994; 24:407-416.

119. Chang C, Kwok SF, Bleecker AB, Meyerowitz EM. *Arabidopsis* ethylene-response gene *ETR1*: similarity to products of two component regulators. Science 1993; 262:539-544.

120. Jonak C, Kiegerl S, Lloyd C, Chan J, Hirt H. MMK2, a novel alfalfa MAP kinase, specifically complements the yeast MPK1 function. Mol Gen Genet 1995; 248:686-694.

121. Mizoguchi T, Irie K, Hirayama T, Hayashida N, Yamaguchi-Shinozaki K, Matsumoto K, Shinozaki K. A gene encoding a mitogen-activated protein kinase kinase kinase is induced simultaneously with genes for a mitogen-activated protein kinase and an S6 ribosomal protein kinase by touch, cold, and water stress in Arabidopsis thaliana. Proc Natl Acad Sci USA 1996; 93:765-769.

122. d'Enfert C, Minet M, Lacroute F. Cloning plant genes by complementation of yeast mutants. Methods Cell Biol 1995; 49:417-430.

123. Davenport KR, Sohaskey M, Kamada Y, Levin DE, Gustin MC. A second osmosensing signal transduction pathway in yeast—Hypotonic shock activates the PKC1 protein kinase-regulated cell integrity pathway. J Biol Chem 1995; 270:30157-30161.

124. Kamada Y, Jung US, Piotrowski J, Levin DE. The protein kinase C-activated MAP kinase pathway of *Saccharomyces cerevisiae* mediates a novel aspect of the heat shock response. Genes Develop 1995; 9:1559-1571.

125. Mazzoni C, Zarov P, Rambourg A, Mann C. The SLT2 (MPK1) MAP kinase homolog is involved in polarized cell growth in *Saccharomyces cerevisiae*. J Cell Biol 1993; 123:1821-1833.

126. Zarzov P, Mazzoni C, Mann C. The SLT2(MPK1) MAP kinase is activated during periods of polarized cell growth in yeast. EMBO J 1996; 15:83-91.

127. Marini NJ, Meldrum E, Buehrer B, Hubberstey AV, Stone DE, Traynor-Kaplan A, Reed SI. A pathway in the yeast cell division cycle linking protein kinase C (Pkc1) to activation of Cdc28 at Start. EMBO J 1996; 15:3040-3052.

128. Costigan C, Snyder M. SLK1, a yeast homolog of MAP kinase activators, has a RAS/cAMP-independent role in nutrient sensing. Mol Gen Genet 1994; 243:286-296.

129. Levin DE, Fields FO, Kunisawa R, Bishop JM, Thorner J. A candidate protein kinase C gene, *PKC1*, is required for the *S. cerevisiae* cell cycle. Cell 1990; 62:213-224.

130. Antonsson B, Montessuit S, Friedli L, Payton MA, Paravicini G. Protein kinase C in yeast. J Biol Chem 1994; 269:16821-16828.

131. Watanabe M, Chen CY, Levin DE. *Saccharomyces cerevisiae PKC1* encodes a protein kinase C (PKC) homolog with a substrate

specificity similar to that of mammalian PKC. J Biol Chem 1994; 269:16829-16836.

132. Levin DE, Bartlett-Heubusch E. Mutants in the *S. cerevisiae PKC1* gene display a cell cycle-specific osmotic stability defect. J Cell Biol 1992; 116:1221-1229.

133. Levin DE, Bowers B, Chen CY, Kamada Y, Watanabe M. Dissecting the protein kinase C/MAP kinase signalling pathway of *Saccharomyces cerevisiae*. Cell Mol Biol Res 1994; 40:229-239.

134. Lee KS, Levin DE. Dominant mutations in a gene encoding a putative protein kinase (*BCK1*) bypass the requirement for a *Saccharomyces cerevisiae* protein kinase C homolog. Mol Cell Biol 1992; 12:172-182.

135. Lee KS, Irie K, Gotoh Y, Watanabe Y, Araki H, Nishida E, Matsumoto K, Levin D. A yeast mitogen-activated protein kinase homolog (Mpk1p) mediates signalling by protein kinase C. Mol Cell Biol 1993; 13:3067-3075.

136. Costigan C, Gehrung S, Snyder M. A synthetic lethal screen identifies *SLK1*, a novel protein kinase homolog implicated in yeast cell morphogenesis and cell growth. Mol Cell Biol 1992; 12:1162-1178.

137. Torres L, Martin H, Garcia-Saez MI, Arroyo J, Molina M, Sanchez M, Nombela C. A protein kinase gene complements the lytic phenotype of *Saccharomyces lyt2* mutants. Mol Microbiol 1991; 5:2845-2854.

138. Martin H, Arroyo J, Sanchez M, Molina M, Nombela C. Activity of the yeast MAP kinase homologue Slt2 is critically required for cell integrity at 37°C. Mol Gen Genet 1993; 241:177-184.

139. Irie K, Takase M, Lee KS, Levin DE, Araki H, Matsumoto K, Oshima Y. *MKK1* and *MKK2*, which encode *Saccharomyces cerevisiae* mitogen-activated protein kinase-kinase homologs, function in the pathway mediated by protein kinase C. Mol Cell Biol 1993; 13:3076-3083.

140. Watanabe Y, Irie K, Matsumoto K. Yeast *RML1* encodes a serum response factor-like protein that may function downstream of the Mpk1 (Slt2) mitogen-activated protein kinase pathway. Mol Cell Biol 1995; 15:5740-5749.

141. Soler M, Plovins A, Martin H, Molina M, Nombela C. Characterization of domains in the yeast MAP kinase Slt2 (Mpk1) required for functional activity and in vivo interaction with protein kinases Mkk1 and Mkk2. Mol Microbiol 1995; 17:833-842.

142. Nickas ME, Yaffe MP. *BRO1*, a novel gene that interacts with components of the pkc1p-mitogen-activated protein kinase pathway in *Saccharomyces cerevisiae*. Mol Cell Biol 1996; 16:2585-2593.

143. Nonaka H, Tanaka K, Hirano H, Fujiwara T, Kohno H, Umikawa M, Mino A, Takai Y. A downstream target of *RHO1* small GTP-binding protein is *PKC1*, a homolog of protein kinase C, which leads to activation of the MAP kinase cascade in *Saccharomyces cerevisiae*. EMBO J 1995; 14:5931-5938.

144. Kamada Y, Qadota H, Python CP, Anraku Y, Ohya Y, Levin DE. Activation of yeast protein kinase C by Rho1 GTPase. J Biol Chem 1996; 271:9193-9196.

145. Qadota H, Python C, Inoue SB, Arisawa M, Anraku Y, Zheng Y, Watanabe T, Levin DE, Ohya Y. Identification of yeast Rho1p GTPase as a regulatory subunit of 1,3-β-glucan synthase. Science 1996; 272:279-281.

146. Ozaki K, Tanaka K, Imamura H, Hilhara T, Kameyama T, Nonaka H, Hirano H, Matsuura Y, Takai Y. Rom1p and Rom2p are GDP/GTP exchange proteins (GEPs) for the Rho1p small GTP-binding protein in *Saccharomyces cerevisiae*. EMBO J 1996; 15:2196-2207.

147. Masuda T, Tanaka K, Nonaka H, Yamochi W, Maeda A, Takai Y. Molecular cloning and characterization of yeast rho GDP dissociation inhibitor. J Biol Chem 1994; 269:19713-19718.

148. Hirano H, Tanaka K, Ozaki K, Imamura H, Kohno H, Hihara T, Kameyama T, Hotta K, Arisawa M, Watanabe T, Qadota H, Ohya Y, Takai Y. *ROM7/BEM4* encodes a novel protein that interacts with the Rho1p small GTP-binding protein in *Saccharomyces cerevisiae*. Mol Cell Biol 1996; 16:4396-4403.

149. Yoshida S, Ikeda E, Uno I, Mitsuzawa H. Characterization of a staurosporine-sensitive and temperature-sensitive mutant, *stt1*, of

Saccharomyces cerevisiae—STT1 is allelic to *PKC1*. Mol Gen Genet 1992; 231:337-344.

150. Yoshida S, Ohya Y, Goebl M, Nakano A, Anraku Y. A novel gene, *STT4*, encodes a phosphatdiylinositol 4-kinase in the *PKC1* protein kinase pathway of *Saccharomyces cerevisiae*. J Biol Chem 1994a; 269:1166-1171.

151. Yoshida Y, Ohya Y, Nakano A, Anraku Y. Genetic interactions among genes involved in the *STT4-PKC1* pathway of *Saccharomyces cerevisiae*. Mol Gen Genet 1994b; 242:631-640.

152. Yoshida S, Ohya Y, Nakano A, Anraku Y. *STT3*, a novel essential gene related to the *PKC1/STT1* protein kinase pathway, is involved in protein glycosylation in yeast. Gene 1995; 164:167-172.

153. Yoshida S, Ohya Y, Hirose R, Nakano A, Anraku Y. *STT10*, a novel class-D VPS yeast gene required for osmotic integrity related to the *PKC1/STT1* protein kinase pathway. Gene 1995; 160:117-122.

154. Garrett-Engele P, Moilanen B, Cyert MS. Calcineurin, the Ca^{2+}/Calmodulin-dependent protein phosphatase, is essential in yeast mutants with cell integrity defects and in mutants that lack a functional vacuolar H^+-ATPase. Mol Cell Biol 1995; 15:4103-4114.

155. Nakamura T, Ohmoto T, Hirata D, Tsuchiya E, Miyakawa T. Genetic evidence for the functional redundancy of the calcineurin and Mpk1-mediated pathways in the regulation of cellular events important for growth in *Saccharomyces cerevisiae*. Mol Gen Genet 1996; 251:211-219.

156. Costigan C, Kolodrubetz D, Snyder M. *NHP6A* and *NHP6B*, which encode HMG1-like proteins, are candidates for downstream components of the yeast *SLT2* mitogen-activated protein kinase pathway. Mol Cell Biol 1994; 14:2391-2403.

157. Shimizu J, Yoda K, Yamasaki M. The hypo-osmolarity-sensitive phenotype of the *Saccharomyces cerevisiae hpo2* mutant is due to a mutation in *PKC1*, which regulates expression of β-glucanase. Mol Gen Genet 1994; 242:641-648.

158. Toda T, Shimanuki M, Yanagida M. Two novel protein kinase C-related genes in fission yeast are essential for cell viability and implicated in cell shape control. EMBO J 1993; 12:1987-1995.

159. Kobori H, Toda T, Yaguchi H, Toya M, Yanagida M, Osumi M. Fission yeast protein kinase C gene homologues are required for protoplast regeneration: a functional link between cell wall formation and cell shape control. J Cell Sci 1994; 107:1131-1136.

160. Blomberg A. Global changes in protein synthesis during osmotic adaption to 0.7 M NaCl of the yeast *Saccharomyces cerevisiae*. J Bacteriol 1995; 177:3563-3572.

161. Miralles VJ, Serrano R. A genomic locus in *Saccharomyces cerevisiae* with four genes up-regulated by osmotic stress. Mol Microbiol 1995; 17:653-662.

162. Wieser R, Adam G, Wagner A, Schüller C, Marchler G, H. R, Krawiec Z, Bilinski T. Heat shock factor-independent heat control of transcription of the *CTT1* gene encoding the cytosolic catalase T of *Saccharomyces cerevisiae*. J Biol Chem 1991; 266:12406-12411.

163. Varela JCS, van Beekvelt C, Planta RJ, Mager WH. Osmostress-induced changes in yeast gene expression. Mol Microbiol 1992; 6:2183-2190.

164. Varela JCS, Praekelt UM, Meacock PA, Planta RJ, Mager WH. The *Saccharomyces cerevisiae HSP12* gene is activated by the high-osmolarity glycerol pathway and negatively regulated by protein kinase A. Mol Cell Biol 1995; 15:6232-6245.

165. Sanchez Y, Taulien J, Borkovich A, Lindquist S. Hsp104 is required for tolerance to many forms of stress. EMBO J 1992; 11:2357-2364.

166. Francois JM, Thompson-Jaeger S, Skroch J, Zellenka U, Spevak W, Tatchell K. *GAC1* may encode a regulatory subunit for protein phosphatase type 1 in *Saccharomyces cerevisiae*. EMBO J 1992; 11:87-96.

167. Fraser J, Luu HA, Neculcea J, Thomas DY, Storms RK. Ubiquitin gene expression: response to environmental changes. Curr Genet 1991; 20:17-23.

168. Arnason T, Ellison MJ. Stress resistance in *Saccharomyces cerevisiae* is strongly correlated with assembly of a novel type of multi-ubiquitin chain. Mol Cell Biol 1994; 14:7876-7883.

169. Kobayashi N, McEntee K. Identification of *cis* and *trans* components of a novel heat shock stress regulatory pathway in *Saccharo-*

myces cerevisiae. Mol Cell Biol 1993; 13:248-256.

170. Evangelista CCJ, Rodriguez Torres AM, Limbach MP, Zitomer RS. Rox3 and Rts1 function in the global stress response pathway in baker's yeast. Genetics 1996; 142:1083-1093.

171. de Virgilio C, Bürckert N, Bell W, Jenö P, Boller T, Wiemken A. Disruption of *TPS2*, the gene encoding the 100-kDa subunit of the trehalose-6-phosphate synthase/phosphatase complex in *Saccharomyces cerevisiae*, causes accumulation of trehalose-6-phosphate and loss of trehalose-6-phosphate phosphatase activity. Eur J Biochem 1993; 212:315-323.

172. Gounalaki N, Thireos G. Yap1p, a yeast transcrptional activator that mediates multidrug resistance, regulates the metabolic stress response. EMBO J 1994; 13:4036-4041.

173. Winderickx J, de Winde JH, Crauwels M, Hino A, Hohmann S, Van Dijck P, Thevelein JM. Expression regulation of genes encoding subunits of the trehalose synthase complex in *Saccharomyces cerevisiae*. Mol Gen Genet 1996; 252:470-482.

174. Mager WH, de Kruijff AJJ. Stress-induced transcription activation. Microbiol Rev 1995; 59:506-531.

175. Ruis H, Schüller C. Stress signaling in yeast. Bioessays 1995; 17:959-965.

176. Boorstein WR, Craig EA. Regulation of yeast *HSP70* gene by a cAMP responsive transcription control element. EMBO J 1990; 9:2543-2553.

177. Belazzi T, Wagner A, Wieser R, Schanz M, Adam G, Hartig A, Ruis H. Negative regulation of transcription of the *Saccharomyces cerevisiae* catalase T (*CTT1*) gene by cAMP is mediated by a positive control element. EMBO J 1991; 10:585-592.

178. Marchler G, Schüller C, Adam G, Ruis H. A *Saccharomyces cerevisiae* UAS element controlled by protein kinase A activates transcription in response to a variety of stress conditions. EMBO J 1993; 12:1997-2003.

179. Martinez-Pastor MT, Marchler G, Schüller C, Marchler-Bauer A, Ruis H, Estruch F. The *Saccharomyces cerevisiae* zinc finger proteins Msn2p and Msn4p are required for

transcriptional induction through the stress-response element (STRE). EMBO J 1996; 15:2227-2235.

180. Schmitt AP, McEntee K. Msn2p, a zinc finger DNA-binding protein, is the transcriptional activator of the multistress response in *Saccharomyces cerevisiae*. Proc Natl Acad Sci USA 1996; 93:5777-5782.

181. Rosenblum-Vos LS, Rhose L, Evangelista CCJ, Boayke KA, Wick P, Zitomer RS, more. The *ROX3* gene encodes an essential nuclear protein involved in *CYC7* expression in *Saccharomyces cerevisiae*. Mol Cell Biol 1991; 11:5639-5647.

182. Thevelein JM. Signal transduction in yeast. Yeast 1994; 10:1753-1790.

183. Albertyn J, Hohmann S, Prior BA. Osmotic stress controls the expression of the *GPD1* gene by different pathways. Submitted for publication.

184. Winderickx J, de Winde JH, Hohmann S, Gustin MC, Thevelein JM. Unpublished results.

185. de Winde JH, Crauwels M, Hohmann S, Thevelein JM, Winderickx J. Differential requirement of the yeast sugar kinases for sugar sensing in the establishment of the catabolite repressed state. Eur J Biochem 1996; (in press).

186. Eriksson P, Påhlmann A-K, Norbeck J, Adler L, Blomberg A. A HOG-pathway independent mechanism controls the *GPD1* promoter in *Saccharomyces cerevisiae* during steady-state growth under osmotic stress. Submitted for publication.

187. Eriksson P, Adler L, Blomberg A. Promoter analysis of the osmotically controlled *GPD1* gene of *Saccharomyces cerevisiae* shows that RAP1 is a major activator and gives evidence for several regulatory elements. Submitted for publication.

188. Kraakman LS, Griffioen G, Zerp S, Groenveld P, Thevelein JM, Mager WH, Planta RJ. Growth-related expression of ribosomal protein genes in *Saccharomyces cerevisiae*. Mol Gen Genet 1993; 239:196-204.

189. Chambers A, Packham EA, Graham IR. Control of glycolytic gene expression in the budding yeast (*Saccharomyces cerevisiae*). Curr Genet 1995; 29:1-9.

190. Marquez JA, Serrano R. Multiple transduction pathways regulate the sodium-extrusion gene *PMR2/ENA1* during salt stress in yeast. FEBS Lett 1996; 382:89-92.

191. Adler L, Blomberg A, Nilsson A. Glycerol metabolism and osmoregulation in the salt-tolerant yeast *Debaryomyces hansenii.* J Bacteriol 1985; 162:300-306.

192. Van Aelst L, Hohmann S, Zimmermann FK, Jans AWH, Thevelein JM. A yeast homologue of the bovine lens fibre MIP gene family complements the growth defect of a *Saccharomyces cerevisiae* mutant on fermentable sugars but not its defect in glucose-induced RAS-mediated cAMP signalling. EMBO J 1991; 10:2095-2104.

193. Thevelein JM, Hohmann S. Trehalose synthase, guard to the gate of glycolysis in yeast? Trends in Biochemical Sciences 1995; 20:3-10.

194. Van Aelst L, Hohmann S, Bulaya B, de Koning W, Sierkstra L, Neves MJ, Luyten K, Alijo R, Ramos J, Coccetti P, Martegani E, de Magelhães NM, Brandão RL, Van Dijck P, Vanhalewyn M, Durnez P, Jans AWH, Thevelein JM. Molecular cloning of a gene involved in glucose sensing. Molecular Microbiology 1993; 8:927-943.

195. Koo SP, Higgins CF, Booth IR. Regulation of compatible solute accumulation in *Salmonella typhimurium*: evidence for a glycine betaine efflux system. J Gen Microbiol 1991; 137:2617-2615.

196. Berrier C, Coulombe A, Szabo I, Zoratti M, Ghazi A. Gadolinium ion inhibits loss of metabolites induced by osmotic shock and large stretch-activated channels in bacteria. Eur J Biochem 1992; 206:559-565.

197. Kwon HM, Handler JS. Cell volume regulated transporters of compatible solutes. Curr Opin Cell Biol 1995; 7:465-471.

198. Jackson PS, Strange K. Volume-sensitive anion channels mediate swelling-activated inositol and taurine efflux. Am J Physiol 1993; 265:C1489-C1500.

199. Strange K, Jackson PS. Swelling-activated organic osmolyte efflux: a new role for anion channels. Kidney Intern 1995; 48:994-1003.

200. Strange K, Emma F, Jackson PS. Cellular and molecular physiology of volume-sensitive anion channels. Amer J Physiol 1996; 39:C711-C730.

201. Paulmichl M, Li Y, Wickman K, Ackerman M, Peralta E, Clapman D. New mammalian chloride channel identified by expression cloning. Nature 1992; 356:238-241.

202. Sprague GFJ, Thorner JW. Pheromone response and signal transduction during the mating process of *Saccharomyces cerevisiae*. In: Jones EW, Pringle JR, Broach JR, eds. The Molecular and Cellular Biology of the Yeast *Saccharomyces*. Gene Expression. Vol. 2. Cold Spring Harbor: Cold Spring Harbor Laboratory Press, 1992:657-744.

203. Reizer J, Reizer A, Saier MHJ. The MIP family of integral membrane channel proteins: sequence comparison, evolutionary relationship, reconstructed pathway of evolution, and proposed functional differentiation of the two repeated halves of the proteins. Crit Rev Biochem Mol Biol 1993; 28:235-257.

204. Yamada S, Katsuhara M, Kelly W, Michalowski CB, Bohnert HJ. A family of transcripts encoding water channel proteins: tissue specific expression in the common ice plant. Plant Cell 1995; 7:1129-1142.

205. Agre P, Brown D, Nielsen S. Aquaporin water channels: unanswered questions and unresolved controversies. Curr Opin Cell Biol 1995; 7:472-483.

206. Engel A, Walz T, Agre P. The aquaporin family of membrane water channels. Curr Opin Struct Biol 1994; 4:545-553.

207. Rao Y, Jan LY, Jan YN. Similarity of the product of the *Drosophila* neurogenic gene *big brain* to transmembrane channel proteins. Nature 1990; 345:163-167.

208. André B. An overview of membrane transport proteins in *Saccharomyces cerevisiae*. Yeast 1995; 11:1575-1611.

209. Gründer S, Thiemann A, Pusch M, Jentsch TJ. Regions involved in the opening of the ClC-2 chloride channel by voltage and cell volume. Nature 1992; 360:759-762.

210. Preston GM, Piazza Carroll T, Guggino WB, Agre P. Appearance of water channels in *Xenopus* oocytes expressing red cell CHIP 28 protein. Science 1992; 256:385-387.

211. van Lieburg AF, Knoers NV, Deen PM. Discovery of aquaporins: a breakthrough in

research on renal transport. Pediatr Nephrol 1995; 9:228-234.

212. Nielsen S, Agre P. The aquaporin family of water channels in kidney. Kidney Intern 1995; 48:1057-1068.

213. Nielsen S, Marples D, Frokiaer J, Knepper M, Agre P. The aquaporin family of water channels in kidney: An update on physiol-ogy and pathophysiology of aquaporin-2. Kidney Intern 1996; 49:1718-1723.

214. King LS, Agre P. Pathophysiology of the aquaporin water channels. Ann Rev Physiol 1996; 58:619-648.

215. Calamita G, Bishai WR, Preston GM, Guggino WB, Agre P. Molecular cloning and characterization of AqpZ, a water channel from *Escherichia coli*. J Biol Chem 1995; 270:29063-29066.

CRUCIAL FACTORS
IN SALT STRESS TOLERANCE

Ramón Serrano, José A. Márquez and Gabino Ríos

1. SALT STRESS AND THE EVOLUTION
OF ION TRANSPORT

The appearance of NaCl as a stress factor is a recent event in the evolution of living organisms. The concentration of NaCl in the ancestral sea where life originated was probably relatively low by present standards (Fig. 5.1). Rocks containing this soluble salt were buried under earth and only exposed to water after emergence to surface during mountain formation. With all the uncertainties in these kind of speculations, we may guess that the first living cells on Earth experienced NaCl concentrations lower than 50 mM. Although we are talking of events occurring 3×10^9 years ago and this plausible assumption should therefore be taken with reservations, we conclude that primitive cells did not need sodium and chloride extrusion systems to prevent the toxic effects of high concentrations of these ions on cellular enzymes. However, as indicated below, a capability for salt extrusion must have been a very early property of cells because of the need to control cellular osmolarity. In this chapter we will try to answer the question of what kind of ion pumps were needed by these primitive cells and how these pumps evolved.

As proposed by Raven and Smith[1] the first ion pump developed by primitive cells was a proton pumping ATPase (H^+-ATPase). This pump aided the export of organic acids generated by the fermentative metabolism that was dominant in this anaerobic world. Later on the proton gradient generated by this pump was utilized to drive the secondary active transport of other molecules. In addition, the combination of the H^+-ATPase with proton-pumping photosynthetic chains gave rise to photosynthetic phosphorylation where the reaction of the pump was reversed for ATP synthesis. After oxygen accumulated as a byproduct of photosynthesis, respiratory proton pumps developed and—together with the

Yeast Stress Responses, edited by Stefan Hohmann and Willem H. Mager.
© 1997 R.G. Landes Company.

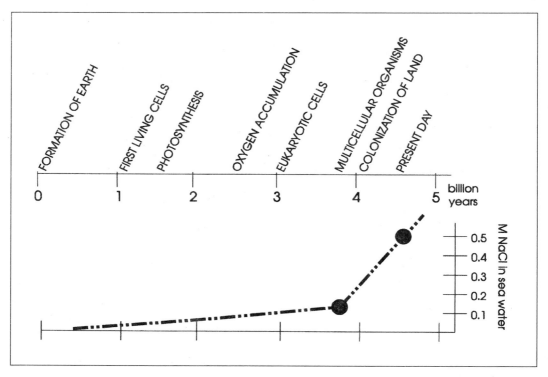

Fig. 5.1. Model about the increase in sea NaCl concentration in the course of biological evolution. The major events are indicated above the time scale. The NaCl line is based on just the present concentration and the one present in the internal fluids of animals, probably corresponding to the sea concentration at the time of land colonization.

H$^+$-ATPase—gave rise to oxidative phosphorylation. The type of ATPase involved in all these activities is the so called "F" or (F$_0$.F$_1$) H$^+$-ATPase, a complex protein machinery made of catalytic and accessory or regulatory subunits. The "F" nomenclature derives from the fact that subunits of this enzyme were identified as "coupling factors" of oxidative phosphorylation. F$_0$ and F$_1$ correspond to the membrane embedded and membrane protruding parts consisting of hydrophobic and hydrophilic subunits, respectively. (F$_0$.F$_1$) H$^+$-ATPases represent one of the most ancient protein machines of living cells and at present are found in bacteria, mitochondria, chloroplasts and in the acidic vacuolar compartment of eukaryotic cells.[2]

The plasmatic and endoplasmatic membranes of present eukaryotic cells contain a different type of pump, the "P" or (E-P) ATPases involved in proton, sodium, potassium and calcium transport.[2] The "P"

nomenclature derives from the fact that these enzymes form a phosphorylated intermediate in the course of the reaction, as implied by the "E-P" symbolism. Its function is devoted to cation extrusion including the protons, but in contrast to the (F$_0$.F$_1$) ATPases they do not participate in ATP synthesis. (E-P) ATPases are also present in bacteria but as minor enzymes involved in transport of potassium and divalent cations they are much less abundant than the major (F$_0$.F$_1$) H$^+$-ATPases devoted to oxidative phosphorylation. The existence of primary pumps for cation transport in addition to proton-cation cotransport systems points to specific advantages of direct ATP-dependent pumping versus coupling to the proton gradient. The appearance of (E-P) ATPases and (F$_0$.F$_1$) ATPases probably represent independent events in the course of evolution based on the large differences in structure and mechanism.

As discussed by Maloney and Wilson[3] ion transport mechanisms solved several basic problems of primitive cells. In addition to pH regulation there was a problem of "colloidal" osmotic swelling determined by impermeant intracellular macromolecules (proteins and nucleic acids). In primitive cells devoid of rigid cell walls, salt extrusion was the only defense against osmotic swelling. Ion extrusion pumps compensate for the osmotic pressure of intracellular macromolecules and prevent cell swelling by generating a gradient with higher salt concentrations outside of the cells. Instead of designing specific pumps for every ion, primitive cells probably utilized the electrochemical proton gradient generated by the H^+-ATPase for salt extrusion. The negative-inside membrane potential drove anion efflux mediated by channels while proton-cation antiports mediated cation efflux. These extrusion systems had to be selective because some ions such as phosphate were needed inside the cells for metabolic purposes and some others such as potassium for electrical balance and pH regulation. Therefore anion extrusion systems were specific for chloride versus phosphate and cation extrusion systems were specific for sodium versus potassium. The important conclusion is that although—as indicated above—the NaCl concentrations in the media of primitive cells were not high enough for toxic effects on enzymes, it is likely that NaCl was a major soluble salt in primeval sea. Therefore some capability for salt extrusion was consubstantial with the origin of cells because of the need of osmotic regulation.

The reason why potassium was selected instead of sodium as the "retained" cation inside the cells was probably a matter of chance. The selection of magnesium instead of calcium as the preferred intracellular divalent cation can be explained by the insolubility of calcium phosphates. But potassium offers no obvious advantage compared to sodium for intracellular systems. However, once this occurred a multitude of intracellular enzymes became "fitted" to

potassium in the course of evolution. These enzymes were activated by binding to this cation and inhibited by sodium which became antagonist to potassium and inhibitory. In addition to blocking potassium binding sites high sodium concentrations could also interfere with magnesium and calcium sites. The case of chloride is simpler probably because no better choice for extruded anion existed. Therefore intracellular enzymes became used to low chloride levels and some of them—mainly those having anionic substrates—could be inhibited by high chloride concentrations. These specific cationic and anionic effects on protein binding sites are more important for salt toxicity than nonspecific ionic strength effects operating at high salt concentrations. Low salt concentrations (50-100 mM) have positive effects on many enzymes because this is the ionic strength of the cytoplasm of normal organisms and protein hydration is maximal under these conditions. High salt concentrations (greater than 0.3-0.5 M) on the other hand are toxic for most enzymes because of the perturbation of the hydrophobic-electrostatic balance between the forces maintaining protein structure.[4,5] The toxicity of salt on cells is observed at intracellular concentrations of Na^+ and Cl^- of 50-100 mM, which is much lower than the levels required for ionic strength effects. Therefore in order to understand salt stress we need to investigate not only ion transport systems but also the intracellular enzyme systems specifically inhibited by Na^+ and Cl^-.[5]

The appearance of a rigid cell wall—due to the complexities of the biosynthetic machinery a relatively late event in cellular evolution—obviated the need for salt extrusion to prevent bursting of cells. Having solved the osmotic problem, wall-containing cells probably maintained the capability for NaCl extrusion to avoid the accumulation inside the cells of toxic concentrations of salt from the surrounding sea. With electrical membrane potentials (negative inside) greater than 60 mV and external sodium concentrations in the order of 0.1 M, the Nernst equation predicts intracellular sodium concentrations greater than 1 M. This high level cannot be tolerated by cellular systems.

It is difficult to ascertain the rate of salinization of the sea at the early times of evolution of living organisms. The present concentration of NaCl in the internal fluids of animals (around 150 mM) is thought to represent the sea concentration at the time multicellular animals evolved and started to colonize the land. This occurred about 0.5×10^9 years ago.[6] If we assume this reasonable speculation to be true then the salt concentration of the sea has increased dramatically during a relatively short period from about 150 (intracellular fluid levels) to 500 mM (present sea levels). This represents a much faster salinization than that occurring during the first 3×10^9 years of evolution (Fig. 5.1). Apparently NaCl-containing rocks have become exposed to water during the last period of the history of our planet and we are confronted with a threat to living organisms. This problem is not only relevant for marine organisms: the dissolved NaCl also concentrates in soils affecting the environment for terrestrial organisms such as plants. Accordingly, the progressive salinization of irrigated land is expected to compromise the future of agriculture in the most productive areas of our planet.[7,8]

A remarkable event occurred at the origin of eukaryotic cells: the $(F_0.F_1)$ H^+-ATPase redistributed from the plasma membrane to the endocytic and vacuolar compartment and (E-P) ATPases remained as the major pumps at the surface of cells.[3] Given the versatility of these pumps it is plausible that in animal cells devoid of a cell wall the indirect sodium extrusion mediated by H^+-Na^+ antiport was replaced by a more efficient Na^+-ATPase. Interestingly, the kind of enzyme which dominates the physiology of animal cells is a (Na^+/K^+)-ATPase which exchanges these two cations. As discussed by Tosteson[9] this is a very efficient system for volume regulation both because of the Na^+/K^+ stoichiometry of 1.5 (3 sodium extruded per 2 potassium taken) and because of the much greater permeability of the plasma membrane to potassium than to sodium. Therefore the sodium gradient (higher concentration outside) is greater than the potassium gradient (higher concentration inside) and compensates for the colloid osmotic pressure. The existence of such an exchange pump as the major transport system of animal cells points to a secondary development at the time animal cells evolved. Transport operations which were mediated before by separate systems—i.e., by a combination of H^+-ATPases, H^+-Na^+ antiports and K^+ channels—was replaced by a superior machinery combining those activities.

Plant cells and eukaryotic microorganisms followed a different specialization at their plasma membrane: a H^+-ATPase of the (E-P) family dominates their physiology. In order to understand this development we must return to a crucial event in evolution: the colonization of land by animals and plants. Animals which adapted to land probably kept the NaCl concentration prevailing in sea water 0.7×10^9 years ago (150 mM, see Fig. 5.1). This allowed osmotic regulation by their osmotically sensitive cells lacking a cell wall. Marine microorganisms and plants adapted to land and fresh water did not need to maintain any capability for sodium transport because the osmotic problem was largely solved by the cell wall and they were transferred to an environment very low in sodium. Accordingly, their capability for sodium transport probably diminished along evolutionary time after they had left the sea. Chloride poses a different problem because it is always present in potassium, magnesium and calcium salts surrounding plants and microorganisms.

The chemiosmotic circuits depicted in Figure 5.2 may represent the prevailing situation of ion transport in animals, microorganisms and plants after colonization of terrestrial habitats. The lack of sodium transporters in fungi and plants is an oversimplification which emphasizes the plausible consequences of more than 0.5×10^9 years of evolution in media without significant sodium concentrations. The

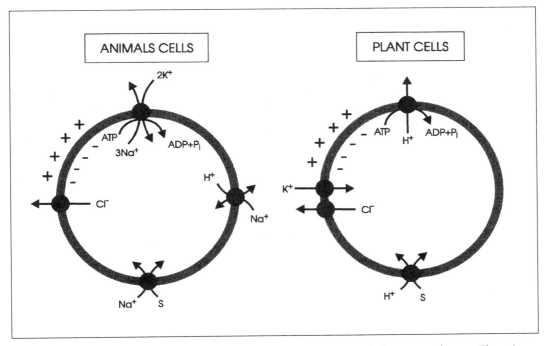

Fig. 5.2. *Different chemiosmotic circuits in animal and plant and fungal plasma membranes. The primary pumps are at the top, the Na^+/K^+-ATPase and the H^+-ATPase, respectively. Ion channels are at the left, a chloride channel in animal cells and potassium and chloride channels in plants and fungi. Finally, at the bottom are the sodium symports for different nutrients (S) in animal cells and the corresponding proton symports of fungi and plants.*

specialization for proton pumping with a powerful (E-P) ATPase with capability to generate pH gradients in excess of 5 pH units[2] must represent an adaptation to acidic medium. Such conditions are highly toxic to animal cells with a Na^+/K^+-ATPase and to bacteria with a less powerful $F_0.F_1$ H^+-ATPase.[10] Apparently the toxicity of protons became predominant in the terrestrial habitats of plants and fungi which concentrated their bioenergetic capabilities on proton transport. A manifestation of the "fitness" of plants to acid pH is the acidification response to the growth hormone auxin exhibited by elongating plant tissues. This response is important for enzymatic "loosening" of the cell wall during growth and is mediated by activation of the plasma membrane (E-P) H^+-ATPase.[11,12]

It is within this scenario of "oblivion" of the sea salt by terrestrial plants and fungi where the significance of salt stress

needs to be situated. The progressive salinization of irrigated land now encounters organisms with almost no defenses against the toxicities of sodium and chloride. An urgent biotechnological challenge is to restore to crop plants the capability for sodium extrusion which their ancestors had. We do not know enough about the sodium transport systems of present marine and terrestrial fungi and plants to design a correct strategy. The only sodium extrusion systems identified so far in nonanimal eukaryotes are the (E-P) ATPase encoded by the *ENA1/PMR2* gene of *Saccharomyces cerevisiae*[13] and the Na^+/H^+ antiporter of *Schizosaccharomyces pombe*.[14] It is not by chance that these two organisms correspond to well-established genetic model systems. The identification of these sodium transport systems has been possible thanks to molecular genetic approaches developed for yeast based on complementation of mutants with plasmid libraries. Biochemical approaches were unsuccessful due to the low abundance of the sodium transport systems. Therefore the significance of salt stress studies performed with yeast is that the results may be of value to the less tractable plant systems where this stress has enormous economical and social relevance.[5,10]

2. OSMOTIC STRESS VS. ION TOXICITIES

NaCl poses two different stress conditions to cells including yeast: osmotic stress and ion toxicities (sodium and chloride toxicities). Osmotic stress is not specific to NaCl and is the subject of chapter 4 of this book. It is important, however, to clarify here which one of these two stress factors limits yeast growth during salt stress. There is no simple answer to this question. In *Saccharomyces cerevisiae* growing on glucose media sodium toxicity is the limiting factor because NaCl is much more toxic than equivalent osmotic concentrations of either sorbitol or KCl. This results from a combination of a high capability for osmotic adjustment and a low capability for sodium extrusion. With other yeast species or with *S. cerevisiae* growing on carbon sources other than glucose the situation is different and osmotic adjustment becomes the limiting factor.[5] As discussed in chapter 4, osmotic adjustment in yeast is based on glycerol synthesis and accumulation. The strategy to achieve high internal glycerol levels during osmotic stress varies between different yeast species and growth conditions. Analogously, the capability for sodium transport differs between yeast species and growth conditions (see below). The balance between these two factors determines the nature of the growth-limiting stress in NaCl containing media: either osmotic stress or sodium toxicity.

Chloride toxicity has not been observed in yeast. We do not know if this is due to either limited uptake of the anion, very active extrusion or chloride-resistant enzyme systems. The most plausible explanation is the presence in yeast of chloride channels which—coupled to the large membrane potential negative inside of this organism[2]—maintain very low intracellular concentrations of this toxic anion. Although there is no information about the molecular basis of chloride transport in yeast, the *GEF1/CLC1* gene encodes a protein similar to animal voltage-gated Cl− channels.[15] A mutational analysis combined with electrophysiological (patch-clamp), physiological and biochemical studies is needed to ascertain whether this gene encodes a chloride extrusion system protecting yeast against the toxicity of this anion.

3. MUTATIONAL ANALYSIS OF SALT TOLERANCE

A mutational analysis has been performed by several laboratories to identify crucial factors in salt tolerance. In order to ensure that sodium toxicity is the growth limiting factor in salt tolerance the mutant screens have been performed in glucose medium with either *S. cerevisiae* or *S. pombe*. Not surprisingly the factors identified have to do with either sodium transport or with sodium-sensitive intracellular enzymes. We will discuss now the methodologies and principle results of such genetic analysis and later on the specific cellular reactions involved in sodium tolerance.

Basically both gain-of-function and loss-of-function approaches have been employed to open the "black box" containing the crucial factors for salt tolerance in yeast. Random overexpression of yeast genes in multi-copy plasmids has previously been utilized to identify the targets of growth inhibitory drugs such as tunicamycin and compactin.[16] We have implemented this approach for the study of salt tolerance and characterized several genes which increase yeast growth under salt stress when overexpressed on multicopy plasmids.[17-19] Disruption of two of them (*HAL1*[17] and *HAL3*[19]) results in decreased salt tolerance. This points to an important role of the encoded proteins as rate-limiting factors for salt tolerance in yeast. The mechanisms of these factors involve the regulation of ion transport systems as discussed below. Disruption of the other gene

(*HAL2*[18]) caused methionine auxotrophy, suggesting that it encoded a salt-sensitive enzyme of this metabolic pathway. The characterization of this enzyme as a sodium-sensitive nucleotidase acting on 3'-phosphoadenosine-5'-phosphate (PAP)[20] is discussed below.

We have recently performed a saturation screening by the random overexpression approach described above and isolated 90 multicopy plasmids with genomic inserts which confer salt tolerance. A preliminary characterization of this collection indicates that our previously identified *HAL1*, *HAL2* and *HAL3* genes are those most frequently isolated being represented each by 8-12 plasmids. Two other genes selected at similar high frequencies are the *ENA1/PMR2* gene encoding a putative sodium-extrusion (E-P) ATPase[13,21] and a novel protein kinase gene originally isolated by S. J. Kron (Whitehead Institute, Cambridge, Massachusetts) and designated *HAL5* (ORF YJL165c; accession number Z49440). All together these five genes account for more than half of the halotolerance genes in our collection. About 20 novel genes which are currently under characterization have been isolated in just 1 to 3 plasmids.

The gain-of-function approach has also been attempted by isolation of salt tolerant yeast mutants. Yeast mutants more tolerant to NaCl are relatively easy to isolate. We have isolated several of these mutants and all corresponded to single point semidominant mutations. Genomic libraries were constructed from DNA of those mutants in yeast centromeric plasmids in order to clone the responsible genes by complementation of the relative salt sensitivity of wild type yeast. In all cases we repeatedly cloned *HAL2* and *HAL3* as extragenic suppressors instead of the gene mutated in the salt tolerant strain. *HAL2* and *HAL3* dramatically improve yeast salt tolerance upon modest overexpression in centromeric plasmids and therefore complicate the cloning by complementation of semidominant mutations. Now that a saturation screening for halotolerance

genes is underway it would be easy to discard in the future extragenic suppressors of salt tolerance mutations by hybridization with known *HAL* genes working in single-copy plasmids.

In a different approach, recessive mutations causing salt sensitivity were complemented by genomic plasmid libraries constructed with wild type yeast DNA. Up to now the only salt sensitive mutant characterized at the molecular level has been the calcineurin mutant isolated by Mendoza et al.[22] Calcineurin or protein phosphatase 2B is a heterodimer that consists of a 60 kDa catalytic subunit (containing a carboxyl-terminal autoinhibitory domain with a Ca^{2+}-calmodulin binding site) stably associated with an essential Ca^{2+}-binding regulatory subunit of about 20 kDa. Calcineurin is activated upon binding of Ca^{2+} to both calmodulin and the regulatory subunit. Ca^{2+}-calmodulin binding to the catalytic subunit displaces the autoinhibitory domain and activates the phosphatase.[23] Null mutants lacking either the two genes encoding calcineurin catalytic subunits—*CNA1* and *CNA2*—or lacking the *CNB1* gene encoding the regulatory subunit each express no measurable calcineurin activity. Such mutants are viable under most conditions but demonstrate slow recovery from α-factor pheromone-mediated arrest[24,25] and increased sensitivity to media with high Na^+ or Li^+.[22,26]

The important role of calcineurin in salt tolerance was also discovered by a pharmacological approach based on the drug FK506. The immunosuppressants cyclosporin A and FK506 act by inhibiting calcineurin-mediated signal transduction in activated T cells. This inhibition involves a complex of the drugs with specific binding proteins the immunophilins. Immunosuppressant-immunophilin complexes bind to calcineurin and thereby inactivate it.[27] In the presence of the immunosuppressant FK506 the growth behavior of wild type cells in high NaCl medium became very similar to those of the calcineurin-deficient mutant.[26]

In addition to these direct genetic approaches to salt tolerance in recent times this phenotype has been investigated after disruption and overexpression of many gene isolates. Overexpression of yeast casein kinase I (encoded by the *YCK1* and *YCK2* genes) increases salt tolerance[28] while mutations on casein kinase II (*CKA1,2* and *CKB1,2* genes for catalytic and regulatory subunits, respectively) results in sodium and lithium sensitivity.[29] Mutations in the yeast calmodulin gene *CMD1* such as *cmd1-3* which cause defective Ca^{2+} binding confer sodium and lithium sensitivity.[19] Overexpression of Ca^{2+}/calmodulin-dependent protein kinases (encoded by the *CMK1,2* genes[30]) improves salt tolerance (unpublished observation of S. J. Kron, Whitehead Institute, Cambridge, Massachusetts).

The results of this preliminary genetic analysis of salt tolerance in *Saccharomyces cerevisiae* define a complex scenario where genes involved in defined transport (*ENA1*) and metabolic (*HAL2*) reactions interplay with a multitude of regulatory components with undefined mechanisms. In the following sections we will try to build a coherent picture from all this dispersed information.

4. ION UPTAKE AND EFFLUX SYSTEMS

The mechanism of cation uptake in yeast is not well understood. Mutations in the *TRK1* gene encoding a plasma membrane protein reduce K^+ uptake and compromise growth in media with relatively low K^+ concentrations.[31,32] Growth is normal at relatively high K^+ concentrations. The concentrations of K^+ required for yeast growth are very dependent on the level of other cations in the medium such as H^+ and NH_4^+ which may compete for K^+ uptake. At normal pH values of 4-5 and with ammonium as nitrogen source 1 mM K^+ is optimal for *TRK1* strains but limiting for the *trk1* mutants which required 10-100 mM. With arginine as nitrogen source the competition of NH_4^+ with K^+ for uptake is avoided and the discriminating concentration is 0.1 mM with the mutant requiring 1-10 mM for optimal growth.[33]

The residual K^+ transport in the *trk1* mutants is due to *TRK2*—a poorly expressed gene related to *TRK1*[34,35]—and to an unidentified system of very low affinity.[34] There is no evidence, however, for a direct transport activity of the *TRK1* and *TRK2* gene products. Therefore they could correspond to membrane regulatory proteins and not to the K^+ transport system itself. The salt tolerance of *trk1* mutants is only slightly decreased as compared to wild type. This suggests that most of the K^+/Na^+ discrimination of yeast cells is achieved at the level of cation efflux and is not due to the major *TRK1* encoded (or regulated) K^+ system.[36] The small decrease in salt tolerance exhibited by the *trk1* mutant is due to a defect in the regulation of the K^+ uptake system during NaCl stress. In *TRK1* cells the cation uptake system experiences during salt stress an increase in affinity for K^+ but not for Na^+. This effect improves the discrimination between K^+ and Na^+. The affinity switch of K^+ transport does not occur in *trk1* cells.[36] The same regulatory change occurs in *TRK1* cells but not in *trk1* cells when the potassium concentration of the medium becomes less than 0.1 mM in ammonium containing media and less than 10 μM in media without ammonium.[33] This explains the inability of *trk1* mutants to grow in low potassium media. NaCl stress must induce potassium starvation due to competition between Na^+ and K^+ for uptake by the same system. Therefore it seems that *TRK1* instead of corresponding to the structural gene for the cation transporter encodes a regulatory component mediating an increase in affinity for K^+ but not for Na^+ when yeast cells experience potassium deficiency. *TRK2* would correspond to an isoform of *TRK1* with reduced expression.[35] The nature of the catalytic component of the cation uptake systems in *S. cerevisiae* is a fundamental gap of knowledge in the field. Recent results with plants have uncovered proteins with homology to animal potassium channels involved in potassium uptake.[37] On the other hand a potassium transport system with homology to the

Kup (TrkD) system of *Escherichia coli* has been identified in the yeast *Schwanniomyces occidentalis*.[38] This system probably operates as a K+/H+ or K+/Na+ symport.[39] A recent report has identified a novel K+ transporter from plants[40] as a Na+/K+ symport,[41] a mechanism originally proposed in *Chara australis* on the basis of electrophysiological evidence.[42] Therefore this type of cotransport mechanism may be widely distributed. These two types of K+ uptake transporters—the channels and the symporters—have not yet been characterized at the molecular level in *S. cerevisiae*. Patch-clamp analysis[44] has demonstrated an inward K+ current activated by high membrane potentials (>0.1 V) which resembles the current mediated by plant K+ uptake channels.[37] The molecular basis for this current is not known. It is not observed in *trk1 trk2* mutants[44] and therefore it could correspond to a K+ uptake channel mediated or modulated by Trk1p and Trk2p.

The important role of cation uptake in salt tolerance is demonstrated by two mutations of *S. cerevisiae* causing increased sodium uptake and decreased salt tolerance. One is in the gene *HOL1* of unknown sequence which probably encodes a cation transporter.[45] The other is in the gene *LIS1/ERG6* encoding a SAM-dependent methyltransferase of the ergosterol pathway. The mutation seems to affect cation transporters indirectly by alteration of the sterol composition of the yeast plasma membrane.[46]

Genetic analysis has established the (E-P)ATPase encoded by the *PMR2/ENA1* gene as the major sodium efflux system of *S. cerevisiae*.[13,36,47] It is surprising, however, that in other yeast species such as *Schizosaccharomyces pombe*[14] and *Zygosaccharomyces rouxii*[48] the major sodium efflux system is a Na+/H+-antiporter and not an ATPase. Proton-sodium antiporters have been described at the physiological level in plasma membrane vesicles from halophytic plants.[49] In *S. cerevisiae* a H+/Li+ antiport has been proposed[50] but it could not be identified in plasma membrane vesicles.[51] Systematic sequencing of the genome has

identified the genes J0909 and D9461.40 with homology to Na+/H+ antiporters from bacteria and animals, respectively.[43] Their physiological role is unknown. An antiporter gene with homology to the Sod2p Na+/H+ antiporter of *Schizosaccharomyces pombe*[14] has recently been characterized in *S. cerevisiae*.[52] Its activity seems to explain the small residual sodium efflux observed in *ena1* mutants.

Potassium efflux could be an important factor in determining the intracellular K+ level and therefore indirectly, salt tolerance. Intracellular systems sensitive to sodium inhibition may be reactivated by potassium and the intracellular Na+/K+ ratio may be the relevant parameter in salt toxicity.[5,7] An outward-rectifying K+ channel has been characterized in *S. cerevisiae* by patch-clamp methodologies.[44,53,54] These channels could participate in balancing charge movements during plasma membrane active transport.[53] Systematic sequencing of the yeast genome has identified a potential K+ channel encoded by the JO911 gene.[43] Patch-clamp analysis in *Xenopus* oocytes[55] and yeast[56] have demonstrated this channel to correspond to the depolarization-activated K+ efflux channel described above. The gene has been renamed as *TOK1*[55] and *YCK1*.[56] This channel seems to be very similar to the one activated by depolarization previously described in plant cells.[57] A mechanosensitive ion channel which is activated by stretching of the membrane, has been described by patch-clamping but its molecular nature is unknown. It may participate in turgor regulation.[58]

In addition to K+-efflux channels, a H+/K+ antiport involved in K+ efflux has been described in yeast plasma membrane vesicles.[51] This system has also been described in plant plasma membrane vesicles[59] and it could be important in determining the intracellular K+ level.

Nothing is known about chloride transport in *S. cerevisiae*. In plant cells uptake of this potentially toxic anion occurs by H+/Cl− symport.[60] Efflux is driven by the negative-inside membrane potential via voltage-regulated anion channels.[61] A chloride channel with homology to animal channels of the CLC-family has recently been cloned.[62] Two yeast genes encoding putative chloride transporters have been sequenced: *CLC1/GEF1*, with homology to animal chloride channels and YBR235w/YBR1601 which has homology to animal Cl⁻/Na⁺ symporters.[43] Future mutational analysis should clarify the physiological role of these putative transporters.

Vacuolar ion transporters in *S. cerevisiae* have only been described at the physiological and electrophysiological levels.[2] They include proton-antiporters for accumulation of protons in the vacuole and channels for efflux. A mutant in vacuolar H⁺-ATPase is much more sensitive to growth inhibition by NaCl and LiCl than wild type[36] pointing to an important role of vacuolar compartmentation in salt tolerance. This was anticipated from the correlation in plants between halophytic features and vacuolar salt accumulation.[5]

A scheme describing the yeast transport systems relevant to salt tolerance is shown in Figure 5.3. The plasma membrane H⁺-ATPase encoded by the *PMA1* gene is the primary chemiosmotic system of the cells. The electrochemical proton gradient generated by the enzyme is the driving force for the uptake of K⁺ and Na⁺ probably through a cation channel modulated by the product of the *TRK1* gene. The nature of this uptake channel is unknown. The alternative possibility that *TRK1* encodes a cation transporter itself seems less likely because of the evidence discussed above. *TRK2* is not represented because it seems to encode an isoform of *TRK1* produced only at low levels. Sodium efflux is mostly dependent on the *ENA1*-encoded ATPase. Other *ENA1*-related ATPase genes present in the *PMR2/ENA1*

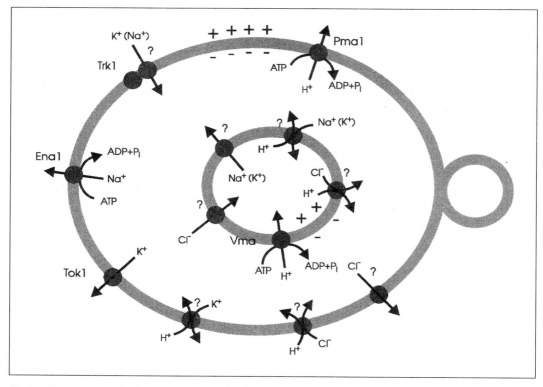

Fig. 5.3. Yeast transport systems relevant to salt tolerance. Question marks indicate transport system identified by physiological and electrophysiological studies but of unknown molecular basis. See text for description of the different systems.

genomic cluster are not indicated because they seem to correspond to low-expression isoforms.[21,47] Potassium efflux could be mediated by either the *TOK1*-encoded channel or by a H^+-K^+ antiport. Chloride movements at the plasma membrane could be mediated by a Cl^--H^+ symport for uptake and by a Cl^- channel for efflux, driven by the negative-inside membrane potential. The absence of significant chloride toxicity in *S. cerevisiae* could be due to either reduced uptake or enhanced efflux in this organism. In some plants, however, chloride is the most toxic component of $NaCl$.[5] The vacuolar membrane (tonoplast) is energized by a complex H^+-ATPase of the $F_0.F_1$ type with multiple subunits (indicated as Vmap). The molecular basis of the different tonoplast proton-cotransports and channels is unknown. The multitude of question marks in this scheme raise a fundamental issue: full understanding of salt tolerance in *S. cerevisiae* requires the molecular characterization of the multitude of ion transport systems in both the plasmatic and vacuolar membranes. As they probably represent minor membrane proteins, the biochemical approaches utilized for major membrane proteins such as Pmalp and Vmap are very difficult to implement. Genetic approaches based on mutant analysis as utilized in the case of the *TRK1* and *ENA1* systems are the best alternatives. In addition the completion of the systematic sequencing of the yeast genome has provided a data base of putative membrane transport proteins[43] which can now be investigated by reverse genetics (generation of mutants with loss and gain of function), as illustrated for the *TOK1/YKC1* encoded channel.

5. REGULATION OF ION HOMEOSTASIS

One important result of the mutational analysis of salt tolerance has been the identification of regulatory components of the ion homeostatic system.[5] In addition to ion transporters such as the *ENA1*-encoded ATPase, several regulatory proteins have been demonstrated to be crucial for salt tolerance. All modulate the *ENA1* sodium ex-

trusion system and some may also affect potassium uptake and efflux systems. Figure 5.4 summarizes current knowledge about salt signal transduction in yeast. There is a major pathway—the HOG pathway—discussed in chapter 4, which is activated by mild, nonspecific osmotic stress. This MAP kinase pathway modulates a multitude of genes whose products are involved in oxidative defenses, heat shock defenses and osmotic adjustment.[63-68]

There is another osmotic pathway which requires higher solute concentrations (1.5 M sorbitol, 0.8 M NaCl) to be activated. This novel pathway has only been identified by a mutation in a gene denominated *EHA1* which has not yet been cloned. Eha1p modulates expression of the *HAL1* and *DDR48* genes (J.A. Márquez and R. Serrano, unpublished). Expression of the *ENA1* gene is regulated by the HOG pathway in response to low osmotic stress. In addition, *ENA1* is specifically induced by high sodium concentrations through a calcineurin-mediated pathway.[69] The fact that the calcineurin pathway is sodium-specific suggest a connection between sodium and calcium during salt stress. This connection between salt stress and calcium has also been proposed in plants.[5] Calcineurin is known to be activated by calcium and since high sodium stimulates the calcineurin pathway it is plausible that sodium increases intracellular free-calcium in yeast. Some preliminary results with the fluorescent calcium indicator fluo-3 indicate that this is the case (A. Ferrando, J.R. Perez-Castiñeira, M. Burgal and R. Serrano; unpublished observations). Calcium homeostasis in *S. cerevisiae* is controlled by several uptake and efflux mechanisms not completely characterized.[70] A nonspecific mechanosensitive plasma membrane channel[54,58] seems to provide the calcium uptake pathway. The activity of this channel could explain an increase in intracellular calcium during osmotic stress. However, as indicated above the calcineurin pathway is sodium-specific and not activated by the nonspecific osmotic stress acting on the mechanosensitive channel. How sodium influences calcium homeostasis can

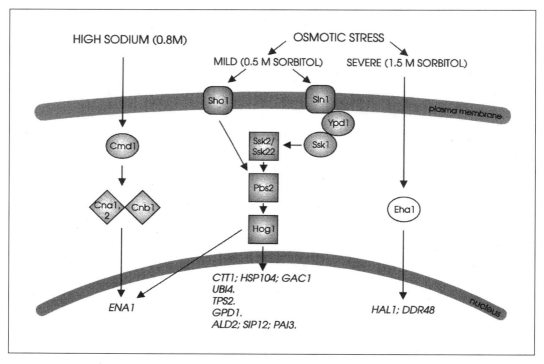

Fig. 5.4. *Multiple pathways control gene expression under salt stress in* Saccharomyces cerevisiae. *Both the HOG and the Eha1p pathways are activated by osmotic stress, but with different sensitivities. The HOG pathway is maximally activated at 0.3 M NaCl (equivalent to 0.5 M sorbitol) while the Eha1p pathway requires higher concentrations of osmolytes (1 M NaCl, equivalent to 1.5 M sorbitol). On the other hand, ENA1 gene expression is controlled by two different pathways, the HOG pathway, responsible for ENA1 induction at low concentrations of osmotica and the calcineurin pathway, which is specifically activated by high sodium concentrations.*

only be a matter of speculation. Calcium efflux from the cytosol is mediated by the Pmr1p ATPase located in the Golgi and by the Pmc1p ATPase and H^+/Ca^{2+} antiporter, both located in the vacuolar membrane. The H^+/Ca^{2+} antiporter is driven by the proton gradient generated by the vacuolar H^+-ATPase. Intracellular sodium could participate in a sodium-calcium antiport activity, releasing calcium from an unknown intracellular compartment. It cannot be excluded, however, that sodium acts from the outside of the cell by opening a calcium-uptake channel.

Known components of the calcium signaling machinery such as calmodulin,[19] calcineurin[22,26,69] and Ca^{2+}/calmodulin-dependent protein kinases[30] (personal communication of S. J. Kron; Whitehead Institute, Cambridge, Massachusetts) participate in the induction of *ENA1* expression by

sodium. On the other hand, additional regulatory components such as Hal1p,[17] Hal3p[19] and the Ppz protein phosphatases[71] determine the global level of *ENA1* expression under both basal and salt stress conditions. Although these regulatory proteins are important for salt tolerance, they do not transduce the sodium-stress signal which seems to have calcium as second messenger. A final aspect of the regulation of Ena1p is that calcineurin—in addition to mediating induction of the *ENA1* gene expression in response to sodium—also directly activates the enzyme.[21]

Some of the components modulating intracellular sodium concentrations via Ena1p such as calcineurin,[22] Hal1p,[17] and Hal3p[19] also modulate intracellular potassium. In the case of calcineurin it seems that the *TRK1*-dependent increase in K^+-affinity occurring during salt stress is dependent on calcineurin. As indicated above for a direct effect of calcineurin on the Ena1p ATPase this activation could be posttranscriptional. In the case of Hal1p (G. Rios and R. Serrano, unpublished data) and Hal3p[19] it seems that these regulatory proteins inhibit potassium efflux during salt stress in a way independent of Ena1p. This effect increases the intracellular K^+ level and therefore the intracellular K^+/Na^+ ratio becomes more favorable and contributes to salt tolerance.[5] Figure 5.5 summarizes the regulatory interactions contributing to ENA1 gene expression in yeast. Also included in Figure 5.5 are the two catabolite repression pathways discussed below.

The important role in salt tolerance of regulatory components in addition to transport systems is demonstrated by the cloning of a putative HMG transcriptional regulator from rats as a suppressor of the potassium transport defect of the S. cerevisiae trk1 mutant.[72] Apparently this heterologous transcription factor when expressed in yeast up-regulates some endogenous K^+ uptake system of trk1 mutant cells such as TRK2 or a potassium uptake channel.

A final point of discussion concerns the lack of specificity of calcineurin action. In addition to the above described effects on ENA1 expression which connect calcineurin to salt tolerance, this calcium-activated protein phosphatase is also involved in a multitude of other cellular functions. In the first place it is required for fast recovery from the transient growth arrest caused by exposure to α-factor pheromone which initiates the mating process.[24,25] Calcineurin modulates calcium homeostasis by inducing the Ca^{2+}-ATPases of the vacuole (Pmc1p) and the Golgi (Pmt1p) and by repressing the H^+/Ca^{2+} antiporter of the vacuole.[73] Calcineurin mutants are very sensitive to manganese, apparently because of enhanced uptake of this divalent cation.[74] Calcineurin also interacts with the MAP kinase pathway involved in cell wall synthesis in S. cerevisiae which includes protein kinase C and the MPK1/SLT2 MAP-kinase (see chapter 4). The osmotic-agent-remedial growth defect of mutants defective in this

pathway is aggravated in the absence of calcineurin.[75] This may be related to the fact that one of the two genes encoding 1,3-β-D-glucan synthase (FKS2) is dependent on calcium and calcineurin for expression.[76] Why the regulation of the salt tolerance gene ENA1 is affected by a pleiotropic regulatory component such as calcineurin is a matter of speculation. It is plausible that all the above responses to calcineurin are due to common primary cellular lesions caused by those different conditions and signaled by increased intracellular calcium. Since the homeostasis of cytoplasmic calcium depends on the integrity of both the plasma membrane and the membranes of internal storage compartments a cytoplasmic calcium increase is a very sensitive indicator of a general membrane-damage. Calcineurin-mediated responses include increased capability for sodium efflux, increased capability for cell cycle start and increased capability for glucan synthesis. Sodium efflux seems to be too specific to fit within a general response of yeast cells to stress because this toxic cation is not usually present in natural yeast habitats. Given the lack of information on the biochemical activities of the Ena1p ATPase, it would be interesting to investigate additional roles of this enzyme not related to sodium transport.

6. CATABOLITE REPRESSION OF SALT TOLERANCE GENES

Genes determining salt tolerance such as ENA1[69] and HAL1 (J.A. Marquez and R. Serrano, unpublished) are not only induced by salt stress but are, on the other hand, repressed by the catabolite repression pathways involved in the response of yeast cells to rapidly fermented sugars such as glucose (see chapter 1). There is a general correlation between conditions favoring fast growth and stress sensitivity.[77,78] In S. cerevisiae fast growth is observed in glucose-rich media and therefore this carbon source has a negative effect on the expression of stress defense genes. Glucose effects in yeast are mediated by two major pathways modulated by at least two protein

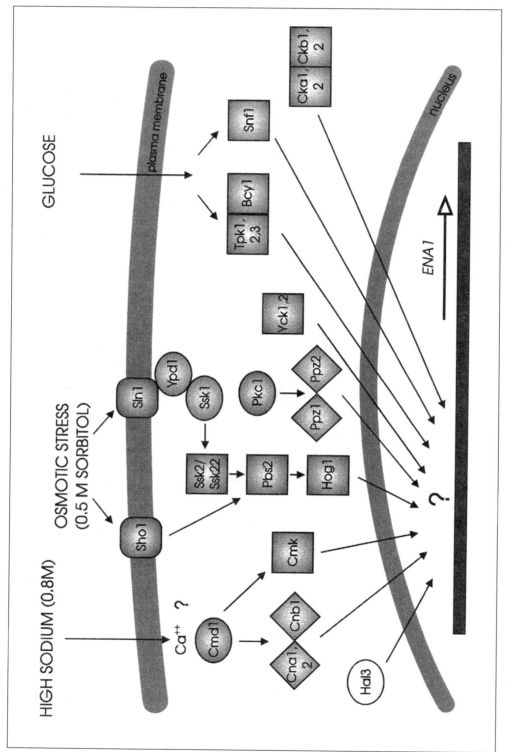

Fig. 5.5. Signal transducing pathways modulating the expression of the ENA1 gene in S. cerevisiae. The mediation of calcium as second messenger of the sodium signal is shown with a question mark because it has not yet been demonstrated.

kinases: the Snf1p protein kinase and protein kinase A (PKA).[79] Fast glucose phosphorylation somehow inactivates the Snf1p protein kinase and increases cAMP levels which then activate protein kinase A. The mechanisms of these effects are still unknown (see chapter 1).[79]

Protein kinase A activity is essential for yeast growth and its activity is modulated by cAMP. When glucose is exhausted from the medium the levels of cAMP decrease (see chapters 1 and 2).[80] The diminished activity of protein kinase A may contribute to the increase in tolerance to several types of stress in stationary phase cultures (see chapter 2).[79] Accordingly, mutants with reduced protein kinase A activity are more stress tolerant and mutants with hyperactive protein kinase A are stress sensitive. These effects are mediated, at least in part, by the negative effect of protein kinase A on the expression of many stress defense genes dependent on the STRE promoter element (see chapter 7).[81,82] Protein kinase A and the HOG protein kinases define parallel pathways converging at an unknown stage with opposite effects on STRE-dependent gene expression.[63] Accordingly, overproduction of the Hog1p protein kinase Pbs2p enhances the stress tolerance of mutants compromised by hyperactivation of protein kinase A.[84] The precise interaction between these two pathways is unknown. The inhibitory effect of protein kinase A on *ENA1* expression[69] could be mediated by this "cross talk" between the HOG protein kinase cascade and protein kinase A. Actually the decrease in cAMP that has been measured during salt stress explains a participation of protein kinase A in Hog1p-mediated induction of *ENA1*.[69] The scenario would be as follows: the HOG-pathway is important for *ENA1* induction and it is inhibited by protein kinase A; salt stress somehow causes a decrease in cAMP levels; lowering of cAMP would inhibit protein kinase A and this would release the inhibition of the HOG pathway. Of course in addition to this indirect mechanism, the HOG pathway would be directly activated by osmotic stress.[63]

The other pathway of glucose signaling in yeast—the Snf1p pathway (see chapter 1)—is also related to stress tolerance. A *snf1* null mutant is sensitive to heat stress and starvation[87], pointing to an important role of this signaling pathway in stress tolerance. However, the stress genes controlled by the Snf1 pathway have not been identified. Preliminary results indicate that *ENA1* may be one of these genes (J.A. Marquez and R. Serrano, unpublished). The stress sensitivity of *snf1* mutants is similar to that of mutants having a hyperactive protein kinase A. Mutations which decrease the activity of protein kinase A moderate the stress sensitivity of *snf1* mutants.[87] This may explain the cAMP-independent effects of glucose exhaustion during yeast growth.[88]

The interaction between the catabolite repression pathways (protein kinases A and Snf1) and the salt-induced (HOG-pathway, calcineurin, calmodulins, Hal1p) and salt-independent (Hal3p, casein kinases, Ppz protein phosphatases) pathways (Fig. 5.5) is not known. A simple hypothesis has been advanced to explain the antagonistic effects of protein kinase A and calcineurin on *ENA1* expression.[89] A transcription factor involved in *ENA1* expression would be inhibited by protein kinase A-mediated phosphorylation and reactivated by calcineurin-mediated dephosphorylation. The identification of the promoter elements and transcription factors involved in *ENA1* expression constitute a goal with top priority for the understanding of salt tolerance in *S. cerevisiae*.

7. CATION-SENSITIVE ENZYME SYSTEMS

Once the intracellular sodium and potassium levels have been adjusted by the activities of influx and efflux systems the final factor determining salt tolerance is the sensitivity of intracellular systems, to the Na^+/K^+ ratio.[5] In *S. cerevisiae* sodium is the toxic component of salt probably because chloride does not reach toxic intracellular levels. In wild type strains growing under standard conditions (76 mM NH_4^+ as nitrogen source and 5 mM K^+) external concentrations of NaCl of around 1 M cause significant growth inhibition which can be explained partially by intracellular sodium toxicity. In addition to this salt effect there is also growth inhibition due to nonspecific osmotic stress which is also observed with equivalent osmotic concentrations of sorbitol or KCl. Under these conditions intracellular sodium levels are in the 100-150 mM range. The intracellular potassium level in the absence of salt stress is 250-300 mM and it decreases to 100-150 mM during salt stress. Therefore growth inhibition occurs when the intracellular Na^+/K^+ ratio is around 1 and the total intracellular monovalent cation level is around 250-300 mM.[17,19] Similar results are obtained in low-potassium media. For example, growth is significantly inhibited and the intracellular Na^+/K^+ ratio is about 1 with external potassium and sodium levels of 0.2 and 50 mM, respectively.[90] These ionic conditions do not have general deleterious effects on cellular proteins and therefore the observed growth inhibition must reflect a specific sensitivity of certain essential cellular systems to the Na^+/K^+ ratio.[5] Ribosomal protein synthesis has been traditionally considered as the crucial cellular system exhibiting this kind of sensitivity.[91] In addition, our mutational analysis in *S. cerevisiae* has uncovered a novel enzyme particularly sensitive to sodium and lithium inhibition which constitutes the relevant target for salt toxicity under in vivo conditions.[18,20] Overexpression of the *HAL2* gene improves growth under sodium and lithium stress without producing any significant alteration in intracellular sodium or lithium. This pointed to a novel mechanism of salt tolerance different from the one involving regulators of *ENA1*-dependent sodium extrusion and potassium transport such as Hal1p, Hal3p, calcineurin and the PPZ protein phosphatases (see above). One logical mechanism was that the *HAL2* product encoded a target of salt toxicity important for cell growth. The clue was provided by the observations that a gene disruption resulted in methionine auxotrophy and that methionine supplementation improved lithium and sodium tolerance.[18] Therefore the product of *HAL2* was postulated to encode a salt-sensitive enzyme of the methionine biosynthetic pathway. A collaboration between our group and that of Y. Surdin-Kerjan at Paris—an expert in sulfur metabolism in yeast—demonstrated that *HAL2* is allelic to *MET22*, a gene encoding an enzyme or a regulator of unknown function within the methionine biosynthetic pathway.[92]

The *HAL2* product has homology to a protein family which include magnesium-dependent lithium-sensitive phosphatases such as inositol monophosphatase, inositol polyphosphate-1-phosphatase and fructose-1,6-bisphosphatase (Fig. 5.6).[93,94] Although overall homology within this family may be relatively low, the most conserved residues constitute the active site of these four enzymes whose three-dimensional structure has been determined.[95-99] The conserved residues make up the motifs described in Table 5.1. They form two adjacent acidic pockets which bind Mg^{2+} ions. These two binding sites include as ligands the hydrolyzable phosphate of the substrate, the first glutamate of motif pre-A and the first aspartate of motif A. In addition, one of the sites includes a water molecule activated for nucleophilic attack by binding to both the magnesium and to the conserved threonine of motif A. The other site also includes the aspartate or glutamate of motif B. The magnesium of this second site probably acts by stabilizing the alcoxi oxyanion. It has been proposed that lithium inhibits these enzymes by binding to this second magnesium site.[96] As this cation binding site is in part formed by the substrate phosphate group lithium inhibition is of the uncompetitive type with respect to the substrate phosphate ester concentration. In this unusual type of inhibition the inhibitor becomes more potent with increasing substrate concentrations. Just the opposite is the case with competitive inhibitors.[100] Although magnesium is essential for

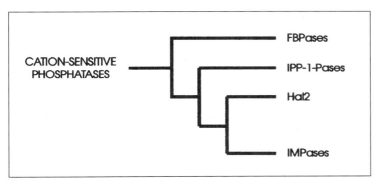

Fig. 5.6. Schematic dendogram of the evolutionary relationships within the family of cation-sensitive phosphatases which includes Hal2p.

Table 5.1. Conserved motifs within the family of cation-sensitive phosphatases which includes Hal2p

Motif	Name[93]
$DX_{22-24}EE$	Motif pre-A
DP(I,L)D(G,S)T	Motif A
(W,Y)(E,D)X_{10}(A,M)GG	Motif B

X represents any amino acid

enzyme activity (optimal concentration 1-5 mM) an excess of this divalent cation (more than 10-20 mM) also produces uncompetitive inhibition. Site-directed mutants of inositol monophosphatase resistant to lithium inhibition are also resistant to high-magnesium inhibition.[101,102] Therefore the inhibitory site for lithium (K_i about 1 mM) can also accommodate magnesium with low affinity (K_i greater than 10 mM). This inhibitory site may correspond to a modification of the magnesium-alcoxi oxyanion site described above after the alcoholic product of the reaction has been released and the free phosphate remains attached to the enzyme.[96]

Since Hal2p belongs to this protein family with cation-sensitive phosphatase activity, we suggested that it could correspond to a salt-sensitive phosphatase.[18] The inositol phosphatases of this family, however, are inhibited by lithium but not by sodium. Actually, Na^+ as well as K^+ and NH_4^+ stimulate activity about 2-fold.[103] Overexpression of Hal2p confers tolerance to both sodium and lithium.[18] Therefore the Hal2p phosphatase should be a special member of the family sensitive to inhibition

by both lithium and sodium. In addition, it was not clear what kind of phosphate ester would constitute the Hal2p substrate.

Old biochemistry textbooks with real biological chemistry are completely forgotten by present day students who are more used to the color picture-full style of modern molecular biology textbooks. However, the clue for Hal2p function was found in one of such biochemistry texbook.[103] Coupling by phosphorylation and subsequent cleavage by a phosphatase is one of the biochemical strategies to drive a synthetic sequence which includes a thermodynamically unfavorable reaction. Hal2p does not correspond to any of the metabolic enzymes involved in methionine biosynthesis but to a side reaction frequently overlooked although essential for this metabolic pathway. Before the inorganic sulfate group can be incorporated into organic molecules it is activated by coupling to ATP (catalyzed by ATP sulfurylase) to produce adenosine-5'-phosphosulfate (APS). A pyrophosphatase then catalyzes the hydrolysis of pyrophosphate resulting in the following overall reaction:

$$ATP + SO_4^{-2} = APS + 2\ P_i$$

This reaction is thermodynamically unfavorable ($\Delta G = + 12$ kJ/mol) and therefore the activated sulfate represented by APS does not accumulate inside the cells in significant amounts. Accumulation of activated sulfate is then based on the further phosphorylation of APS by ATP catalyzed by APS kinase. The product of this reaction is 3'-phosphoadenosine-5'-phosphosulfate (PAPS):

$$APS + ATP = PAPS + ADP$$

This additional phosphorylation of the adenosine of APS drives the accumulation of activated sulfate in the form of PAPS. However, when this activated sulfate is utilized by either reduction to sulfite or transfer to other molecules an unusual nucleotide is formed: 3'-phosphoadenosine-5'-phosphate (PAP). In order to recycle adenosine PAP must be hydrolyzed to AMP. The specific nucleotidase catalyzing this reaction is encoded by the yeast *HAL2* gene and is very sensitive to inhibition by lithium (K_i = 0.1 mM) and sodium (K_i = 20 mM). Potassium activates at high concentrations (>0.1 M) and counteracts the inhibition caused by lithium and sodium.[20] During salt stress the enzyme is inhibited "in vivo" and PAP accumulates inside the cells. This results in product inhibition of enzymatic reactions, generating PAP from PAPS. In yeast PAPS reduction to sulfite during methionine biosynthesis seems the most important reaction affected because growth under salt stress is ameliorated by methionine supplementation.[18] PAPS-dependent sulfate transfer reactions are significant both in animal cells for sulfoprotein and sulfopolysaccharide formation as well as in plant cells for sulfolipid production.

It remains to be investigated if the homologous Hal2p phosphatases from other organisms are also sensitive to sodium in addition to lithium. Preliminary results indicate that rat liver Hal2p is sensitive to lithium but not to sodium (J.R. Murguía and R. Serrano, unpublished). Therefore a plausible scenario is that enzymes of the Hal2p family (see Fig. 5.6) in the animal world are sensitive to lithium but not to sodium. This is probably due to the evolutionary pressure posed by the high intracellular sodium levels (around 20-50 mM) present in animal cells. On the other hand, enzymes of this phosphatase family in terrestrial plants and fungi have not experienced such sodium levels and evolved into sodium-sensitive enzymes. Structural studies should clarify the molecular basis for this subtle change restricting sodium entry into the inhibitory cation site of the enzyme.

If our hypothesis is true we could expect that other phosphatases of the family from fungi and plants would be sensitive to sodium. Given the crucial role of inositol phosphates in cellular regulation[100] sodium inhibition of inositol monophosphatase and inositol polyphosphate-1-phosphatase may constitute an important target for sodium toxicity in terrestrial fungi and plants.

8. PERSPECTIVES

The discussion above suggests that engineering salt transport and salt toxicity targets to improve salt tolerance in an organisms such as yeast is a goal which can be reached in a reasonable time. We have identified some of the crucial factors in salt tolerance operating at the cellular level and by proceeding step by step we can dramatically improve this phenotype.

Two important gaps of knowledge for the understanding of the physiology of salt tolerance in yeast are the roles played by chloride transport through the plasma membrane and by compartmentation of sodium and chloride in the vacuole. Also, the primary targets for the toxicity of intracellular chloride should be investigated. Concerning sodium toxicity, secondary targets after the Hal2p phosphatase step is bypassed should be studied. Signal transduction pathways modulating the expression of salt-induced genes have started to be elucidated (Figs. 5.4 and 5.5) but nothing is known about primary sodium sensors and about transcription factors acting on salt-induced genes and their regulation at the molecular level.

It is hoped that this basic knowledge obtained with yeast will provide not only biotechnological tools for generation of salt tolerant crop plants but also principles to start the mutational analysis of salt tolerance in model plants such as *Arabidopsis thaliana*. In the end the satisfaction remains that after understanding basic principles we can manipulate a model organism such as yeast in a predictable way.

ACKNOWLEDGMENTS

Work from the authors' laboratory has been supported by a grant from the European Union (Brussels) within the Project of Technological Priority (BIO2-CT93-0400).

REFERENCES

1. Raven JA, Smith FA. The evolution of chemiosmotic energy coupling. J Theor Biol 1976; 57:301-310.

2. Serrano R. Transport across yeast vacuolar and plasma membranes. In: Broach JR, Pringle JR, Jones EW, eds. The Molecular and Cellular Biology of the Yeast *Saccharomyces*: Genome Dynamics, Protein Synthesis, and Energetics. Cold Spring Harbor Laboratory Press, 1991:523-585.

3. Maloney PC, Wilson TH. The evolution of ion pumps. Bioscience 1985; 35:43-48.

4. Wyn Jones RG, Pollard A. Proteins, enzymes and inorganic ions. In: Laüchli A, Pirson, A, eds. Encyclopedia of Plant Physiology. New Series. Berlin: Springer-Verlag, 1983:528-562.

5. Serrano R. Salt tolerance in plants and microorganisms: toxicity targets and defense responses. Int Rev Cytol 1996; 165:1-52.

6. Valentine JW. The evolution of multicellular plants and animals. Sci Am 1978; 239:140-158.

7. Serrano R, Gaxiola R. Microbial models and salt stress tolerance in plants. Crit Rev Plant Sci 1994; 13:121-138.

8. Ashraf M, Breeding for salinity tolerance in plants. Crit Rev Plant Sci 1994; 13:17-42.

9. Tosteson DC. Regulation of cell volume by sodium and potassium transport. In: Hoffman JF, ed. The Cellular Functions of Membrane Transport. Englewood Cliffs, New Jersey: Prentice Hall Inc, 1964:3-22.

10. Serrano R: Recent molecular approaches to the physiology of the plasma membrane proton pump. Bot Acta 1990; 103:230-234.

11. Hager A, Menzel H, Krauss A. Versuche und Hypothese zur Primärwirkung des Auxins beim Streckungswachstum. Planta 1971; 100:47-75.

12. Hager A, Debus G, Edel H-G, Stransky H, Serrano R. Auxin induces exocytosis and the rapid synthesis of a high-turnover pool of plasma membrane H+-ATPase. Planta 1991; 185:527-537.

13. Haro R, Garciadeblas B, Rodriguez-Navarro A. A novel P-type ATPase from yeast involved in sodium transport. FEBS Lett 1991; 291:189-191.

14. Jia Z-P, McCullough N, Martel R, Hemmingsen S, Young P.G. Gene amplification at a locus encoding a putative Na^+/H^+ antiporter confers sodium and lithium tolerance in fission yeast. EMBO J 1992; 11:1631-1640.

15. André B. An overview of membrane transport proteins in *Saccharomyces cerevisiae*. Yeast 1995; 11:1575-1611.

16. Rine J, Hansen W, Hardeman E, Davis RW. Targeted selection of recombinant clones through gene dosage effects. Proc Natl Acad Sci USA 1983; 80:6750-6754.

17. Gaxiola R, Larrinoa IF, Villalba JM, Serrano R. A novel and conserved salt-induced protein is an important determinant of salt tolerance in yeast. EMBO J 1992; 11: 3157-3164.

18. Gläser HU, Thomas D, Gaxiola R, Montrichard F, Surdin-Kerjan Y, Serrano R. Salt tolerance and methionine biosynthesis in *Saccharomyces cerevisiae* involve a putative phosphatase gene. EMBO J 1993; 12:3105-3110.

19. Ferrando A, Kron SJ, Rios G, Fink GR, Serrano R. Regulation of cation transport in *Saccharomyces cerevisiae* by the salt tolerance gene *HAL3*. Mol Cell Biol 1995; 15:5470-5481.

20. Murguia JR, Bellés JM, Serrano R. A salt-sensitive 3'(2'),5'-bisphosphate nucleotidase involved in sulfate activation. Science 1995; 267:232-234.

21. Wieland J, Nitsche AM, Strayle J, Steiner H, Rudolph HK. The *PMR2* gene cluster encodes functionally distinct isoforms of a putative Na^+ pump in the yeast plasma membrane. EMBO J 1995; 14:3870-3882.

22. Mendoza I, Rubio F, Rodriguez-Navarro A, Pardo JM. The protein phosphatase calcineurin is essential for NaCl tolerance of *Saccharomyces cerevisiae*. J Biol Chem 1994; 269:8792-8796.

23. Shenolikar S. Protein serine/threonine phosphatases. New avenues for cell regulation. Annu Rev Cell Biol 1994; 10:55-86.

24. Cyert MS, Thorner J. Regulatory subunit (CNB1 gene product) of yeast Ca^{2+}/calmodulin-dependent phosphoprotein phosphatases is required for adaptation to pheromone. Mol Cell Biol 1992; 12: 3460-3469.

25. Foor F, Parent SA, Morin N, Dahl AM, Ramadan N, Chrebet G, Bostian KA, Nielsen JB. Calcineurin mediates inhibition by FK506 and cyclosporin of recovery from α-factor arrest in yeast. Nature 1992; 360:682-684.

26. Nakamura T, Liu Y, Hirata D, Namba H, Harada S, Hirokawa T, Miyakawa T. Protein phosphatase type 2B (calcineurin)-mediated, FK506-sensitive regulation of intracellular ions in yeast is an important determinant for adaptation to high salt stress conditions. EMBO J 1993; 11:4063-4071.

27. Kunz J, Hall MN. Cyclosporin A, FK506 and rapamycin: more than just immunosuppression. Trend Biochem Sci 1993; 18:334-338.

28. Robinson LC, Hubbard EJA, Graves PR, DePaoli-Roach AA, Roach PJ, Kung C, Haas DW, Hagedorn CH, Goebl M, Culbertson MR, Carlson M. Yeast casein kinase I homologues: an essential gene pair. Proc Natl Acad Sci USA 1992; 89:28-32.

29. Bidwai A, Reed JC, Glover CVC. Cloning and disruption of CKB1, the gene encoding the 38-kDa b subunit of Saccharomyces cerevisiae casein kinase II (CKII). Deletion of CKII regulatory subunits elicits a salt-sensitive phenotype. J Biol Chem 1995; 270:10395-10404.

30. Pausch MH, Kaim D, Kunisawa R, Admon A, Thorner J. Multiple Ca^{2+}/calmodulin-dependent protein kinase genes in a unicellular eukaryote. EMBO J 1991; 10: 1511-1522.

31. Ramos J, Contreras P, Rodriguez-Navarro A. A potassium transport mutant of Saccharomyces cerevisiae. Arch Microbiol 1985; 143:88-93.

32. Gaber RF, Styles CA, Fink GR. TRK1 encodes a plasma membrane protein required for high-affinity potassium transport in Saccharomyces cerevisiae. Mol Cell Biol 1988; 8:2848-2859.

33. Rodriguez-Navarro A, Ramos J. Dual system for potassium transport in Saccharomyces cerevisiae. J Bacteriol 1984; 159:940-945.

34. Ko CH, Gaber RF. TRK1 and TRK2 encode structurally related K+ transporters in Saccharomyces cerevisiae. Mol Cell Biol 1991; 11:4266-4273.

35. Ramos J, Alijo R, Haro R, Rodriguez-Navarro A. TRK2 is not a low-affinity potassium transporter in Saccharomyces cerevisiae. J Bacteriol 1994; 176:249-252.

36. Haro R, Bañuelos MA, Quintero FJ, RubioF, Rodriguez-Navarro A. Genetic basis of sodium exclusion and sodium tolerance in yeast. A model for plants. Physiol Plant 1993; 89:868-874.

37. Schroeder JI, Ward JM, Gassmann W. Perspectives on the physiology and structure of inward-rectifying K$^+$ channels in higher plants: biophysical implications for K$^+$ uptake. Annu Rev Biophys Biomol Struct 1994; 23:441-471.

38. Bañuelos MA, Klein RD, Alexander-Bowman SJ, Rodriguez-Navarro A. A potassium transporter of the yeast Schwanniomyces occidentalis homologous to the Kup system of Escherichia coli has a high concentrative capacity. EMBO J 1995; 14:3021-3027.

39. Bakker EP. Low-affinity K+ uptake systems. In:Bakker EP, ed. Alkali Cation Transport Systems in Prokaryotes. Boca Raton: CRC Press, 1993:253-276.

40. Schachtman DP, Schroeder JI. Structure and transport mechanism of a high-affinity potassium uptake transporter from higher plants. Nature 1994; 370:655-658.

41. Rubio F, Gassmann W, Schroeder JI. Na$^+$-driven K$^+$ uptake by the plant K$^+$ transporter HKT1 and mutations conferring salt tolerance. Science 1995; 270:1660-1663.

42. Smith FA, Walker NA. Transport of potassium in Chara australis: 1. a symport with sodium. J Membrane Biol 1989; 108:125-137.

43. André B. An overview of membrane transport proteins in Saccharomyces cerevisiae. Yeast 1995; 11:1575-1611.

44. Bertl A, Anderson JA, Slayman CL, Gaber RF. Use of Saccharomyces cerevisiae for patch-clamp analysis of heterologous membrane proteins: characterization of Kat1, an inward-rectifying K$^+$ channel from Arabidopsis thaliana, and comparison with endogenous yeast channels and carriers. Proc Natl Acad Sci USA 1995; 92:2701-2705.

45. Gaber RF, Kielland-Brandt MC, Fink GR. HOL1 mutations confer novel ion transport

in *Saccharomyces cerevisiae*. Mol Cell Biol 1990; 10:643-652.

46. Welihinda AA, Beavis AD, Trumbly RJ. Mutations in Lis1 (Erg6) confer increased sodium and lithium uptake in *Saccharomyces cerevisiae*. Biochim Biophys Acta 1994; 1193:107-117.

47. Garciadeblas B, Rubio F, Quintero FJ, Bañuelos MA, Haro R, Rodriguez-Navarro A. Differential expression of two genes encoding isoforms of the ATPase involved in sodium efflux in *Saccharomyces cerevisiae*. Mol Gen Genet 1993; 236:363-368.

48. Watanabe Y, Miwa S, Tamai Y. Characterization of Na$^+$/H$^+$-antiporter gene closely related to the salt-tolerance of yeast *Zygosaccharomyces rouxii*. Yeast 1995; 11:829-838.

49. Braun Y, Hassidim M, Lerner HR, Reinhold L. Evidence for a Na$^+$/H$^+$ antiporter in membrane vesicles isolated from roots of the halophyte *Atriplex nummularia*. Plant Physiol 1988:87:104-108.

50. Rodriguez-Navarro A, Sancho ED, Perez-Lloveres C. Energy source for lithium efflux in yeast. Biochim Biophys Acta 1981; 640:352-358.

51. Camarasa C, Prieto S, Ros R, Salmon JM, Barre P. Evidence for a selective and electroneutral K$^+$/H$^+$ exchange in *Saccharomyces cerevisiae* using plasma membrane vesicles. Yeast 1996; (in press).

52. Prior C, Potier S, Souciet, J-L, Sychrova H. Characterization of the *NHA1* gene encoding a Na$^+$/H$^+$-antiporter of the yeast *Saccharomyces cerevisiae*. FEBS Lett 1996; (in press).

53. Bertl A, Slayman CL, Gradmann D. Gating and conductance in an outward-rectifying K$^+$ channel from the plasma membrane of *Saccharomyces cerevisiae*. J Membrane Biol 1993; 132:183-199.

54. Gustin MC, Martinac B, Saimi Y, Culbertson MR, Kung C. Ion channels in yeast. Science 1986; 233:1195-1197.

55. Ketchum KA, Joiner WJ, Sellers AJ, Kaczmarek LK, Goldstein AN. A new family of outwardly rectifying potassium channel proteins with two pore domain in tandem. Nature 1995; 376:690-695.

56. Zhou X-L, Vaillant B, Loukin SH, Kung C, Saimi Y. YKC1 encodes the depolariza-tion-activated K+ channel in the plasma membrane of yeast. FEBS Lett 1995; 373:170-176.

57. Blatt MR, Thiel G. Hormonal control of ion channel gating. Annu Rev Plant Physiol Plant Mol Biol 1993; 44:543-567.

58. Gustin MC, Zhou X-L, Martinac B, Kung C. A mechanosensitive ion channel in the yeast plasma membrane. Science 1988; 242:72-765.

59. Cooper S, Lerner HR, Reinhold L. Evidence for a highly specific K$^+$/H$^+$ antiporter in membrane vesicles from oil-seed rape hypocotyls. Plant Physiol 1991; 97:1212-1220.

60. Felle HH. The H$^+$/Cl$^-$ symporter in root-hair cells of *Sinapis alba*. An electrophysiological study using ion-selective microelectrodes. Plant Physiol 1994; 106:1131-1136.

61. Hedrich R. Voltage-dependent chloride channels in plant cells: identification, characterization, and regulation of a guard cell anion channel. Curr Top Membranes 1994; 42:1-33.

62. Lurin C, Geelen D, Barbier-Brygoo H, Guern J, Maurel C. Cloning and functional expression of a plant voltage-dependent chloride channel. Plant Cell 1996; 8:701-711.

63. Schüller C, Brewster JL, Alexander MR, Gustin MC, Ruis H. The HOG pathway controls osmotic regulation of transcription via the stress response element (STRE) of the *Saccharomyces cerevisiae CTT1* gene. EMBO J 1994; 13:4382-4389.

64. Gounalaki N, Thireos G. Yap1, a yeast transcriptional activator that mediates multidrug resistance, regulates the metabolic stress response. EMBO J 1994; 13:4036-4041.

65. Miralles V, Serrano R. A genomic locus in *Saccharomyces cerevisiae* with four genes up regulated by osmoticstress. Mol Microbiol 1995; 17:653-662.

66. Brewster JL, de Valoir T, Dwyer ND, Winter E, Gustin MC. An osmosensing signal transduction pathway in yeast. Science 1993; 259:1760-1763.

67. Maeda T, Wurgler-Murphy SM, Saito H. A two-component system that regulates an osmosensing MAP kinase cascade in yeast. Nature 1994; 369:242-245.

68. Maeda T, Takekawa M, Saito H. Activation of yeast PBS2 MAPKK by MAPKKKs or by binding of an SH3-containing osmosensor. Science 1995; 269:554-557.

69. Marquez JA, Serrano R. Multiple transduction pathways regulate the sodium-extrusion gene *PMR2/ENA1* during salt stress in yeast. FEBS Lett 1996; 382:89-92.

70. Davis TN. Calcium in *Saccharomyces cerevisiae*. In: Means AR, ed. Advances in Second Messenger and Phosphoprotein Research. Vol 30. New York: Raven Press, 1995: 339-358.

71. Posas F, Camps M, Ariño J. The PPZ protein phosphatases are important determinants of salt tolerance in yeast cells. J Biol Chem 1995; 270:13036-13041.

72. Lesage F, Hugnot J-P, Amri E-Z, Grimaldi P, Barhanin J, Lazdunski M. Expression cloning in K+ transport defective yeast and dustribution of HBP1, a new putative HMG transcriptional regulator. Nucl Acids Res 1994; 22:3685-3688.

73. Tanida I, Hasegawa A, Iida H, Ohya Y, Anraku Y. Cooperation of calcineurin and vacuolar H+-ATPase in intracellular Ca^{2+} homeostasis of yeast cells. J Biol Chem 1995; 270:10113-10119.

74. Farcasanu IC, Hirata D, Tsuchiya E, Nishiyama F, Miyakawa T. Protein phosphatase 2B of *Saccharomyces cerevisiae* is required for tolerance to manganese, in blocking the entry of ions into the cells. Eur J Biochem 1995; 232:712-717.

75. Garrett-Engele P, Moilanen B, Cyert MS. Calcineurin, the Ca^{2+}/calmodulin-dependent protein phosphatase, is essential in yeast mutants with cell integrity defects and in mutants that lack a functional vacuolar H+-ATPase. Mol Cell Biol 1995; 15:4103-4114.

76. Mazur P, Morin N, Baginsky W, Elsherbeini M, Clemas JA, Nielsen JB, Foor F. Differential expression and function of 2 homologous subunits of yeast 1,3-beta-D-glucan synthase. Mol Cell Biol 1995; 15:5671-5681.

77. Piper PW. Molecular events associated with acquisition of heat tolerance by the yeast *Saccharomyces cerevisiae*. FEMS Microbiol Rev 1993; 11:339-356.

78. Mager WH, Varela JC. Osmostress response of the yeast *Saccharomyces*. Mol Microbiol 1993; 10:253-258.

79. Thevelein JM. Signal transduction in yeast. Yeast 1994; 10:1753-1790.

80. François J, Eraso P, Gancedo C. Changes in the concentration of cAMP, fructose-2,6-bisphosphate and related metabolites and enzymes in *Saccharomyces cerevisiae* during growth on glucose. Eur J Biochem 1987; 164:369-373.

81. Marchler G, Schüller C, Adam G, Ruis H. A *Saccharomyces cerevisiae* UAS element controlled by protein kinase A activates transcription in response to a variety of stress conditions. EMBO J 1993; 12:1997-2003.

82. Belazzi T, Wagner A, Wieser R, Schanz M, Adam G, Hartig A, Ruis H. Negative regulation of transcription of the *Saccharomyces cerevisiae* catalase T (*CTT1*) gene by cAMP is mediated by a positive control element. EMBO J 1991; 10:585-592.

83. Cherry JR, Johnson TR, Dollard C, Schuster JR, Denis CL. Cyclic AMP-dependent protein kinase phosphorylates and inactivates the yeast transcriptional activator ADR1. Cell 1989; 56:409-419.

84. Boguslawski G. PBS2, a yeast gene encoding a putative protein kinase, interacts with the *RAS2* pathway and affects osmotic sensitivity of *Saccharomyces cerevisiae*. J Gen Microbiol 1992; 138:2425-2432.

85. Keleher CA, Redd MJ, Schultz J, Carlson M, Johnson AD. Ssn6-Tup1 is a general repressor of transcription in yeast. Cell 1992; 68:709-719.

86. Gancedo JM. Carbon catabolite repression in yeast. EurJ Biochem 1992; 206:297-313.

87. Thompson-Jaeger S, François J, Gaughran JP, Tatchell K. Deletion of *SNF1* affects the nutrient response of yeast and resembles mutations which activate the adenylate cyclase pathway. Genetics 1991; 129:697-706.

88. Cameron S, Levin L, Zoller M, Wigler M. cAMP-independent control of sporulation, glycogen metabolism, and heat shock resistance in *S. cerevisiae*. Cell 1988; 53:555-566.

89. Hirata D, Harada S, Namba H, Miyakawa T. Adaptation to high-salt stress in *Saccharomyces cerevisiae* is regulated by Ca^{2+}/calmodulin-dependent phosphoprotein phos-

phatase (calcineurin) and cAMP-dependent protein kinase. Mol Gen Genet 1995; 249:257-264.

90. Camacho M, Ramos J, Rodriguez-Navarro A. Potassium requirements of *Saccharomyces cerevisiae*. Curr Microbiol 1981; 6:295-299.

91. Lubin M. Cell potassium and the regulation of protein synthesis. In:Hoffman JF, ed. The Cellular Functions of Membrane Transport. Englewood Cliffs: Prentice-Hall, 1964:193-211.

92. Thomas D, Barbey R, Henry D, Surdin-Kerjan Y. Physiological analysis of mutants of *Saccharomyces cerevisiae* impaired in sulphate assimilation. J Gen Microbiol 1991; 138:2021-2028.

93. Neuwald AF, York JD, Majerus PW. Diverse proteins homologous to inositol monophosphatase. FEBS Lett 1991; 294:16-18.

94. York JD, Ponder JW, Majerus PW. Definition of a metal-dependent/Li$^+$-inhibited phosphomonoesterase protein family based upon a conserved three-dimensional core structure. Proc Natl Acad Sci USA 1995; 92:5149-5153.

95. Bone R, Springer JP, Atack JR. Structure of inositol monophosphatase, the putative target of lithium therapy. Proc Natl Acad Sci USA 1992; 89:10031-10035.

96. Pollack SJ, Atack JR, Knowles MR, McAllister G, Ragan CI, Baker R, Fletcher SR, Iversen LL, Broughton HB. Mechanism of inositol monophosphatase, the putative target of lithium therapy. Proc Natl Acad Sci USA 1994; 91:5766-5770.

97. York JD, Ponder JW, Chen ZW, Mathews FS, Majerus PW. Crystal structure of inositol polyphosphate 1-phosphatase at 2.3-Angstrom resolution. Biochemistry 1994; 33:13164-13171.

98. Zhang Y, Liang J-Y, Huang S, Ke H, Lipscomb WN. Crystallographic studies of the catalytic mechanism of the neutral form of fructose-1,6-bisphosphatase. Biochemistry 1993; 32:1844-185.

99. Xue Y, Huang S, Liang J-Y, Zhang Y, Lipscomb WN. Crystal structure of fructose-1,6-bisphosphatase compexed with fructose 2,6-bisphosphate, AMP, and Zn^{2+} at 2.0-Å resolution: aspects of synergism between inhibitors. Proc Natl Acad Sci USA 1994; 91:12482-12486.

100. Nahorski SR, Ragan CI, Challiss RAJ. Lithgium and the phosphoinositide cycle: an example of uncompetitive inhibition and its pharmacological consequences. Trends Pharmacol Sci 1991; 12:297-303.

101. Pollack SJ, Knowles MR, Atack JR, Broughton HB, Ragan CI, Osborne SA, McAllister G. Probing the role of metal ions in the mechanism of inositol monophosphate by site-directed mutagenesis. Eur J Biochem 1993; 217:281-287.

102. Gore MG, Greasley P, McAllister G, Ragan CI. Mammalian inositol monophosphatase: the identification of residues important for the binding of Mg2+ and Li+ ions using fluorescence spectroscopy and site-directed mutagenesis. Biochem J 1993; 296: 811-815.

103. Metzler DE. Biochemistry. New York: Academic Press, 1977:636.

===== CHAPTER 6 =====

OXIDATIVE STRESS RESPONSES IN THE YEAST *SACCHAROMYCES CEREVISIAE*

Nicholas Santoro and Dennis J. Thiele

1. INTRODUCTION

Stress is defined as a force or system of forces which tends to strain or deform a body or that which has a disquieting influence.[1] Indeed, all living organisms actively engaged in movement, growth, metabolism, flight or exposure to the environment are simultaneously subjected to a wide array of stresses of either a physical or chemical nature. In fact survival in the face of stress is more the rule than the exception to the rule. The exposure to and metabolism of oxygen is both essential and dangerous for organisms which live in an aerobic environment. Much of what we have learned about oxidative stress responses is patterned after a number of elegant investigations of prokaryotic systems relevant to oxidative stress protection. Several comprehensive reviews are available to summarize this area of investigation.[2-6] Much of what we learn using microorganisms as model systems to study oxidative stress responses will no doubt have an important influence on our understanding of the mechanisms by which mammals and other higher eukaryotes cope with oxidative stress. This is of critical importance in biomedical science since oxidative damage has been linked to a number of neurodegenerative diseases, circulatory defects, cancers and the process of aging.[7-9] In this chapter we will summarize our current understanding of how cells generate, sense, respond to and mount a protective mechanism toward the damaging consequences of oxidative stress using baker's yeast cells as a eukaryotic model system.

Yeast Stress Responses, edited by Stefan Hohmann and Willem H. Mager.
© 1997 R.G. Landes Company.

2. HOW IS OXIDATIVE STRESS GENERATED IN LIVING CELLS AND WHAT DAMAGE IS DONE?

Oxidative stress is a generic term for the stress associated with cells or organisms sensing, responding to and protecting themselves from the generation of reactive oxygen species (ROSs). It is important at the outset to summarize the nature of reactive oxygen species, the mechanisms by which they are generated and the damage at the biochemical and cellular level which ensues as a consequence of elevated levels of ROSs. Reactive oxygen species (or free radicals) are highly unstable molecules which have one or more unpaired electrons. As such these chemical entities may be stabilized through the attack and removal of electrons from other sources with the concomitant destabilization of the donor molecule.[10,11] The major ROSs with respect to reactive oxygen species and yeast cells include superoxide anion ($O_2^-\cdot$), hydroxyl radical ($HO\cdot$), singlet oxygen (O), and hydrogen peroxide (H_2O_2). Dioxygen (O_2)—with two unpaired electrons in the π^* orbital—is also a radical species. As Halliwell and Gutteridge point out the two unpaired electrons have the same quantum number. Thus were O_2 to oxidize another species, the two needed electrons must have parallel spin to occupy the π^* orbitals of dioxygen.[10,11] Since the spin requirement places constraints on the incoming electrons dioxygen tends to accept electrons consecutively, thereby slowing the reaction rate of dioxygen with nonradical molecules. Singlet oxygen is a more reactive form of O_2 which is generated by the movement of one unpaired electron to complete one shell which leaves the other orbital empty. The generation of singlet oxygen can occur in biological systems but requires energy input such as through photo-excitation.

One of the most common and abundant reactive oxygen species is the superoxide radical (O_2^-). In nature superoxide anion is generated via the one electron reduction of dioxygen during mitochondrial respiration or during other electron transport events which occur in plant chloro-

plasts or in microsomes of the endoplasmic reticulum, in the respiratory burst of macrophages and neutrophils (using an NADPH-dependent oxidase complex) and through a number of other processes which occur during aerobic conditions.[12] In mitochondria superoxide anion can either be generated at complex I (NADH-Ubiquinone) or complex III (cytochrome c) of the electron transport chain (Fig. 6.1). Superoxide anion can also be generated through auto-oxidation via interactions with cellular reductants (such as glutathione, NADH and others) and via the action of a number of chemical agents such as paraquat and menadione, which cross biological membranes with relative ease and engage in redox chemistry under aerobic conditions.[12] Although superoxide anion itself is not a highly reactive radical species, superoxide gives rise to other ROSs which are highly reactive and damaging. Therefore it is critical that cells possess the ability to remove this abundant ROS to preclude the formation of increased levels of the more damaging free radical species such as hydroxyl radical. The destruction of superoxide anion is carried out by a class of enzymes known as superoxide dismutases which will be described in more detail elsewhere in this chapter.

The disproportionation of superoxide via the catalytic activity of superoxide dismutase gives rise to hydrogen peroxide as one product with the reaction given in Table 6.1. Furthermore, several oxidases are known to produce hydrogen peroxide via their normal reaction mechanisms.[10,11] Catalases, which will be discussed later, carry out the destruction of hydrogen peroxide with the generation of water (Table 6.1). Although hydrogen peroxide has no unpaired electrons (and as such cannot be considered a free radical species) it efficiently crosses biological membranes and serves as a key chemical reactant in the generation of the highly reactive and damaging free hydroxyl radical ($HO\cdot$) via several mechanisms.[10,11] First, H_2O_2 undergoes a slow spontaneous decomposition to yield $HO\cdot$. Secondly, the metal-dependent de-

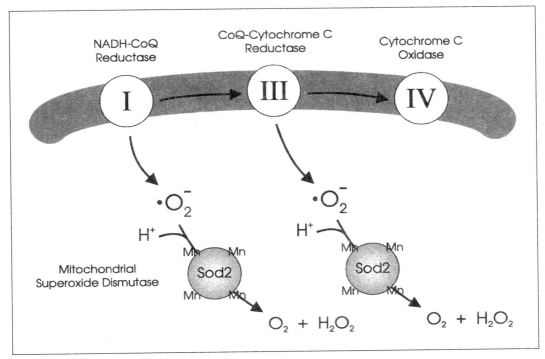

Fig. 6.1. The generation of superoxide anion in the respiratory chain. Superoxide anion is generated during normal cellular respiration at complex I (NADH-CoQ reductase) and complex III (CoQ-Cytochrome C reductase). The tetrameric Mn superoxide dismutase (Sod2p), located in the mitochondrial lumen disproportionates the superoxide anion generated during respiration. Yeast cells that are respiratory deficient can dispense with Sod2p.

Table 6.1. Reactions for the generation or destruction of reactive oxygen species

Superoxide Dismutase	$2O_2^{-\cdot} + 2H^+ \longrightarrow O_2 + H_2O_2$
Fenton Reaction	$Metal^{(n+1)} + H_2O_2 \longrightarrow Metal^{(n+1)+} + HO\cdot + H_2O$
Haber-Weiss Reaction	$Metal^{(n+1)+} + O_2^{-\cdot} \longrightarrow Metal^{(n+1)} + O_2$
Catalase	$2H_2O_2 \longrightarrow 2H_2O + O_2$

composition of hydrogen peroxide via the Fenton reaction (shown in Table 6.1) is a highly efficient means of generating hydroxyl radical from hydrogen peroxide. Metals which may participate in these reactions include the reduced forms of the redox active metals Iron (Fe II), Copper (Cu I), Titanium (Ti III) and perhaps others, although Fe (III) can also generate HO· in the appropriate chelate complex.[10,11] Furthermore, the Haber-Weiss reaction which involves the oxidized form of Fe (Fe III) or Cu (Cu II) and superoxide anion generates the reduced form of either metal which can be coupled to Fenton chemistry

for the generation of hydroxyl radical. It is clear that any elevations in the levels of superoxide anion, hydrogen peroxide or the redox active metals Fe and Cu are likely to lead to the formation of high levels of hydroxyl radical by the chemical mechanisms listed in Table 6.1. Therefore the valence state and bioavailability of these redox active metals is a key determinant in their ability to participate in the generation of reactive oxygen species. Later in this chapter we summarize the homeostatic mechanisms utilized by yeast cells to govern the reduction, transport, distribution and sequestration of redox active metals.

It is important to note that hydroxyl radical is also efficiently generated in vivo via the homolytic fission of O-H bonds in water from exposure to ionizing radiation.[10,11]

Once generated, what conditions determine the efficacy with which ROSs generate cellular damage and what types of cellular damage ensues? Several features of reactive oxygen species contribute to their toxicity, including the site(s) of generation, rates of chemical reactivity, abundance and diffusability. Furthermore, the physiological state of the cell is important with respect to how rapidly the cells are dividing or carrying out biosynthetic reactions and what additional stresses the cells are being exposed to which may exacerbate the oxidative stress. Although it is not the purpose of this chapter to exhaustively cover each of these features for all ROSs, these

are important considerations when one studies oxidative stress-mediated cellular damage, repair and protection. A number of excellent reviews in this area are available for more details.[7,13-15]

Reactive oxygen species are capable of generating cellular damage at a number of levels, including direct protein or enzyme inactivation, membrane damage due to lipid peroxidation and damage to DNA (Fig. 6.2). Proteins and enzymes are damaged by hydroxyl radical via preferential oxidation of the amino acids tyrosine, phenylalanine, tryptophan, histidine, methionine and cysteine. Indeed, the oxidation of some amino acids to carbonyl derivatives is thought to play a role in flagging these proteins for turnover and there is an increase in the level of these carbonyl derivatives during aging.[7] Hydroxyl radical

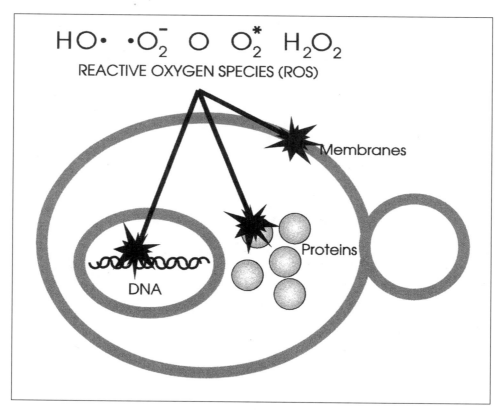

Fig. 6.2. Reactive oxygen species (ROSs) are extremely damaging to biological macromolecules. Simplified overview of each reactive oxygen species which is generated and the primary macromolecules that are damaged. All cells must be able to sense and respond to ROS in order to repair macromolecules and to inactivate ROSs.

mediates extensive lipid peroxidation, especially with polyunsaturated fatty acids which are particularly susceptible to hydrogen abstraction due to the presence of double bonds. The carbon radical formed in the fatty acid subsequently forms a conjugated diene capable of cross-linking fatty acids within cellular membranes.[16] Hydroxyl radical is a key culprit in DNA damage, generating several base lesions in double-stranded DNA including thymine glycol, 8-oxoguanine and formamidopyrimidine as well as other base oxidation products.[13] The double-stranded nature of genomic DNA renders it much less susceptible to oxidative damage than single-stranded DNA, however, oxidative damage to double-stranded DNA is very common.[13] Not surprisingly, studies have demonstrated that eukaryotic DNA packaged into chromatin is significantly more resistant to oxidative damage than naked DNA, underscoring the role of DNA packaging as a protective mechanism as well as playing a key role in compartmentalization and gene regulation.[17] In addition to the bases of DNA, the phosphodiester sugar backbone is susceptible to oxidative damage, resulting in strand gaps or double-stranded breaks.[13] It is clear that although other ROSs generate macromolecular damage in biological systems, the hydroxyl radical is a key causative agent in oxidative damage.

3. YEAST: AN EXCELLENT MODEL FOR STUDIES OF OXIDATIVE STRESS RESPONSES

As described elsewhere in this volume, the baker's yeast *Saccharomyces cerevisiae* is truly an outstanding experimental model to dissect and understand the biochemical mechanisms by which cells sense and respond to stress (introduction and chapters 7, 8). *S. cerevisiae* is a eukaryotic microorganism with sixteen chromosomes and a mitochondrial genome, and the DNA of both has been completely sequenced. The availability of the yeast genome sequence has provided a remarkable resource for the yeast biologist. The genome sequence allows rapid gene identification, strain to strain comparisons, the identification of putative structural and functional homologues to known proteins from yeast or other organisms and a variety of other uses which enhance the utility of this model system and accelerate experimental progress. Although the entire sequence of the yeast genome is known, this information is most elegantly utilized when combined with classical yeast genetics for the isolation and analysis of loci involved in a given pathway of interest. Classical yeast genetics is indispensable for the isolation and characterization of genes encoding structural, catalytic or regulatory proteins or RNAs and for identifying genes encoding interacting components in a pathway or complex by extragenic suppression analysis.

As detailed in a variety of sources,[18,19] yeast biologists have available a wide array of plasmids for the introduction of genes which are episomally maintained at either low or high copy number or which can be targeted to integrate at chromosomal loci homologous to plasmid-borne sequences. Therefore plasmid-borne copies of genes and promoters can be mutated and re-introduced into yeast cells to firmly establish structure-function relationships. The efficiency with which yeast cells carry out homologous recombination has been utilized to generate chromosomal mutant alleles, harboring either a point mutation or a complete deletion simply by transformation with DNA fragments genetically altered by standard in vitro recombinant DNA techniques. The ability to generate chromosomal "gene knockouts" with ease has greatly facilitated our ability to observe phenotypes which can be directly associated with the loss of function of genes of interest. Alternatively, dominant "gain of function" alleles can be introduced into the yeast genome for structure-function analysis. We describe a number of examples in which yeast genetics and molecular biology have been combined to identify components and their mechanisms of action in oxidative stress responses.

The ability to grow yeast cells both aerobically during which oxidative phosphorylation (respiration) is the primary source of energy (with a smaller contribution from glycolysis) or anaerobically (with glycolysis as the sole source of ATP) makes this an experimental system particularly suited for oxidative stress studies. Since normal respiration is an important physiological source of reactive oxygen species, enforced respiration by the growth of yeast cells on nonfermentable carbon sources is a facile means of generating chronic, yet nonlethal oxidative stress in the laboratory. Furthermore, since yeast strains which harbor deletions of important oxidative stress protection genes are sensitive even to atmospheric oxygen concentrations, the ability to propagate such strains anaerobically or under extremely low oxygen tensions is advantageous.

Although there is an abundance of features that make yeast cells first rate experimental models for studies of oxidative stress, the ability to grow yeast under highly defined conditions and in quantities sufficient for biochemical analysis and protein purification are advantages which cannot be understated. Indeed, given the mechanistic similarity with which yeast and humans are known to carry out a wide range of physiological processes,[20] the ability to combine the power of molecular biology, classical genetic analysis, biochemistry and cell biology and to manipulate the yeast milieu at will all serve to make *S. cerevisiae* an ideal system to study oxidative stress. In the following sections we describe the results from a number of studies which have capitalized on these approaches to begin to define the molecular mechanisms by which yeast cells defend themselves from oxidative stress.

4. METAL HOMEOSTASIS AND OXIDATIVE STRESS

As described above, the generation of reactive oxygen species is catalyzed by a number of redox active metals through well characterized chemical reactions (Table 6.1). Although a number of metal ions play essential roles as structural or catalytic cofactors in biology, the chemical reactions

in which these metals participate foster severe oxidative stress conditions. Therefore given the link between metals and oxidative stress, it is important to consider the mechanisms by which yeast cells acquire and maintain homeostatic control of redox active metals. Two metals found throughout biology are the prime culprits for the generation of free radical species through intracellular chemical reactions. Copper (Cu) and Iron (Fe) are metals which are essential for life due to their involvement as enzyme cofactors in a wide variety of biochemical reactions. However, these same two metals are highly toxic due to their proclivity to engage in the redox reactions previously discussed, which result in the generation of reactive oxygen species. These two facts dictate that all cells must establish fine-tuned mechanisms which allow cells to accumulate sufficient levels of Cu and Fe for normal biochemical reactions, yet prevent the accumulation of these metals to levels which unleash their toxic effects. Although questions of Cu and Fe homeostasis have been investigated for decades—in part due to the existence of human diseases caused by imbalances in Cu or Fe levels (see below)—a detailed understanding of Cu and Fe homeostatic mechanisms has not been forthcoming in prokaryotes nor in higher eukaryotic systems. In contrast, recent work in the baker's yeast has broken new ground in beginning to formulate a comprehensive understanding of Cu and Fe homeostatic mechanisms and furthermore, to demonstrate that these two metals are physiologically intimately intertwined (Figs. 6.3, 6.4).

4.1. IRON TRANSPORT IN *S. CEREVISIAE*

A major question relevant to the accumulation of Cu and Fe in yeast cells is: how are these metals transported into cells and what cellular factors are involved? A number of elegant studies in *S. cerevisiae* have demonstrated the existence of a surprising array of proteins involved in Cu or Fe transport. Furthermore, additional work has demonstrated that yeast vacuoles are

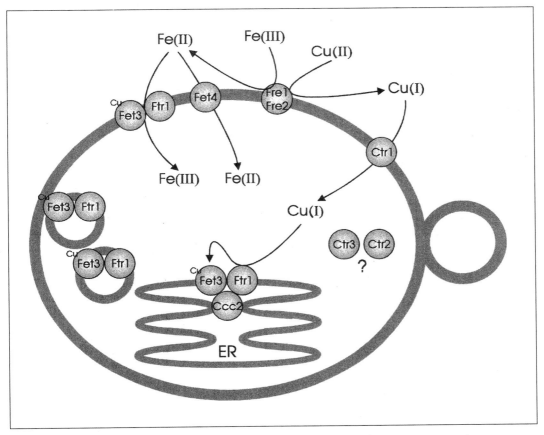

Fig. 6.3. Copper and iron transport in S. cerevisiae. The close association between copper and iron transport is shown. Both systems rely on the Fre1p/Fre2p reductases to catalyze the reduction of either copper or iron prior to transport. The iron transporters consist of a high affinity system (Fet3p/Ftr1p) and a low affinity system (Fet4p). The components of the copper transport system are Ctr1p, a plasma membrane copper transporter, and Ccc2p, an intracellular copper transporter. The Fet3p/Ftr1p complex is supplied with copper in the endoplasmic reticulum (ER) through the action of Ccc2p before vesicular transport to the plasma membrane. The exact location(s) and function of the other components of the copper transport system, Ctr2p and Ctr3p, are currently under investigation.

important for metal homeostasis, suggesting that intracellular metal compartmentalization is important for metal homeostasis. These systems have been comprehensively reviewed in detail elsewhere.[21] However, we will briefly summarize these systems to place them in the context of metal homeostasis and oxidative stress (Fig. 6.3). We shall first consider Fe transport mechanisms of which there are several in yeast cells. A number of studies in yeast have demonstrated that the high affinity transport of Fe requires the reduction of Fe(III) to Fe(II) via an extracellular Fe reductase activity.[22] The rationale for the

requirement for this reduction is that Fe(III) is extremely insoluble at physiological pH while Fe(II) is much more soluble. This Fe reductase activity is derived from two distinct genes—*FRE1* and *FRE2*—which encode membrane proteins that are 24.5% identical.[22,23] Both *FRE1* and *FRE2* mRNA levels are elevated during Fe starvation and repressed during Fe repletion. Interestingly, however, *FRE1* appears to be responsible for the bulk of reductase activity during log phase while *FRE2* predominates during stationary phase.[23] *S. cerevisiae* cells bearing disruptions of both the *FRE1* and *FRE2* genes are unable to grow on Fe

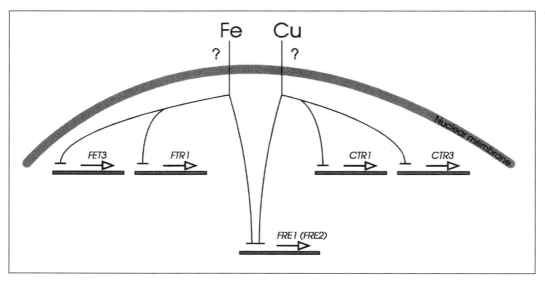

Fig. 6.4. Genes whose protein products are involved in copper and iron transport are themselves regulated at the level of transcription by the availability of these same metal ions. The FET3 and FTR1 genes are induced by Fe starvation and repressed by Fe repletion. CTR1 and CTR3 are induced by Cu starvation and repressed by Cu repletion. Expression of FRE1 and FRE2 is induced by either Fe or Cu starvation and repressed by either Fe or Cu repletion. The mechanism by which these metal ions are transported from the cytoplasm across the nuclear membrane as well as their chemical form is unknown.

limiting medium, thereby demonstrating that the Fe reductase activity is critical for high affinity Fe acquisition.[23] This same reductase activity also plays an important role in high affinity Cu transport as discussed below.

The actual transport of Fe into yeast cells appears to largely occur via two systems, one with high affinity (Km ~ less than 1 mM) and one with low affinity (Km ~ 40 mM). A high affinity system for Fe transport has been elucidated at least in part by taking advantage of the genetics of yeast combined with cunningly designed selection schemes. Kaplan and colleagues utilized the well established damage mediated by free radicals via a complex of the antibiotic streptonigrin (SNG) and Fe as a means to enrich for yeast mutants defective in high affinity Fe transport.[24] Under conditions of Fe starvation in which the Fe transport system is induced wild type cells are killed in the presence of SNG whereas Fe transport mutants survive. As a result an important component of the high affinity Fe transport system—*FET3*—

has been identified.[24] The *FET3* gene encodes a glycosylated plasma membrane protein of 72 kDa which harbors an extracellular domain with Fe reductase activity. Interestingly, Fet3p contains clear sequence homology to a group of proteins known as multicopper oxidases with compelling homology to the human serum ferroxidase protein ceruloplasmin. The latter is a copper binding protein long known to be an important component of the Cu and Fe homeostatic machinery in higher organisms.[25,26] Consistent with this notion Fet3p-dependent Fe transport absolutely requires Cu since cells bearing mutations in components involved in either Cu transport or distribution are defective in Fe transport.[27] A number of investigations served as the basis for a model in which Fet3p provides a Cu-dependent oxidation of Fe(II) to Fe(III) during transport. In this model Fet3p is functionally and perhaps physically coupled to a plasma membrane bound high affinity Fe transporter.[21] Additional genetic studies have identified a Fet3p-dependent transporter as the product of the *FTR1* gene.[28] The Ftr1 protein is a plasma membrane protein thought to bind Fe atoms via glutamate carboxylate groups. In addition, it plays an essential role in the loading of Fet3p with Cu in the endoplasmic reticulum-Golgi secretory compartments.[28] Subsequently the Ftr1p-Fet3p complex moves to the plasma membrane perhaps in proximity to the Fre1p/Fre2p reductases where Fe transport into the cell ensues. Consistent with these proteins playing critical roles in high affinity Fe transport both the *FET3* and *FTR1* genes are repressed by Fe and induced by growth under Fe limiting conditions. In addition to the high affinity Fe transport system encoded by the *FET3* and *FTR1* genes yeast cells also harbor a low affinity Fe transporter encoded by—*FET4*—which was identified by suppression of the Fe transport defects in a *fet3Δ* background by high levels of Fet4p expression. Like the high affinity Fe transport machinery the Fet4p Fe transport pathway depends upon the Fe reductase Fre1p/Fre2p for transport (Fig. 6.3). However, the Fet4p system will also transport other metals such as Cadmium and Cobalt.[29]

4.2. COPPER TRANSPORT AND HOMEOSTASIS

Surprisingly, utilizing a powerful genetic selection system to identify Fe homeostatic genes, Dancis and co-workers isolated a yeast mutant defective in high affinity Cu transport and isolated the corresponding wild type gene denoted *CTR1* (Cu transporter 1).[27] *CTR1* deletion mutants are defective in high affinity Cu transport and as a consequence suffer from a variety of phenotypes which can all be biochemically ascribed to Cu insufficiency since exogenous Cu restores the wild type phenotypes. Deletion of the *CTR1* gene also results in poor growth on Cu deficient medium, defective respiration (presumably due to the Cu and Fe requirement in cytochrome oxidase) and defective high affinity Fe transport (due to the requirement for Cu by the Fet3p Cu-dependent ferroxidase). Furthermore, and of great relevance

to oxidative stress protection *ctr1Δ* strains are highly sensitive to redox cycling drugs such as menadione and have little or no Cu,Zn superoxide dismutase (SOD) activity.[30] This latter observation is in keeping with the absolute requirement of Cu,Zn SOD—a major defense against superoxide anion—for Cu as a catalytic cofactor.[12] This observation also underscores the fine balance of Cu and Fe levels cells must maintain since too much copper results in the generation of reactive oxygen species. Yet too little Cu results in an inability to deal with reactive oxygen species produced by normal aerobic metabolism or other nonmetal catalyzed mechanisms.

Ctr1p is a plasma membrane glycoprotein (Fig. 6.3) bearing in its amino terminus several repeats of the presumptive Cu binding domain Met-X-X-Met which resembles a domain found in several prokaryotic Cu homeostatic proteins.[27] Although Ctr1p is capable of homo-oligomerization, it is currently unknown which proteins Ctr1p interacts with in vivo to carry out high affinity transport. Interestingly, two additional yeast genes have been identified which also appear to play a role in Cu transport. Kampfenkel et al demonstrated that an *Arabidopsis thaliana* cDNA when produced in a *ctr1Δ* strain could partially suppress some of the growth defects associated with *ctr1Δ*.[31] A yeast homologue was identified through the yeast genome data base—denoted *CTR2*—which had similar protein sequence and functional properties to the plant gene. As yet, however, no evidence exists for the direct involvement of Ctr2p in Cu transport. Recently the use of extragenic suppression studies identified a new yeast gene involved in high affinity Cu transport. Using a *ctr1Δ* mutant strain, a strain harboring a dominant mutation was isolated which allowed cells to dispense with *CTR1* yet utilize respiratory carbon sources for growth.[32] This mutant allele denoted *CTR3ˢ* (Copper Transporter 3 suppressor) allows the expression of the frequently quiescent *CTR3* gene encoding an integral membrane protein which restores high affinity Cu transport and other normal phenotypes to strains harboring a *ctr1Δ*.[32] It is interesting to note that *CTR3* expression is normally shut down in the vast majority of laboratory yeast strains analyzed via the insertion of a transposon (Ty2) in the promoter region. The *CTR3'* allele harbors only a remnant of the excised 6 kilobase Ty2 element, restoring promoter function and *CTR3* gene expression. Ctr3p is a small 27.5 kDa protein which exhibits significant sequence similarity to the Ctr2p carboxyl terminus. Furthermore, similar to the Fet3p/Ftr1p and Fet4p Fe transport systems in yeast, Ctr1p and Ctr3p require the action of the *FRE1*-encoded Cu reductase for normal high affinity Cu transport.[32] However, the precise mechanisms by which Ctr1p, Ctr2p or Ctr3p function in Cu transport remain to be elucidated (Fig. 6.3).

The *FRE1, FRE2, FET3, FTR1, CTR1* and *CTR3* genes are all regulated by metal ion availability (Fig. 6.4). *FET3* and *FTR1* are induced by Fe starvation and repressed by Fe repletion but not regulated by Cu.[21] *CTR1* and *CTR3* are induced by Cu starvation and repressed by Cu repletion but not regulated by Fe. As one might predict based on the involvement of the *FRE1* and *FRE2*—encoded Fe/Cu reductases in both Fe and Cu transport, the expression of the *FRE1* and *FRE2* genes is repressed by either Fe or Cu and induced by either Fe or Cu starvation. Therefore distinct sensors must exist in yeast cells to sense and respond to changes in the levels of Fe or Cu and act on distinct and overlapping target genes. Perhaps some of the sensing molecules in this regulatory network detect metals directly or the reactive oxygen intermediates which result from metal-based chemistry. We discuss these possibilities elsewhere in this review. Recently it has been shown that the Aft1 protein directly regulates these Fe-responsive genes via Fe alterations of Aft1 DNA binding activity.[33]

Once Cu enters yeast cells it must be distributed appropriately to Cu-requiring proteins such as Cu,Zn superoxide dismutase, cytochrome oxidase and Fet3p. Inappropriate distribution may result in ab-

normally high intracellular local Cu concentrations which may be available for the generation of reactive oxygen species. Indeed, the two severe human genetic diseases, Menkes syndrome and Wilsons disease, have been demonstrated to be due to defects in Cu distribution in mammalian cells.[34,35] In Menkes disease copper export from cells is defective in most tissues with the exception of the liver. Therefore Cu cannot be appropriately distributed to Cu requiring enzymes. In Wilsons disease Cu is not incorporated into ceruloplasmin—a plasma membrane ferroxidase protein similar to Fet3p—and Cu fails to be excreted from the liver into the bile, causing Cu toxicity. It is likely that a significant source of the Cu toxicity is due to the generation of ROSs by the reactions described in Table 6.1. The human genes which when mutated cause Menkes syndrome and Wilsons disease have recently been isolated and sequenced.[34,35] Indeed, the proteins encoded by the Menkes and Wilsons genes respectively are surprisingly similar (approximately 60%). Both proteins harbor six copies of an amino-terminal putative metal binding motif, a proposed cation channel, a large hydrophilic ATP-binding domain and a domain which resembles the P-type ATPase-phosphorylated-phosphatase region. In the absence of any biochemical data at present both the Mnk and Wnd proteins resemble bacterial cation-transporting P-type ATPases and therefore are likely to be involved in Cu transport. Interestingly, in a rat model for Wilsons disease—the Long-Evans Cinnamon rat—expression of the Wilsons disease gene is either absent or extremely low.[36,37] This is highly significant with respect to oxidative stress since Long-Evans Cinnamon rats accumulate high levels of Cu in the liver and suffer from hepatocellular carcinoma at an incidence of greater than 90%.[38] This cancer is potentially due to damage caused by the generation of reactive oxygen species from the high levels of hepatic Cu since early chelation therapy in these rats dramatically lowers the incidence of hepatocellular carcinoma.[39,40]

A particularly striking example of the conservation between yeast and human biology with respect to Cu homeostasis is exemplified by the yeast *CCC2* gene encoding a structural and presumably functional homologue of the human Menkes/Wilsons disease gene products. The *CCC2* gene was isolated independently by two groups. One group used degenerate oligonucleotides encompassing sequences conserved in the Mnk and Wdn proteins in PCR reactions. The other group isolated *CCC2* by virtue of suppression of a defect in calcium homeostasis via an unknown mechanism.[41,42] Yeast cells bearing a disruption of the endogenous *CCC2* gene grow poorly in low Cu medium and are defective in high affinity Fe transport. This Fe transport defect has been demonstrated to be due to the inability to assemble Fet3 apo-protein with Cu and therefore Fet3p in *ccc2Δ* strains has defective ferroxidase activity.[41] This is consistent with the observation that some Wilsons disease patients exhibit Fe homeostasis defects.[34,35] Ccc2p is likely to function in an intracellular secretory compartment such as the endoplasmic reticulum-Golgi (Fig. 6.3) since Fet3p is located in the plasma membrane.

4.3. PROTECTION FROM CU TOXICITY

When *S. cerevisiae* cells encounter high Cu concentrations they respond by activating the transcription of two genes encoding isoforms of a metal binding protein known as metallothionein (MT). These protein isoforms—encoded by the *CUP1* and *CRS5* genes—respectively, function to protect yeast cells from the elevated Cu levels by binding and sequestering Cu atoms through cysteine thiolates with high affinity (Fig. 6.5A).[43,44] The yeast MT proteins resemble those from higher eukaryotic organisms in that they are small proteins and highly enriched in cysteine residues. The presence of 12 cysteine residues in the *CUP1* encoded MT is critical for both protection from Cu toxicity in vivo and the binding of 6 to 7 Cu atoms per polypeptide in vitro. The sequestration of Cu

by MT thereby renders this metal unavailable for engaging in ROS generating redox chemistry. Furthermore, as described elsewhere in this review, MTs of both yeast and mammalian origin harbor potent antioxidant activity.[45] Whereas deletion of the *CUP1* gene renders cells highly sensitive to Cu, deletion of the *CRS5* gene causes only a very modest shift in Cu sensitivity compared to that observed in *cup1Δ* strains.[46]

How do *S. cerevisiae* cells sense elevations in Cu levels which may approach toxic conditions and respond by activating *CUP1* and *CRS5* gene transcription? The answer to this question was investigated based on the known dramatic Cu sensitivity phenotype of *cup1Δ* mutants which provided a clear indication that the expression of the *CUP1* gene is important for Cu detoxification.[47] First using *CUP1* as a model Cu-inducible promoter, a series of promoter deletion and synthetic fusions were constructed to identify the cis-acting elements required for Cu-responsive gene transcription.[48] These experiments and subsequent work revealed that a binding site with the consensus: 5'-TTTTGCTG-3' plays a critical role in the induction of *CUP1* by copper (Fig. 6.5A). In concert with this work genetic studies ensued which on the basis of the importance of the *CUP1* gene in Cu resistance identified a class of yeast mutants which rendered cells both defective in *CUP1* Cu-responsive transcription and Cu detoxification. These original allelic mutations—designated *ace1-1* or *cup2*—were demonstrated to be recessive and act *in trans* to *CUP1*.[49,50] The cloning of the wild type allele (heretofore referred to as *ACE1* for consistency), in vitro biochemical experiments and in vivo gene expression studies together provided a basic framework for understanding how yeast cells sense increases in Cu and respond by activating MT gene transcription.[51] Ace1p is a 225 amino acid protein with a highly basic amino terminal region (110 amino acids) which encompasses a Cu-activated DNA binding domain. In short, the Ace1 DNA binding domain undergoes a Cu-dependent conformational change

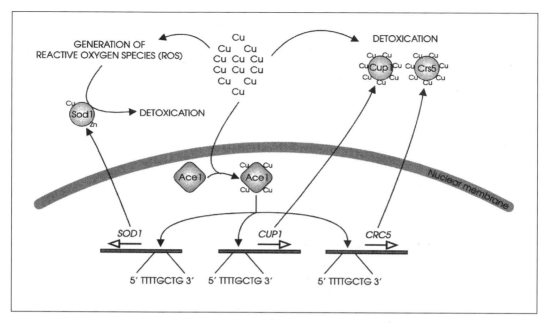

Fig. 6.5A. Sod1p (Cu,Zn Sod), Cup1p and Crs5p are involved in cellular copper detoxification in the cytosol either by directly binding copper or by catalyzing the breakdown of ROSs generated from copper or iron. The genes encoding these three proteins are themselves regulated by copper availability through the transcription factor Ace1p. Ace1p binds copper and is then able to interact with a consensus site found in the promoter of these genes to activate transcription. Thus, the cytosolic availability of copper regulates the expression of proteins that participate in detoxification in the cytoplasm.

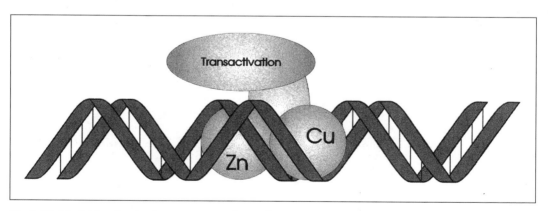

Fig. 6.5B. Model for the Ace1p copper metalloregulatory transcription factor interaction with DNA of the metal responsive element. A monomeric Ace1p molecule is shown interacting with a copper responsive cis-acting DNA element such as those found in the CUP1, CRS5 or SOD1 promoters. In this model two sub-domains of the DNA binding domain are formed, either around the tetrameric Cu (I) cluster and a single Zn atom (as shown) or around two independent Cu (I) sites. One sub-domain makes contacts with the major groove of the DNA and the other sub-domain makes contacts with the DNA minor groove. The binding of Ace1p to the DNA in a copper-dependent manner brings a potent transcriptional activation domain to the DNA which interacts with the yeast cellular transcriptional machinery.

which drives the folding of this domain to allow Ace1p to bind to the consensus sequence described above (Fig. 6.5A). Within the amino terminus lie 12 cysteine residues, 11 of which are essential for the coordination of 4 to 6 Cu atoms in its reduced form (Cu I). Cu(I) coordination was demonstrated to occur via trigonal coordination geometry through the cysteine thiols. This would thereby drive Ace1p folding differently than Cu (II) coordination which occurs via a distinct coordination geometry. Neither Ace1p-mediated DNA binding nor target gene transcription is induced by either Zn or Cd. These metals, however, prefer tetrahedral coordination via thiols. Thus it is clear that differences in coordination chemistry dictate the metal specificity for gene activation in response to the redox-active metal copper. Recent work also suggests that Ace1p binds to a single Zn atom but the biological significance of this remains to be demonstrated.[52] It is interesting that although Ace1p binds to DNA as a monomer the protein contacts both the major and minor groove of the binding site, suggesting that the amino-terminal DNA binding domain may have two subdomains.[53,54] Perhaps one subdomain is formed around the Cu coordination sites and the other around the single zinc site (Fig. 6.5B).

The binding of Cu to Ace1p is thought to occur in a highly cooperative fashion, thereby ensuring rapid responses to small fluctuations in Cu concentrations. Once bound to DNA Ace1p rapidly and potently activates target gene transcription through an as yet poorly defined trans-activation domain located in the acidic carboxyl-terminal 115 amino acid residues.[51] In addition to activating the expression of *CUP1* approximately 20- to 50-fold, Ace1p has been shown genetically to be necessary for the modest (3- to 5-fold) induction of *CRS5* mRNA and an Ace1p binding site consensus lies within the *CRS5* promoter (Fig. 6.5A).[46] Furthermore, in response to elevated Cu concentrations Ace1p activates the transcription of the yeast *SOD1* gene—

encoding Cu,Zn superoxide dismutase—again via a single consensus Ace1 binding site in the *SOD1* promoter (Fig. 6.5A).[55] Although the rationale for Ace1p-mediated Cu-inducible transcription of the *SOD1* gene is uncertain, two possibilities have been put forth.[55] First, since Cu,Zn SOD absolutely requires Cu as a catalytic cofactor, yeast cells may signal the presence of elevated levels of Cu so that apo SOD is not fully produced under conditions of insufficient Cu. Secondly, since Cu readily generates harmful reactive oxygen species through interactions with superoxide anion it would be beneficial to the cell to produce additional Cu,Zn superoxide dismutase under conditions of Cu excess. In either or both cases regulation of the yeast Cu,Zn SOD by Ace1 represents a clear case in which there is an interdigitation of redox active metals and reactive oxygen species in cell signaling.

5. YEAST DEFENSE MECHANISMS AGAINST OXIDATIVE STRESS

It is clear that cells either generate or are exposed to a number of distinct oxygen-derived free radicals that cause many types of cellular damage. Cells have evolved a number of antioxidant activities to protect themselves from the oxidative insults via enzymatic activities, low molecular weight stoichiometric antioxidants, redox metal sequestration and repair activities (Table 6.2). A subset of these activities are constitutively present while others are biosynthetically induced in response to the presence of oxidative stress. Furthermore, the subcellular location of antioxidant activities plays a critical role in the sensing, removal and repair of damage from oxidative stress. Here we describe a number of mechanisms by which yeast cells protect themselves from oxidative stress through either the removal of free radicals, the repair of free radical-mediated cellular damage or through the biosynthetic induction of antioxidant defense components.

5.1. SUPEROXIDE DISMUTASES (SODs): SOD-DEFICIENT YEAST AS AN EXPERIMENTAL TOOL

From bacteria to humans a primary cellular defense against oxygen toxicity involves one or more forms of the enzyme superoxide dismutase (SOD). Superoxide dismutases (SODs) are a group of metalloenzymes that catalyze the conversion of superoxide anion to hydrogen peroxide and dioxygen.[56] Eukaryotic cells possess a copper/zinc-containing SOD (Cu,Zn SOD) that is predominantly cytosolic and a manganese-containing SOD that is mitochondrial (Fig. 6.6).[12] Consistent with an important role in oxidative stress protection, exposure of yeast cells to hyperoxia or paraquat has been shown to induce the synthesis of the mRNA and protein for both SODs.[57-59] Bacteria typically have both iron and man-

Table 6.2. Yeast oxidative stress protection gene products

Gene	Protein	Activity in oxidative stress defense
SOD1	Cu,Zn SOD	cytosolic, superoxide anion dismutation
SOD2	Mn SOD	mitochondrial, superoxide anion dismutation
SPE2	S-adenosylmethionine decarboxylase	polyamine biosynthesis
CUP1	metallothionein	sequestering copper and free radicals
ATX1	unknown	sequence homology to bacterial metal transporters; Cu accumulation
TSA	thiol-specific antioxidant	sequence homology to bacterial alkyl hydroperoxide reductase; catalyzes removal of thiyl radicals
CTT1	catalase T	cytosolic, hydrogen peroxide dismutation
CTA1	catalase A	peroxisomal, hydrogen peroxide dismutation
APN1	AP endonuclease	repair of abasic and 3' fragmented DNA generated by oxidative damage
IMP2	multifunctional	required for repair of oxidatively damaged DNA
GSH1	g-glutamylcysteine synthetase	catalyzes first and rate-limiting step in glutathione biosynthesis
YAP1/YAP2	B-Zip transcription factors	overexpression confers pleiotropic drug resistance, regulates expression of several oxidative stress genes
TRX2	thioredoxin	abundant cellular antioxidant
HAP1	heme-responsive transcription factor	activates transcription of genes regulated by oxygen and heme
MAC1	nuclear protein	affects expression of *CTT1*, *FRE1*, *CTR1* and genes required for respiratory growth
POS9	putative response regulator protein	involved in cellular resistance to oxidative stress
HSF	heat shock transcription factor	activates expression of *CUP1* by oxidative stress

ganese-containing SODs, however, a recent
report has demonstrated the presence of a
Cu,Zn SOD in prokaryotes.[60] The *S. cerevisiae* Cu,Zn SOD encoded by the *SOD1*
gene is a homodimer of two 16 kDa subunits. Each subunit binds to a single Cu
and Zn atom which are absolutely required
for catalysis.[12] Superoxide is ushered into
the catalytic site via a positively charged
electrostatic channel, thereby rendering this
a highly efficient enzyme with a catalytic
rate constant of 10^9 M^{-1} s^{-1}.[12] The yeast
manganese SOD encoded by the *SOD2*
gene is tetrameric, having four 23 kDa
subunits, each of which binds to a single
Mn ion.[12] As for the Cu,Zn SOD an electrostatic channel appears to promote interactions between the MnSOD enzyme and
the superoxide radical substrate. Although
the yeast MnSOD is localized to the mito-

chondria, it is encoded by a nuclear gene
and the protein is imported into the lumen of the mitochondria via a 27 amino
acid amino-terminal mitochondrial targeting sequence.[61] Therefore—like humans—
yeast contain a predominantly cytosolic and
a mitochondrial SOD. The distinct locations of these enzymes are a key determinant in their cellular role in oxidative stress
protection. The importance of superoxide
dismutase activity in cellular protection is
underscored by the finding that the
MnSOD gene is deleted in some forms of
human melanoma and the expression of exogenously introduced MnSOD in these cells
can suppress the malignant phenotype.[62]

The exploration of SOD function in
oxidative stress protection in yeast is
greatly facilitated by the ability to selectively delete genes from the yeast genome

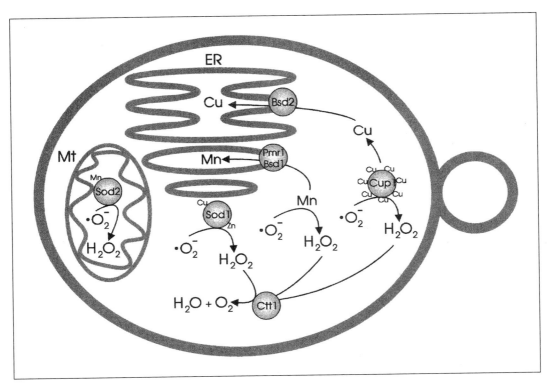

Fig. 6.6. Yeast proteins involved in extranuclear antioxidant defense. Cytosolic oxidative stress defenses include cytosolic catalase T (Ctt1p), metallothionein (Cup1p), and Cu,Zn Sod (Sod1p). Yeast antioxidant defenses which are found in cellular compartments include MnSod (Sod2p) which is found in the mitochondrion, Bsd2p which is in the endoplasmic reticulum and Pmr1p which is in the Golgi. The influence of metal ions is again underscored since all of these proteins—except Ctt1p—utilize metals for either catalytic activity, transport or sequestration.

or to reintroduce mutant alleles of SOD genes. Cells harboring inactivating mutations in either the *SOD1* or *SOD2* genes have been shown to have general oxygen intolerance and sensitivity to free-radical generating compounds in both prokaryotes and eukaryotes.[12] Furthermore, yeast cells overexpressing Cu,Zn SOD have been shown to suppress the oxygen sensitivity of *spe2* mutants which are deficient in S-Adenosyl-methionine decarboxylase activity.[63] Yeast strains containing mutations in the Cu,Zn SOD (*SOD1*) exhibit a variety of growth abnormalities including auxotrophies for lysine, cysteine, and methionine when grown aerobically and sensitivity to atmospheric levels of oxygen or hyperoxia and to superoxide-generating drugs such as paraquat and menadione.[64] Furthermore, consistent with the generation of hydroxyl radical from superoxide, cells harboring an *sod1Δ* undergo dramatically elevated spontaneous mutagenesis presumably due to the oxidative damage to DNA.[64] Although the biochemical basis for each of the oxygen-dependent amino acid auxotrophies is unclear recent work suggests that the lysine auxotrophy may be due to oxygen-mediated inactivation of alpha-amino adipate transaminase, a key enzyme in the lysine biosynthetic pathway.[65] *SOD1* deletion strains also fail to grow on nonfermentable carbon sources. It is unclear whether this is due to the inability of cells to cope with superoxide generated in the respiratory chain or whether there is one or more component of the chain which is sensitive to superoxide radical generated through respiration-independent mechanisms.

Yeast strains lacking the mitochondrial Mn SOD due to inactivation of the *SOD2* gene are slightly sensitive to normal oxygen concentrations when grown fermentatively. However, these strains are sensitive to hyperoxia and growth on respiratory carbon sources in atmospheric oxygen levels.[59,66] In support of the mitochondrial electron transport chain as an important source of superoxide it has been demonstrated that *sod2Δ* cells grow as well as wild type cells in hyperoxia if the function of

the mitochondrial electron transport chain was abolished using rho° mutants.[67] Recent work expressing heterologous Mn SODs in yeast cells lacking the *SOD2* gene has defined some of the minimal structural requirements for the proper function and localization of the heterologous Mn SOD. For example, *E. coli* FeSOD fused to the yeast *SOD2* mitochondrial targeting presequence protected yeast deficient in MnSOD from oxidative stress. On the other hand, expression of *E. coli* FeSOD without the mitochondrial targeting presequence did not confer protection.[68] Similarly, the maize *Sod3* gene complemented the paraquat-induced oxidative stress sensitivity of an *sod2Δ* strain.[69] The first 8 amino acids of the maize *Sod3* transit peptide were required to target the protein to the yeast mitochondria in vivo.

5.2. THE BSD GENES AS SUPPRESSORS OF SOD-DEFICIENT YEAST

By capitalizing on yeast genetics and the pleiotropic phenotypes associated with superoxide dismutase-deficient yeast cells, extragenic suppression studies have identified additional components of the oxidative stress response. The aerobic defects and lysine and methionine auxtrophies of *sod1Δ* mutants have been used as the selection basis to isolate recessive extragenic suppressor mutations. Either one of the two genes identified as *BSD1* or *BSD2* (bypass SOD defects) appeared to suppress the methionine and lysine auxotrophies and oxygen sensitivity of *sod1Δ sod2Δ* double mutants.[65,70,71] The function of the *BSD* gene products again provides one point of convergence between oxygen toxicity and metal homeostasis since both gene products are thought to participate in metal ion accumulation and or distribution (Fig. 6.6).

The *BSD1* gene was cloned by complementation of a recessive nuclear gene mutation which reversed many of the oxygen-dependent phenotypes associated with an *sod1Δ sod2Δ* double deletion strain.[65,71] The *BSD1* gene is identical to *PMR1* encoding a 104 kDa P-type ATPase homologue that localizes to the Golgi and is proposed to function in both calcium ion metabolism and in secretory protein transport and processing.[72,73] Mutants lacking *PMR1* accumulate elevated levels of cytosolic calcium and display calcium-dependent perturbations in the protein secretory pathway.[72,74] Recently, however, it has been shown that these calcium related alterations are not responsible for the suppression of aerobic defects by *pmr1* mutants in SOD deletion strains.[71] The oxygen resistance of strains lacking *SOD1*, *SOD2* and *PMR1* is dependent on extracellular manganese.[71] Strains lacking *PMR1* accumulate elevated levels of manganese and are extremely sensitive to manganese ion toxicity. The suppression phenotypes of *pmr1* mutants can be reproduced in *PMR1* wild-type yeast by elevating intracellular manganese concentrations via exogenously added manganese. Based on these observations it has been proposed that *BSD1/PMR1* (localized in the Golgi) functions in maintaining low cytosolic levels of both calcium and manganese (Fig. 6.6).[71] *PMR1* would thus serve a physiological role in supplying manganese ions to activate various manganese-dependent enzymes involved in protein processing and secretion. Mutants in which *PMR1* is inactivated may suppress the oxygen toxicity of strains lacking SOD as a result of the hyper-accumulation of manganese in the cytosol where this ion acts by a free-radical scavenging mechanism to reverse oxidative damage in the absence of Cu,Zn SOD.[71] There are three lines of evidence in support of this mechanism. First, mutants lacking SOD can be suppressed for the lysine auxotrophy by addition of manganese to the medium. Secondly, *pmr1* mutants increase the resistance of yeast lacking SOD to paraquat approximately 10-fold.[71] Finally, it has been well established that Mn is an excellent superoxide dismutase mimetic.[56,75] Although the loss of *PMR1* functions to protect cells from oxidative stress in strains that are SOD-deficient, it is unknown if there are conditions under which changes in *PMR1* function or expression may play a promi-

nent role in oxidative stress protection in wild type yeast cells.

The *BSD2* gene encodes a 37.5 kDa transmembrane protein that also appears to participate in the homeostasis of heavy metal ions (Fig. 6.6). *BSD2* mutants not only reverse the aerobic defects of yeast lacking SOD but also display an increased sensitivity to copper and cadmium toxicity and an increased accumulation of these metal ions.[70] Although the *CUP1*-encoded metallothionein is also able to suppress *sod1Δ* defects (see below), the level of *CUP1* gene expression is not increased in *bsd2* mutants and *bsd2* mutants are able to suppress SOD defects in either *cup1* or *ace1* deletion strains. Furthermore, *bsd2* mutations suppress both the lysine and methionine auxotrophies of *sod1* mutants. Overexpression of the *CUP1* metallothionein gene suppresses only the methionine but not lysine auxotrophy.[45] These data suggest that *bsd2* mutations do not reverse oxidative damage through a *CUP1*-dependent mechanism.[70] Again since Cu also has known SOD mimetic activity perhaps alterations in intracellular Cu levels or distribution caused by *BSD2* mutations generate bioavailable Cu to serve in this capacity.

The Bsd2 protein does not contain the metal-binding motifs found in yeast metallothioneins or other known copper transport proteins.[70] If Bsd2p binds copper or cadmium directly, it may contain a metal binding motif not commonly used by MTs or related metal binding proteins. Site-directed mutagenesis has been used to determine how *BSD2* functions in metal homeostasis.[76] The Bsd2 protein contains several Tyr-Tyr pairs and a single Cys residue; these sites may serve as potential metal binding ligands. Interestingly, mutation of the Tyr[131]-Tyr[132] pair to Phe-Phe abolishes the ability of the *bsd2* mutant to suppress SOD defects but mutation of the single cysteine has no effect.[76] The observation that Bsd2p participates in copper ion accumulation and the presence of the three potential transmembrane domains has led to the suggestion that *BSD2* functions

in the intracellular transport or sequestering of copper and cadmium into compartments such as the vacuole or Golgi
(Fig. 6.6).[70] Additional support for this
hypothesis comes from the recent findings
that Bsd2p is heavily glycosylated, contains
an endoplasmic reticulum (ER) localization
signal and in fact localizes to the ER by
subcellular fractionation and indirect
immunofluorescence.[76]

Therefore, the *SOD1, SOD2, BSD1,
BSD2* and *CUP1* genes are intimately
linked to both the oxidative stress response
and cellular transition metal homeostasis
(Fig. 6.6). These studies have established
an important biological link between the
homeostasis of redox-active transition metals and oxygen radical metabolism in aerobic cells. They also suggest that oxidative
damage in eukaryotic cells can be prevented through alterations in the homeostasis of critical redox-active transition
metals, a theme which pervades this
review, as discussed below.

5.3. METALLOTHIONEIN

Metallothioneins (MTs) are an important class of eukaryotic stress-responsive
proteins whose biosynthesis is induced by
a variety of environmental and physiological stresses.[43,44] MTs are small cysteine-rich
metal-binding proteins that play a central
role in metal detoxification and correspondingly, production of MT is transcriptionally induced by metals.[77] It has also been
shown that MTs are able to protect cells
from the toxicity of a number of xeno-
biotics such as chemotherapeutic alkylating agents and from the deleterious effects
of radiation and chemicals that can generate free radicals.[78] Mouse metallothionein
was recently shown to protect cells against
the cytotoxic and DNA-damaging effects
of nitrous oxide.[79] In support of MTs having a role in protecting cells against oxidative stress, MT transcription in mammals
is induced by a variety of chemicals that
generate reactive oxygen species or which
induce the inflammatory response.[80]

Recent experiments have demonstrated
that expression of the *CUP1* encoded MT

or mammalian MTs suppresses a number
of oxidative-stress induced growth defects
of yeast lacking Cu,Zn superoxide dismutase.[45] For example, expression of the
metallothionein gene *CUP1* in the presence
of copper restored growth on lactate and
suppressed the requirement of *sod1Δ* mutants for cysteine and methionine during
aerobic growth. Furthermore, the expression of the *CUP1* MT increased the cellular resistance to paraquat. In vitro assays
measuring the inhibition of spontaneous
auto-oxidation of 6-hydroxy dopamine
demonstrated that Cup1p (purified or
from whole-cell extracts) exhibits antioxidant activity.[45] How might this high affinity Cu(I) binding protein function as an
antioxidant? Pulse radiolysis experiments
have indicated that the Cu(I) MT protein
scavenges both superoxide anion and hydroxyl radical with reaction rate constants
of 6.5×10^6 and 2.2×10^{11} per mole per
second, respectively.[81] Therefore unlike
the catalytic activity of SODs the *CUP1*
encoded Cu(I) MT is a stoichiometric
antioxidant which is biosynthetically
induced to high intracellular levels
(Figs. 6.5A and 6.6).

Consistent with a role in oxidative
stress protection, the yeast *CUP1* gene is
transcriptionally activated when cells are
grown in the presence of high oxygen tension, during respiration or exposure to redox cycling drugs.[45,82,83] Since each of these
conditions is known to generate superoxide anion, this has led to the suggestion
that *CUP1* activation occurs in response to
superoxide radical or a metabolic deriva-
tive. The mechanisms involved in oxidative stress sensing and *CUP1* gene activation are discussed later in this review. Thus
both yeast and mammalian MT proteins are
an important line of defense against oxidative
stress.

5.4. THE *ATX1* GENE

Recently the *ATX1* gene was isolated
in a screen for genes which when present
in multiple copies, suppress the lysine auxotrophy of aerobically growing yeast lacking both *SOD1* and *SOD2*.[84] Furthermore,

sod mutants lacking a functional *ATX1* gene display an increased sensitivity to atmospheric levels of dioxygen. Since wild-type and *sod* mutants deleted for the *ATX1* gene are hypersensitive toward paraquat and display increased sensitivity toward hydrogen peroxide, *ATX1* appears to be involved in protecting cells against the toxicity of both superoxide anion and hydrogen peroxide. Like the *BSD1* and *BSD2* genes *ATX1* is also involved in metal homeostasis. *ATX1* encodes a 8.2 kDa polypeptide that has significant homology to some bacterial metal transporters.[84] The Atx1 protein exhibits a high degree of identity to the bacterial periplasmic mercury transporters comprising the MerP family. Consistent with a role in metal metabolism, Atx1p contains the highly conserved metal-binding sequence, MTCXXC, found in bacterial mercury transporters and the putative copper-binding sites for several yeast and the human putative copper transporters, Mnk and Wnd.[34,35,84] Interestingly, overexpression of *ATX1* causes cells to accumulate copper. However, these cells also have lower basal levels of *CUP1* expression, suggesting an increase in the compartmentalization or sequestration of the elevated Cu.[84] A role for *ATX1* in copper metabolism is also suggested by the observation that *ATX1* is unable to suppress SOD defects when cells are depleted of copper using the highly specific Cu chelator bathocuproine disulfonic acid (BCS). Thus *ATX1* has been proposed to be involved in the intracellular transport or sequestration of copper. A role for *ATX1* in oxidative stress protection is further supported by the finding that *ATX1* gene expression is strongly induced by oxygen.[84]

5.5. THE THIOL-SPECIFIC ANTIOXIDANT GENE TSA

Based upon an activity which inhibits oxygen radical-dependent enzyme inactivation, a 25 kDa protein denoted thiol-specific antioxidant (TSA) was purified from yeast cells.[85] TSA homologues are also known to exist in bovine, rat and humans.[86] Interestingly, the Tsa protein shows no significant homology to any of the standard antioxidant enzymes such as catalase, superoxide dismutase, or glutathione peroxidase, nor does Tsap possess activities characteristic of these three enzymes.[87] Tsa is an enzyme that protects cellular components against oxidation systems in which a thiol functions as a reducing equivalent (for example, $DTT/Fe^{3+}/O_2$) but not against an oxidation system without thiol.[85] Tsa activity is measured in vitro by assaying the level of protection against oxidative inactivation of glutamine synthase by $DTT/Fe^{3+}/O_2$.[85] Tsa is an abundant protein that was shown to constitute approximately 0.7% of the total soluble protein of yeast grown aerobically.[88] Exposure of yeast to conditions that generate oxidative stress, including exposure to 100% O_2 or Fe^{3+}, results in an increase in the synthesis of Tsa.[89] Haploid *tsa* disruption mutants are viable under air and the growth rates of *tsa* mutants and wild-type strains are identical under anaerobic conditions. However, under aerobic conditions the *tsa* mutant grows more slowly than wild-type cells and this difference is more pronounced under oxidative stress generated by peroxides or the redox cycling drug paraquat.[88] In addition, the *tsa* mutant is more sensitive than the wild-type strain to the pro-oxidants t-butyl hydroperoxide, H_2O_2, and cumyl hydroperoxide.[88] These results clearly demonstrate that Tsap is a physiologically important antioxidant activity.

Although the biochemical mechanism by which Tsap functions in oxidative stress protection is currently under study, recent investigations have begun to outline its mechanism of action. Sequence analysis of the cloned rat and yeast Tsa coding regions revealed significant homology to a component of the *Salmonella* alkyl hydroperoxide reductase AhpC.[87] The AhpC component was shown to have Tsa activity and it was suggested that the AhpC/TSA proteins represent a large family of related antioxidant enzymes present in organisms from all kingdoms.[86] Therefore a comprehensive understanding of the mechanism by which Tsa functions is likely to emanate from studies of many of the AhpC/TSA family members. Tsap does not contain redox cofactors such as heme, flavin or metal ions but it does contain two cysteine residues at positions 47 and 170.[87,90] These two cysteines are found in a majority of the members of the AhpC/Tsa family and although both cysteines were found to be necessary to maintain the dimeric structure of oxidized Tsa, only Cys-47 was found to be essential for the in vitro antioxidant activity of Tsa. This cysteine is present in all of the AhpC/Tsa family proteins and since Tsa does not contain redox cofactors, cys-47 is likely to be the site of oxidation by oxidized cellular substrates. Additional evidence suggests that Tsa may protect cells from oxidative damage by catalyzing the removal of thiyl radicals.[91] Since Tsa is an abundant cytosolic protein, it is likely to play a major role in the protection of cytosolic enzymes from direct oxygen radical inactivation.

5.6. YEAST CATALASES

The hydrogen peroxide formed by superoxide dismutase, a number of enzymatic activities and by the nonenzymatic reaction of hydroperoxy radicals is scavenged by catalases, ubiquitous heme proteins that catalyze the dismutation of hydrogen peroxide into water and molecular oxygen.[92] There are two known catalase genes in *S. cerevisiae*—*CTA1* and *CTT1*—both of which are hemoproteins and encode the peroxisomal catalase A and the cytosolic catalase T enzymes, respectively.[92] Catalases together with glutathione peroxidase are responsible for removing hydrogen peroxide formed in cells via the reaction shown in Table 6.1. Catalases are known to be tetrameric hemoproteins possessing four identical subunits.[92] The iron cofactor in the heme group is in the Fe^{3+} state and during catalysis the iron ion is reduced to Fe^{2+} and reoxidized. Therefore as with superoxide dismutases a redox active metal which generates ROSs plays a critical role in the ability of catalases to protect the cell from oxidative damage. Catalases are extremely efficient, possessing an unusually high turnover rate and apparent second order rate constant (the latter usually approaches the diffusion rate).[93] To our knowledge it is not known whether this extremely high catalytic efficiency is required for preventing oxidative stress damage in aerobically growing wild-type yeast cells. However, using yeast genetics and molecular biology tools it should be possible to address this question by replacing the wild-type catalase genes with mutant catalases possessing reduced catalytic efficiency. Using isogenic wild-type and *CTT1* deficient yeast strains, catalase T has been shown to protect yeast from heat and oxidative stress.[94] In these experiments *CTT1* deficient yeast strains were more sensitive than wild-type cells to exposure to either 50°C or to hydrogen peroxide exposure following a 37°C pretreatment. It has been suggested by a number of investigations that elevated temperatures exacerbate the generation of reactive oxygen species.[95] Perhaps this occurs in part through temperature dependent increases in metabolic processes such as respiration or enzymatic activities which generate ROIs as reaction byproducts.

5.7. OXIDATIVE DNA DAMAGE: PROTECTION AND REPAIR

As previously mentioned, reactive oxygen species such as superoxide anion and the hydroxyl radical are formed in cells by normal oxidative metabolism, ionizing radiation, near-UV light and certain chemical agents.[96] It has been proposed that the oxidative damage to DNA caused by reactive oxygen species contributes to aging and diseases such as cancer.[9,97] It has been estimated that the number of oxidative lesions to DNA per cell per day is about 100,000 in the rat and by the time the rat is 2 years old it has about 2 million lesions per cell.[9] Reactive oxygen species produce a wide variety of lesions, including strand breaks, base loss or damage and fragmentation of the deoxyribose moiety.[98,99] Oxidizing agents produce a spectrum of DNA base damage, including ring saturation, fragmentation, and contraction.[98] In addition, both pyrimidines and purines are modified to form a variety of products such as hypoxanthine.[98] The bulk of hydrogen peroxide-mediated cell toxicity is due to DNA damage via the generation of hydroxyl radical through the Fenton reaction (see Table 6.1). Indeed, the generation of hydroxyl radical in proximity to DNA is highly likely to lead to damage, considering the estimated half life for hydroxyl radical of 10^{-9} seconds and the estimated diffusion distance in this time frame of between 3 and 10 nm.[100] Although much of what this section will focus on involves proteins that either sense or repair damaged DNA, oxidative agents can also cause gross rearrangement of the DNA. Recently the effect of five oxidative mutagens on the frequency of both intrachromosomal and interchromosomal recombination in S. cerevisiae was investigated.[101] Two of these agents—paraquat and hydrogen peroxide—are commonly used to generate an oxidative stress response in yeast. All five of the agents significantly increased the frequency of intrachromosomal recombination in a dose dependent-manner; only hydrogen peroxide increased the frequency of interchromosomal recombination. It is suggested that hydroxyl radical causes the increased recombination because the radical scavenger DMSO significantly inhibited the induction of both intrachromosomal and interchromosomal recombination by hydrogen peroxide.[101] Several recent reviews regarding oxidative DNA damage describe this area in more detail.[13,96,102,103]

Much of the work which has been done studying DNA damage and repair has used E. coli. Hence many of the genes encoding proteins involved in oxidative DNA damage protection and repair which have been studied are E. coli genes. In contrast to E. coli, much less work has been done using eukaryotic cells as an experimental system to study oxidative DNA damage. S. cerevisiae is an excellent model system for studying DNA repair in eukaryotic cells for several reasons: its low DNA content, powerful genetics and the availability of a large number of well-studied radiation-sen-

sitive (rad) DNA repair mutants.[102] In addition, there is a high degree of functional conservation between S. cerevisiae and human DNA repair genes.[102,104] This section will summarize the mechanisms used to repair and protect yeast cells from oxidative DNA damage.

Several DNA damaging agents such as ionizing radiation, bleomycin or peroxides exert their effects via the formation of reactive oxygen species which generate apurinic/apyrimidinic (AP) sites and single strand breaks with 3'-deoxyribose fragments (Fig. 6.7).[99] The alkylating agent methyl methane sulfonate (MMS), which will be discussed below, indirectly generates AP sites through intermediates in the excision of alkylated bases by DNA glycosylases.[98,99] Class II AP endonucleases are the major cellular enzymes responsible for initiating the repair of AP sites.[13] These enzymes also possessrepair diesterase activity for several 3'-damages in DNA. Since spontaneous base loss occurs at an estimated frequency of between 1,000 and 10,000 per cell per day if left unrepaired, AP sites and 3'-damages are potentially lethal and mutagenic.[105,106]

The major class II AP endonuclease/3'-diesterase of S. cerevisiae is encoded by APN1. Apn1p is a homologue of E. coli endonuclease IV, one of two enzymes in E. coli involved in repairing AP sites and 3'-DNA fragments (Fig. 6.7).[107] The major human class II AP endonuclease, Ape, possesses weak 3'-diesterase activity but is a potent hydrolytic AP endonuclease.[13] Like sod1Δ strains, yeast strains containing mutations in the APN1 gene display an elevated spontaneous mutation rate.[64,107] Genetic studies indicate that this increase in the rate of spontaneous mutations correlates with endogenously generated AP sites.[108] However, these same studies could not rule out the possibility that oxidative strand breaks requiring a 3'-repair diesterase were contributing to the increased spontaneous mutation rate in apn1 cells. Since the human Ape possesses very weak 3'-diesterase activity relative to its hydrolytic AP endonuclease activity, it was used

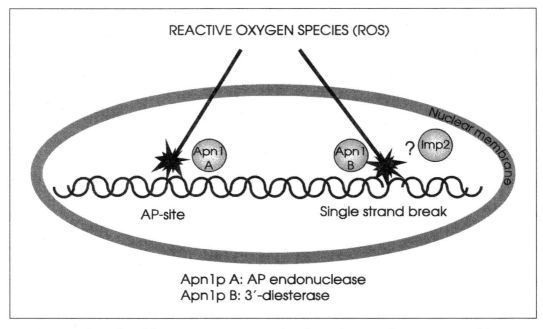

Fig 6.7. ROSs that either diffuse into or are generated in the nucleus can damage DNA and thus repair mechanisms must exist to repair damaged DNA. Two common forms of oxidative DNA damage are apurinic/ apyrimidinic (AP) sites and single-strand breaks with 3'-deoxyribose fragments. These two types of DNA damage are repaired by class II AP endonuclease. Apn1p is the major class II AP endonuclease in S. cerevisiae. Like other class II endonucleases, Apn1p possesses repair activity for both AP sites (depicted as Apn1p A) and DNA strand breaks (depicted as Apn1p B). Imp2p is also involved in the repair of oxidatively damaged DNA, perhaps as a transcription factor important for the expression of oxidative protection genes.

in *apn1* mutant cells to determine the relative contribution of these two enzymatic activities to the *apn1* mutator phenotype.[109] Expression of Ape in *apn1* yeast mutants similar to levels that are normal for Apn1 restored resistance to MMS to near wild-type levels but provided little protection against hydrogen peroxide. Hydrogen peroxide can generate both AP sites and 3' fragmented strands and, as previously mentioned, Ape is relatively specific for AP sites. Consistent with these results expression in yeast of *E. coli* endonuclease IV, which has AP endonuclease and a very potent 3'-diesterase activity, provides resistance to both MMS and hydrogen peroxide.[109] Thus it is proposed that the weak 3'-repair diesterase activity of Ape is responsible for the inability of this enzyme to protect against 3'-damage in *apn1* cells. Furthermore, Ape expression at levels 3- to 4-fold lower than those described above

are sufficient to reduce the spontaneous mutation rate of *apn1* mutants to wild-type levels. Taken together these results suggest that AP sites generated in vivo are responsible for the increased spontaneous mutation rate in *apn1* cells.[109]

Bleomycin is a glycopeptide antitumor antibiotic that is used in the treatment of a variety of human cancers.[110,111] Bleomycin is biologically active in part due to its ability to bind simultaneously to DNA and metal ions.[110,111] Therefore the proximity of bleomycin and metals to the DNA induces DNA damage such as the formation of AP sites and 3' strand fragments through the generation of oxygen derived free radicals.[98,99] Recently the high spontaneous mutation rate of *apn1* yeast strains was used to generate mutants that are hypersensitive to bleomycin.[112] Two DNA repair proteins that are involved in the repair of bleomycin-induced DNA damage are encoded by the *RAD6* and *RAD52* genes;[113] however, the *rad6* and *rad52* mutants are not yet well characterized with respect to this function. These mutants might help to better understand the mechanisms involved in repairing bleomycin-induced oxidative DNA damage in eukaryotic cells.[112]

An *apn1* yeast strain was used to isolate mutants that exhibit hypersensitivity to bleomycin to identify other components involved in protection of yeast cells from oxidative damage to DNA.[112] The most sensitive one of three mutants isolated displaying different sensitivities to bleomycin was chosen for further study. This strain is also hypersensitive to chemical oxidants such as hydrogen peroxide and *tert*-butyl-hydroperoxide, but is no more sensitive to, other DNA-damaging agents such as the alkylating agent MMS, UV light, and 4-NQO.[112] Therefore the mutation appears to be defective specifically in the response to oxidative DNA-damaging agents. In contrast, the *rad6* and *rad52* mutants are hypersensitive to a wide variety of DNA damaging agents including MMS, gamma rays and 4-NQO. Similar phenotypic differences were observed when the drug sensitivities of a larger group of DNA repair-defective mutants in nucleotide excision, postreplication and recombinational repair pathways were compared to the bleomycin sensitive mutant denoted *imp2*.[112] Using this bleomycin hypersensitive mutant the *IMP2* gene was cloned from a yeast single copy library by its ability to restore wild-type levels of bleomycin resistance. Interestingly, the *IMP2* gene was previously independently isolated for its ability to complement a yeast mutant which is unable to grow on galactose, maltose and raffinose in the absence of functional mitochondria.[114] The *IMP2* gene encodes a 312 amino acid protein containing a putative nuclear localization signal sequence and a highly acidic domain.[112] Consistent with *IMP2* playing an important role in protecting cells from oxidative DNA damage, yeast cells carrying a disruption of the *IMP2* gene are hypersensitive to the oxidative agents hydrogen peroxide, *tert*-butyl-

hydroperoxide, paraquat, menadione and diamide. These cells, however, are as sensitive as the wild-type to agents which do not use oxidative mechanisms to damage DNA such as, MMS, UV, or 4-NQO.[112]

Using assays to detect chromosomal DNA fragmentation due to strand breaks resulting from bleomycin treatment the *imp2* mutant was shown to be unable to repair DNA strand breaks even after 6 hours of incubation in fresh growth medium following bleomycin treatment and removal.[112] An assay detecting the repair of strand breaks containing blocked 3' termini showed that the *imp2* deletion mutant when compared with the wild-type cells contained roughly the same level of DNA strand breaks and was as effective in removing blocked 3' termini. However, one striking difference is that although the wild-type cells efficiently repaired most of the strand breaks within 5 hours following bleomycin treatment, the *imp2* deletion mutant was still actively engaged in strand break repair at this time. These data suggest that the hypersensitivity of the *imp2* mutants to oxidants results from a defect in the repair of the oxidatively damaged DNA.[112] However, it is also possible that the product of the *IMP2* gene may be acting in a pathway that removes ROSs and the disruption of *IMP2* leaves the cell unable to inactivate ROSs. Thus the *imp2* mutant even 6 hours after bleomycin treatment may continue to experience ROS-mediated DNA damage.

The presence of a highly acidic amino-terminal domain in the Imp2 protein led to the proposal that this protein might function as a transcriptional activator.[112] Fusions of Imp2p to the DNA binding domain of the *lexA* gene were constructed in a *CYC1* reporter plasmid since Imp2p lacks a known DNA binding motif. In this plasmid the endogenous *CYC1* activating sequences were replaced with the lexA DNA binding sequence. Fusion of the amino-terminal 116 amino acids of Imp2p to lexA powerfully activated transcription from the reporter plasmid. A fusion containing the carboxy-terminal amino acids

117 through 312 and lexA was unable to activate transcription. Fusion of the full length Imp2p to lexA poorly activated transcription.[112] Imp2p lacks a recognizable DNA binding motif. It has, therfore been proposed that the single leucine-rich repeat region in the carboxy-terminus of Imp2 may be responsible for bringing this protein into close proximity to DNA through protein-protein interactions. *IMP2* gene expression is not induced at the mRNA level by oxidative agents. Further studies will undoubtedly unravel the biochemical role the Imp2 protein plays in the protection of yeast cells from oxidative DNA damage.

6. OXIDATIVE STRESS SENSORS

The exposure of cells to oxidative insults demands that cells mount an instantaneous response to protect themselves or to repair damage to macromolecules. As such yeast cells contain relatively high steady state levels of a number of antioxidants such as Cu,Zn superoxide dismutase and glutathione. However, the chronic exposure to oxidative stress additionally requires rapid adaptive responses to either synthesize increased levels of existing antioxidants or to call into play new lines of oxidative stress protection via biosynthetic induction mechanisms. This requires that cells be equipped with regulatory molecules to rapidly sense and respond to oxidative stress. Since a number of yeast genes are transcriptionally induced upon exposure to oxidative stresses of distinct nature, there exist distinct oxidative stress-responsive transcriptional regulatory molecules. In this section we summarize our current understanding of the yeast global adaptive response to oxidative stress, the identity of oxidative stress responsive transcription factors, their known target genes and their mechanisms of sensing oxidative stress to elicit transcriptional responses.

6.1. THE ADAPTIVE RESPONSE TO OXIDATIVE STRESS

Much work on adaptive responses to oxidative stress has focused on the adaptive oxidative stress responses of bacteria,

especially *Escherichia coli* and *Salmonella typhimurium*.[2,6,115] Few studies have been done on the adaptation to oxidative stress in yeast. The adaptive response occurs when cells are pretreated with sublethal concentrations of oxidant (e.g., hydrogen peroxide or menadione) whereupon the induction of a protective response occurs that allows the cells to become resistant to subsequent treatments with higher normally lethal concentrations of the oxidant. Initial studies using yeast showed that cells treated with nonlethal concentrations of hydrogen peroxide or menadione did indeed adapt to become resistant to the lethal effects of higher doses of the oxidant.[116-118] Whether pretreatment with distinct oxidative stress agents confers protection against subsequent challenge with heterologous oxidative stress agents in yeast is currently unclear.

In order to address whether different oxidants may induce a distinct adaptive stress response, two-dimensional gel electrophoretic analysis of proteins prepared from cells induced by either hydrogen peroxide or menadione was performed.[119] Protein extracts were analyzed from cells treated with sublethal concentrations of either oxidant, i.e. concentrations that induce the adaptive response. Analysis of the gels showed that the proteins that were induced fell into three groups: those induced by both menadione and hydrogen peroxide and those induced solely by either hydrogen peroxide or menadione. In a similar study using two-dimensional gel analysis at least 21 proteins were shown to exhibit increased expression following hydrogen peroxide adaptation.[120] This adaptation required de novo protein synthesis since 5 µg/ml cycloheximide inhibited the adaptive response and the increased expression of these 21 proteins. This study also showed the transient nature of the adaptive response. At 45 minutes after pretreatment cells were strongly resistant to 0.8 mM hydrogen peroxide. However, by 4 hours after pretreatment and removal of hydrogen peroxide the adaptive protective response was totally lost. These de-adapted cells, however, could be given a second pre-treatment and again mount an adaptive response.[120] Thus the adaptive response to oxidative stress in yeast is rapid, regenerable and occurs in parallel with the induction of a defined set of proteins, many of which are known to serve protective roles under oxidant exposure.

Recently the role of intracellular glutathione levels on the adaptation to hydrogen peroxide was examined.[121] First, cells depleted of glutathione via the inactivation of the chromosomal gene encoding g-glutamylcysteine synthetase (*GSH1*) were shown to be exquisitely more sensitive to hydrogen peroxide than the isogenic wild type cells. Secondly, wild type but not *gsh1Δ* cells pretreated with H_2O_2 adapted to lethal concentrations of peroxide.[121] The sublethal hydrogen peroxide concentrations used in this study (0.2 mM) were sufficient to significantly raise the enzyme activities (1.5-fold each) of glutathione reductase and glucose-6-phosphate dehydrogenase, responsible for the recycling of glutathione. Consistent with the involvement of glutathione incubation of cells with the three amino acids that constitute glutathione, glycine, cysteine and glutamine, increased the intracellular glutathione level and these cells were more resistant to hydrogen peroxide.[121] Thus intracellular glutathione and increases in the level of glutathione synthesis play an important role in the adaptive response of yeast to hydrogen peroxide.

6.2. YAP1 AND YAP2,
YEAST TRANSCRIPTION FACTORS
OF THE B-ZIP FAMILY

The *S. cerevisiae YAP1*[122-126] and *YAP2*[127,128] gene products are proteins containing a basic-leucine zipper (B-ZIP) domain similar to that found in the c-Jun family of transcriptional activators in mammalian cells, though they are not necessarily functionally homologous. Within these proteins a leucine zipper dimerization domain is adjacent to a basic region (B-ZIP) that interacts directly with DNA. The yeast B-ZIP family consists of the transcription factors Gcn4p, Yap1p, Yap2p and

other related family members recently revealed from the analysis of the completed yeast genome DNA sequence data base. It is clear from numerous examples in higher eukaryotes that B-ZIP proteins may either homo- or hetero-oligomerize, resulting in distinct DNA binding specificities and biological functions.[129] Indeed, the members of the yeast B-ZIP family have been identified on the basis of different functions. Gcn4p—the yeast prototype in this family—is a transcriptional activator of genes encoding amino acid biosynthetic enzymes (see chapter 2).[130] Several groups have independently identified the *YAP1* gene due to its ability in high copy number plasmids to confer pleiotropic drug resistance. The gene was variously named *PDR4* (PDR, pleiotropic drug resistance) based on its ability to confer resistance to sulfometuron methyl and cycloheximide,[123] *PAR1* by conferring resistance to the iron chelators 1,10-phenanthroline and 1-nitroso-2-napthol[125] and *SNQ3* by giving resistance to 4-nitroquinoline-N-oxide, trenimon and MNNG.[124] Additionally, the *YAP1* gene was cloned based on its ability to bind to a human *c-jun* DNA binding site (AP-1 recognition element or ARE) without consideration of function.[122,126] For clarity we will refer to this gene product as Yap1p.

Like *YAP1*, *YAP2* was cloned independently by several labs.[127,128] The metal chelator 1,10-phenanthroline is able to inhibit cell proliferation and causes cells to accumulate in the G1 stage as stationary phase-like cells. *YAP2* was isolated in a genetic screen to identify genes that when present in high copy would overcome yeast cell growth arrest induced by low concentrations of 1,10-phenanthroline.[128] Moreover, overexpression of either *YAP1* or *YAP2* allows cells to tolerate normally toxic levels of iron chelators and zinc, suggesting that both of these genes are involved in iron and zinc homeostasis either directly or indirectly.[127] Overexpression of *YAP2* also gives rise to pleiotropic drug resistance, however, the expression of *YAP2* at either physiological or supraphysiological levels reduce acquired thermotolerance of yeast cells.[127,128]

The role of the *YAP1* and *YAP2* genes in the regulation of the adaptive response to oxidative stress in yeast was recently investigated (Fig. 6.8).[131] Using deletion strains both *YAP1* and *YAP2* were shown to be critical for the induction of the adaptive stress response to hydrogen peroxide. However, both single mutants were able to induce a normal adaptive protective stress response toward superoxide anion-generating drug menadione.[131] Previous work from the same lab showed that stationary phase cultures are considerably more resistant to oxidants than exponentially growing cultures. Correspondingly, stationary phase cultures of either the *yap1* or *yap2* single mutant were more sensitive toward hydrogen peroxide than the wild-type. However, neither single mutation had a significant effect on the resistance of stationary phase cultures towards menadione. Similarly, either single mutation reduces the resistance of respiring cells towards hydrogen peroxide but has little effect on the resistance of these cells towards menadione.[131]

Like many of the other genes involved in the response to oxidative stress described in this report, *YAP1* and *YAP2* affect heavy metal sensitivity/resistance specifically for the metal cadmium.[127,132-134] Disruption mutants such as *yap1* and to a greater extent the *yap1 yap2* double disruption strains are hypersensitive to cadmium. Either *YAP1* or *YAP2* in high copy can suppress the cadmium sensitivity of the disruption strains as well as offer increased levels of resistance over the wild-type strain.[127,134] Therefore the Yap1 and Yap2 proteins play an important role in protecting yeast cells from a variety of conditions compatible with the generation of oxidative stress (Fig. 6.8).

What are the target genes for the action of Yap1p and Yap2p? Although the direct targets for Yap2p have not yet been reported there are likely to be many based on the pleiotropic phenotypes associated with Yap2p. However, a number of studies have established that the Yap1p protein plays a direct role in the activation of antioxidant gene expression in response to oxidative stress (Fig. 6.8).

First *yap1* null mutants display increased sensitivity to hydrogen peroxide and chemicals which generate superoxide anion.[125,135] The specific activities of several enzymes involved in the oxidative stress response including glutathione reductase, total superoxide dismutase activity and glucose 6-phosphate dehydrogenase were shown to be decreased in *yap1* disruption mutants and increased in *YAP1* overexpression strains.[135] Glutathione levels were similarly affected in *yap1* disruption mutants and *YAP1* overexpression strains. No significant differences between the disruption mutants, *YAP1* high-copy transformants and wild-type strains were reported for the activities of several enzymes not involved in oxygen metabolism.[135]

A link between the Cd sensitivity of *yap1* deletion strains and a putative transcriptional target for *YAP1* was forged by the isolation of the yeast *YCF1* (Yeast Cadmium Factor 1) gene. *YCF1* was isolated based upon its ability (when present on a multicopy plasmid) to confer highly elevated Cd resistance to wild type yeast strains.[136] Furthermore, inactivation of the chromosomal *YCF1* gene renders cells hypersensitive to Cd toxicity. Based upon primary structural features, YCF1 protein is a member of a subfamily of ATP binding cassette (ABC) transporter proteins typified by the Cystic Fibrosis Transmembrane Conductance Regulator (hCFTR) and

Fig. 6.8 (opposite). The yeast B-ZIP transcription factors Yap1p and Yap2p regulate the expression of genes which effect the cellular response to oxidative stress. Both transcription factors bind to AP-1 recognition elements (ARE) found in the promoters of target genes. The yeast gene TRX2 encodes thioredoxin while the GSH1 gene product catalyzes the first step in the biosynthesis of glutathione. Both of these gene products are able to remove ROSs. Glutathione reductase (Glr1p) is required to regenerate reduced glutathione. Excess Cd may generate oxidative stress by depleting cellular glutathione which occurs due to the transport of Cd/glutathione complexes into the vacuole by Ycf1p.

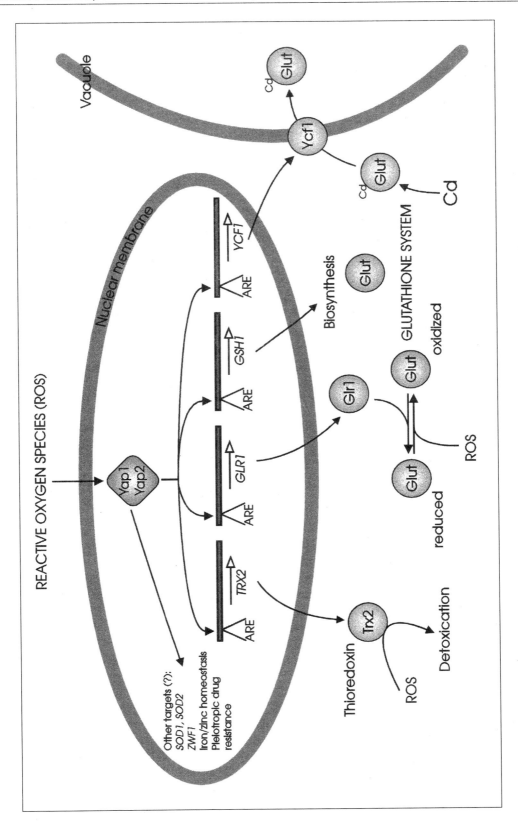

the human Multidrug Resistance-associated Protein (MRP). Subsequently, the YCF1 protein was demonstrated to be a vacuolar membrane-associated Mg-ATP-dependent transporter of glutathione-conjugated moieties, suggesting the possibility that YCF1 mediates Cd resistance by transporting Cd-glutathione conjugates into the vacuole (Fig. 6.8).[137] Recently the *YCF1* gene has been shown to be an important physiological target gene for Yap1p-mediated transcriptional activation (Fig. 6.8).[132] The epistatic relationship of *YAP1* to *YCF1* was displayed by the failure of overexpression of *YAP1* in a *ycf1* disruption background to produce elevated cadmium resistance. DNase I footprinting of a segment of the *YCF1* promoter showed that recombinant Yap1p binds a novel ARE which overlaps a dyad symmetry element and shows only limited homology to the SV40 AP-1 recognition element. A comparison of the AREs found in yeast promoters suggests that there may be considerable variation in the consensus ARE sequence.[132] Although as yet unproven, the ability of Cd to generate an oxidative stress in yeast cells may be due to the depletion of glutathione, via Cd coordination at the expense of its use as an antioxidant.

Two other genes in addition to *YCF1* have been shown to represent in vivo targets of *YAP1* transcriptional control:[138,139] *GSH1* encoding g-glutamylcysteine synthetase gene and *TRX2* encoding one of the two yeast thioredoxins (Fig. 6.8). Both of these gene products play important roles in oxidative stress protection by providing nonenzymatic cellular defenses against reactive oxygen species (Fig. 6.8). The low molecular weight dithiols glutathione and thioredoxin are two of the most important and abundant cellular antioxidants which maintain a strong reducing environment in the cell. In yeast the *GSH1* gene product catalyzes the first and rate-limiting step in the biosynthesis of glutathione.[140,141] The finding that *GSH1* is transcriptionally controlled by Yap1p provides a molecular basis for two observations: increases in levels of g-glutamylcysteine synthetase activity increased glutathione production and elevation of *YAP1* gene dosage leads to increased intracellular glutathione levels.[138] The promoter of the *GSH1* gene contains an AP-1-like response element which matches the SV40 ARE in 11 of 12 positions. Yap1p protein binds to the *GSH1* ARE in vitro and in addition, in vivo expression analysis showed that *GSH1* expression via the wild type element but not a mutated form is responsive to *YAP1* gene dosage.[138] Furthermore, strains carrying a mutant *GSH1* gene lacking the *YAP1* ARE are more sensitive to cadmium than wild-type cells.

Using a genetic screen the *TRX2* gene was shown to be a direct target of the Yap1p transcription factor.[139] In this same study cells carrying a disruption of *yap1* were shown in agreement with previous results to be hypersensitive to hydroperoxides and thioloxidants. These treatments also resulted in the stimulation of both Yap1p DNA binding activity and Yap1p-dependent transcription that required functional Yap1p DNA binding sites. This stimulation was not due to an increase in the synthesis of Yap1p but rather an apparent modification of preexisting protein.[139] The increased expression of the *TRX2* gene conferred hyper-resistance to H_2O_2 and the alkyl hydroperoxide, *t*BOOH. In addition, deletion of *TRX2* resulted in cells that are hypersensitive to these same two compounds, establishing a clear functional link between oxidative stress resistance via *TRX2* and Yap1p activation of *TRX2* gene transcription. In this same study the *GSH1* gene was shown to be induced by both menadione and hydrogen peroxide, however, the *TRX2* gene was shown to not be inducible by menadione. Deletion of *YAP1* abolished the peroxide-mediated induction of *GSH1* but the menadione inducibility of *GSH1* was not dependent on *YAP1*.[139] Very recently the yeast glutathione reductase gene—*GLR1*—was also demonstrated to be directly regulated by Yap1p (Fig. 6.8).[142]

6.3. CATALASE T ACTIVATION THROUGH AN STRE AND MAC1P

Catalase A (*CTA1*) gene expression is induced by fatty acids and growth on nonfermentable carbon sources and strongly repressed by glucose.[92] Expression of the *CTT1* gene is negatively regulated by cAMP and has been shown to be induced by oxygen, heme, nutrient starvation, copper, heat shock, osmostress and oxidative stress (Fig. 6.9).[92] Expression of both genes is induced by oxygen via positive regulation by heme (Fig. 6.9).[92] In yeast heme has been shown to have a major role in signaling the presence of oxygen to regulate the expression of catalase genes and additional genes such as those encoding cytochrome c oxidase that are regulated by oxygen or required during aerobic growth.[143] The regulation of *CTT1* and *CTA1* by heme occurs through the transcription activator Hap1p and binding sites

for Hap1p are present in the promoter regions of both catalase genes (Fig. 6.9).[92] Hap1p is a heme-dependent transcriptional activator protein which harbors a repeated 10 amino acid heme binding site which is conserved in several enzymes. The binding of heme somehow unmasks the activity of the Hap1p DNA binding domain and the trans-activation domain.[144]

In addition to the control via heme, exposure of yeast cells to copper modestly induces the mRNA level of *CTT1* but not that of *CTA1*.[145] Furthermore, this Cu-dependent induction of *CTT1* is independent of the Ace1p Cu metalloregulatory factor. This suggests the presence of either a distinct Cu responsive transcription factor for *CTT1* or that *CTT1* may be induced due to an oxidative stress which occurs as a consequence of increases in Cu levels (Fig. 6.9).

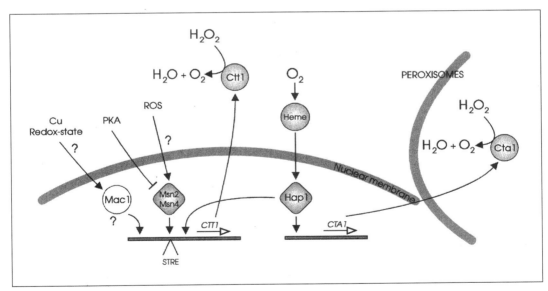

Fig. 6.9. Model for the transcriptional regulation of the yeast catalase genes. The regulation of both catalase genes is quite complex, each involving multiple metabolic signals as well as transcription factor binding sites. Hap1p binding sites are present in the promoter of both the peroxisomal catalase Cta1p and the cytosolic catalase Ctt1p. Ctt1p is also regulated through a stress response element (STRE). Protein kinase A (PKA) is shown to negatively regulate Ctt1p. Msn2p and Msn4p have been shown to bind STREs and possess transactivation domains, however, their biochemical mode of action remains to be elucidated. Although Mac1p is required for the hydrogen peroxide-induced expression of Ctt1p, the biochemical basis for this is not clearly understood at present.

As described above, *CTT1* mRNA levels are induced by a variety of stresses including heat shock. The *CTT1* promoter has an upstream activating sequence element that is distinct from the heat shock element (HSE) and activates transcription of *CTT1* independent of HSF.[146] A 13 bp element in the upstream region of the promoter has been shown to be sufficient for the regulation of expression of a *CTT1*-reporter gene fusion in response to changes in cAMP levels, nitrogen starvation and heat shock. This DNA element contains a sequence, designated stress response element (STRE, consensus 5'-CCCCT-3'[146]) that is under both negative and positive control. It is negatively regulated by protein kinase A and activated by nitrogen starvation, increases in osmolarity, heat shock and oxidative stress.[146,147] Although the STRE of *CTT1* is activated by a variety of stresses, only the response to osmotic stress is dependent on the *HOG1* MAP kinase pathway (see chapter 8 for details).[147]

Recently two structurally related proteins encoded by the *MSN2* and *MSN4* genes have been shown to bind specifically to oligonucleotides containing STREs from the *HSP12*, *CTT1*, and *DDR2* genes, all of which are regulated by multiple stresses through an STRE (Fig. 6.9).[148] In addition, *msn2 msn4* double mutants are unable to activate transcription of these genes in response to a variety of stresses, including carbon source starvation, heat, high salt and 7% ethanol. The *MSN2* gene was previously isolated in a genetic screen for multicopy plasmids which restore the growth of a temperature sensitive *SNF1* protein kinase mutant.[149] *MSN4* was then isolated by sequence homology to *MSN2*.[149] Both proteins contain Cys_2His_2 Zinc finger motifs and fusions of either LexA-Msn2p or LexA-Msn4p proteins were shown to be potent transcriptional activators.[149] The link between the *MSN2* and *MSN4* genes and oxidative stress protection is demonstrated in the approximately 50-fold reduction in the viable cell numbers of *msn2 msn4* mutants compared to wild-type cells following exposure to hy-

drogen peroxide.[148] Why should *CTT1* be activated by heat and other stresses? One possibility is that these other stresses increase either the generation of or sensitivity to H_2O_2. Several studies have reported increased ROS generation when cells are grown at higher than normal temperatures.[95] The enhanced protection against hydrogen peroxide exposure provided by *CTT1* requires pretreatment at 37°C.[94] This observation exemplifies the synergism seen in the induction of *CTT1* by heme, nutrient starvation and heat.[92]

Expression of the *CTT1* gene is also intertwined with a protein that plays a critical role in the expression of genes intimately involved in Cu and Fe transport. This evidence stems from functional studies of the Mac1 protein, a nuclear regulatory protein which is important for the expression of the Fre1p and Ctr1p activities.[150,151] The amino-terminal region of Mac1p is highly similar to the amino-terminal regions of the Ace1p and Amt1 Cu metalloregulatory factors of *S. cerevisiae* and *Candida glabrata*, respectively.[150] This domain of Ace1p has been shown to be required for both DNA and copper binding and in studies of Amt1 it is proposed that this domain harbors a minor groove DNA binding motif. Therefore Mac1p is related to a family of copper-dependent transcription factors.[77] Although Mac1p has only 3 of the 11 cysteine residues required for Ace1p function in its amino terminus, there are 16 additional cysteine residues in the carboxyl-terminal region of Mac1p which may be involved in metal coordination, redox regulation or play another role. Therefore the *MAC1* gene is proposed to encode a transcription factor (Fig. 6.9) based on several results: the sequence similarity to Ace1p and Amt1p; immunofluorescence data which show that Mac1p is a nuclear protein; and two target genes in addition to *CTT1* (*FRE1* and *CTR1*) that have been identified that depend upon a functional *MAC1* gene for normal expression.[150]

Yeast cells suffering from a mutation in the *MAC1* gene exhibit several phenotypes: they grow slowly on YPD plates;

they are respiratory deficient as observed by the inability to grow on nonfermentable carbon sources and the severe reduction in oxygen uptake; and they are hypersensitive to a variety of stresses including heat, exposure to cadmium, zinc, lead and hydrogen peroxide.[150] It is intriguing that addition of copper can rescue the slow growth, respiratory deficiency, and the stress phenotypes of *mac1* cells and is consistent with the observation that Mac1p is required for high affinity Cu transport and *CTR1* expression.[151] Although *mac1* mutants are not hypersensitive to copper, a dominant *MAC1* allele designated *MAC^{up1}* confers Cu sensitivity. This mutation leads to elevated levels of basal and induced expression of *FRE1* mRNA and Fre1p reductase activity. However, *MAC^{up1}* mutant does not lead to an elevated level of either basal or induced expression of *CTT1*, suggesting that *MAC1* is required but not sufficient for the hydrogen peroxide-induced transcription of *CTT1*.[150] Thus *MAC1* has been proposed to be a sensor for the intracellular Cu ion concentration or the cellular redox state (Fig. 6.9).[150,151] It will be important to further study *MAC1* to identify additional target genes and determine the role of *MAC1* in regulating metal metabolism and the oxidative stress response.

6.4. YEAST TWO COMPONENT-LIKE SYSTEMS

Using a genetic screen for hydrogen peroxide sensitivity 34 recessive mutants representing 16 complementation groups (*pos1* through *pos16*, for peroxide sensitivity) were recently isolated that have growth phenotypes similar to *yap1Δ* mutants.[152] Mutants in the *pos9* complementation group were similar to *yap1* mutants in that both grow like wild-type under standard laboratory conditions. Yet they are highly sensitive to hydrogen peroxide and show no increased sensitivity to osmotic stress. However, unlike *yap1* mutants which have decreased levels of glutathione and enzymes involved in oxidative stress protection (see above) *pos9* mutants have wild-type levels

of glutathione and several enzymes involved in oxidative stress protection. Mutants in other complementation groups showed significantly elevated levels of enzymes involved in oxidative stress. Genetic crosses of all *pos* mutants with *yap1* mutants except *pos9* led to increased sensitivity to hydrogen peroxide.[152] However, *pos9 yap1* double mutants showed the same sensitivity as either single mutant to hydrogen peroxide. This observation led the authors to suggest that the function of Pos9p in oxidative stress protection is closely linked to that of Yap1p.[152]

The isolation and sequencing of the *POS9* gene revealed that Pos9p is identical to the putative response regulator type protein Skn7p.[153,154] The response regulator protein and a membrane-bound sensing kinase make up the prokaryotic "two-component signaling systems." Recently genes encoding proteins homologous to two component regulatory system proteins have been found in yeast, denoted *SLN1* and *SSK1*.[155,156] Both of these genes function in the HOG-pathway which is involved in the cellular response to osmotic stress (chapter 4). Currently, however, very little is known about the role and mechanism of action of the Pos9 protein in oxidative stress responses in yeast cells. It will be very interesting to determine whether some of the oxidative stress sensing and response mechanisms in yeast are also of the two component type.

6.5. HEAT SHOCK TRANSCRIPTION FACTOR

As described earlier in this review, expression of either the yeast *CUP1*-encoded metallothionein or monkey MT isoforms can suppress many of the oxidative stress phenotypes of cells bearing an *sod1Δ* allele.[45] Accordingly, both the *CUP1* gene in yeast and the mouse MT genes are known to be transcriptionally induced by oxidative stress.[45,78,82,83] In yeast cells the *CUP1* mRNA levels were strongly induced by either growth on nonfermentable carbon sources or by the administration of menadione (a pro-oxidant which generates superoxide anion through redox cycling). This induction of *CUP1* mRNA has been recently shown to occur independently of the Ace1p Cu metalloregulatory factor but to require the yeast Heat Shock Transcription Factor (Hsf1p) and a functional *CUP1* promoter heat shock element (Fig. 6.10).[82,83] Although this was initially somewhat surprising, a number of previous studies demonstrated that the mammalian HSF proteins are activated by conditions or chemicals that generate oxidative stress such as electron transport uncouplers and redox cycling drugs.[157,158] Several points are noteworthy for the activation of *CUP1* by Hsf1p in response to oxidative stress. In response to heat Hsf1p-mediated *CUP1*, transcription and Hsf1p phosphorylation are transient while both *CUP1* gene expression and Hsf1p phosphorylation are sustained in response to menadione-induced oxidative stress.[83] Furthermore, the pattern of tryptic phosphopeptides resolved from Hsf1p derived from cells subjected to heat shock or oxidative stress are distinct. Moreover, the activation of *CUP1* by heat and oxidative stress requires a carboxyl-terminal Hsf1p trans-activation domain whereas heat shock activation of classical

heat shock responsive genes such as those in the *HSP70* family is independent of this domain. Interestingly, the *HSP70* gene family is poorly induced by menadione administration. Taken together these results suggest that Hsf1p has a critical role in eukaryotic oxidative stress protection and that heat and oxidative stress signals are differentially communicated to Hsf1p to activate gene transcription.[83] A biological link between oxidative stress-dependent HSF-mediated *CUP1* activation was clearly established by the observation that a single functional copy of the *CUP1* gene protected cells from menadione toxicity in an HSE-dependent manner.[83] Therefore Hsf1p may be a key component of the oxidative stress responsive machinery, however, the precise mechanisms whereby this factor (or a putative protein kinase) senses the oxidative stress and activates Hsf1p remain to be investigated. Furthermore, the identification of additional target genes which respond to oxidative stress via HSF is an interesting avenue for experimentation.

One potential mechanism that yeast cells may use to sense oxidative stress is through the use of metal sulfur clusters. A number of enzymes and regulatory mol-

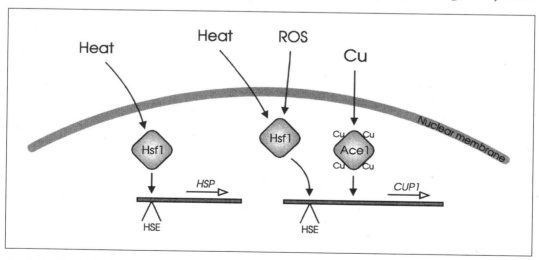

Fig. 6.10. Model for the transcriptional regulation of yeast stress genes by heat shock transcription factor (Hsf1p). Hsf1p is constitutively bound to some heat shock elements (HSE) and heat inducibly bound to other HSEs in the promoters of stress genes. Exposure of yeast to 36°C or higher temperatures results in the activation of classic heat shock protein genes (HSP). Expression of the stress gene CUP1, however, is activated by Hsf1p in response to heat and ROSs. CUP1 expression is activated independently by Ace1 in response to elevated copper levels.

ecules utilize a switch based upon the formation or destruction of Iron-sulfur clusters.[159,160] It has been proposed that metal-sulfur clusters may be a widely used means of sensing changes in the levels of ROSs since cellular oxidants such as the ROSs described in this review are capable of destroying such clusters. Perhaps transcription factors, protein kinases or other regulatory molecules in yeast cells have adapted to use this type of regulatory mechanism for sensing some forms of oxidative stress.

7. SUMMARY AND PERSPECTIVES

What can studies of oxidative stress in yeast tell us about the etiology of human disease states associated with oxidative stress? The power and versatility of yeast genetics coupled with the vast body of knowledge concerning the physiological impact of antioxidant defenses such as superoxide dismutase has made a significant impact on our understanding of human diseases related to oxidative stress. Amyotrophic lateral sclerosis (ALS, Lou Gehrig's Disease) is a motor neuron degenerative disease of which approximately 10-15% of the cases are familial (familial amyotrophic lateral sclerosis, FALS) and about 20-25% of FALS cases are correlated with dominantly inherited mutations in SOD1, the human gene encoding Cu,Zn SOD.[161] Indeed, the expression in yeast $sod1\Delta$ strains of FALS-associated SOD1 proteins from humans clearly established that these enzymes displayed no reduction in superoxide anion disproportionation activity.[162] A number of published and ongoing studies aimed at gaining a mechanistic understanding of the biochemical basis for FALS strongly suggest that mutations in Cu,Zn SOD increase the rate of a side reaction which is responsible for initiating the pathologic changes in FALS.[162-165] This is more consistent with the known dominant nature of FALS-associated mutations in Cu,Zn SOD. In addition to superoxide dismutase activity, Cu,Zn SOD also has two potentially toxic non-SOD activities,

a nonspecific peroxidase activity and catalysis of the nitration of tyrosyl residues by peroxynitrite.[56] Recently evidence was reported to suggest that these two side reactions are increased in the mutant FALS-associated Cu,Zn, SOD.[161] Two mutated Cu,Zn SOD enzymes associated with FALS expressed in and purified from $sod1\Delta$ yeast cells were shown to catalyze the oxidation of a model substrate at an increased rate compared to the wild-type enzyme. Previous studies demonstrated that the reaction in which Cu,Zn SOD oxidizes substrates by H_2O_2 occurs at the Cu ion bound in the active site of the enzymes subunit.[56] Addition of copper chelators to the in vitro reaction significantly reduced the formation of the oxidized substrate by both mutant Cu,Zn SODs.[161] In a neural cell culture model for FALS overexpression of wild-type Cu,Zn SOD inhibited apoptosis whereas similar levels of expression of the FALS-associated mutant enzymes induced apoptosis. Interestingly, in this model cell culture system addition of copper chelators also inhibited the induction of apoptosis in the cell lines expressing the FALS forms of Cu,Zn SOD.[161] These observations suggest that the modulation of bioavailable Cu levels in FALS patients could play a therapeutic role in disease treatment.

It is clear from information provided in this review that even a single-celled eukaryotic organism such as yeast has dedicated a vast array of structural molecules, enzymatic activities and regulatory factors to the sensing, protection and repair of damage via oxidative stress. Furthermore, the interdigitation of a number of pathways, including metal homeostatic mechanisms, normal physiological processes and other cellular stresses all impinge on the ability of yeast cells to cope with oxidative stress. The value of yeast as an experimental system in studies of oxidative stress lies in our ability to understand what processes play a key role in oxidative stress protection and the precise nature of the biochemical mechanisms used for these processes.

ACKNOWLEDGMENTS

The work in our laboratory on oxidative stress in yeast is supported by a predoctoral fellowship from the United States Environmental Protection Agency to N.S., a grant from the National Institutes of Health, RO1 GM41840 and a Taisho Excellence in Research Program Award from Taisho Pharmaceuticals Co., Ltd. to D.J.T. D.J.T. is a Burroughs Wellcome Toxicology Scholar.

REFERENCES

1. Webster. Webster's Contemporary American Dictionary of the English Language. Woodbury: Bobley Publishing Corp., 1979.
2. Demple B, Amabile-Cuevas CF. Redox redux: The control of oxidative stress responses. Cell 1991; 67:837-839.
3. Baeuerle PA, Pahl HL. Oxygen and the control of gene expression. BioEssays 1994; 16:497-502.
4. Jamieson DJ. Oxidative stress response of *Saccharomyces cerevisiae*. Redox Report 1995; 1:89-95.
5. Moradas-Ferreira P, Costa V, Piper P, Mager W. The molecular defences against reactive oxygen species in yeast. Mol Microbiol 1996; 19:651-658.
6. Storz G, Tartaglia LA, Farr SB, Ames BN. Bacterial defences against oxidative stress. Trends Genet 1990; 6:363-368.
7. Stadtman ER. Protein oxidation and aging. Science 1992; 257:1220-1224.
8. Halliwell B. Free radicals and antioxidants a personal view. Nutr Rev 1994; 52:253-265.
9. Ames BN, Shigenaga MK, Hagen TM. Oxidants, antioxidants, and the degenerate diseases of aging. Proc Natl Acad Sci USA 1993; 90:8013-8017.
10. Halliwell B, Gutteridge JMC. Oxygen toxicity, oxygen radicals, transition metals and disease. Biochem J 1984; 219:1-14.
11. Halliwell B, Gutteridge JMC. Free radicals in Biology and Medicine. Clarendon Press, Oxford, 1989.
12. Gralla EB, Kosman DJ. Molecular genetics of superoxide dismutases in yeast and related fungi. Adv Genet 1992; 30:251-319.
13. Demple B, Harrison L. Repair of oxidative damage to DNA: enzymology and biology. Ann Rev Biochem 1994; 63:915-948.
14. Davies KJA, Wiese AG, Sevanian A, Kim EH. Repair systems in oxidative stress. Molecular Biology of Aging. Alan R. Liss, Inc., 1990:123-141
15. Packer L, Glazer AN. Oxygen radicals in biological systems. Methods in Enzymology. Vol. 186, Part B. San Diego, CA: Academic Press, Inc., 1990.
16. Coyle JT, Puttfarcken P. Oxidative stress, glutamate, and neurodegenerative disorders. Science 1993; 262:689-695.
17. Ljungman M, Hanawalt PC. Efficient protection against oxidative damage in chromatin. Mol Carcinog 1992; 5:264-269.
18. Ausubel FM, Brent R, Kingston RE, Moore DD, Seidman JG, Smith JA, Struhl K, eds. Current Protocols in Molecular Biology. New York: John Wiley & Sons, 1987.
19. Guthrie C, Fink GR, eds. Guide to Yeast Genetics and Molecular Biology. Methods in Enzymology. Vol. 194. San Diego, CA: Academic Press, Inc., 1991.
20. Bassett DEJ, Boguski MS, Hieter P. Yeast genes and human disease. Nature 1996; 379:589-590.
21. Askwith CC, de Silva D, Kaplan J. Molecular biology of iron acquisition in *Saccharomyces cerevisiae*. Mol Microbiol 1996; 20:27-34.
22. Dancis AK, RD, Hinnebusch AG, Barriocanal JG. Genetic evidence that ferric reductase is required for iron uptake in *Sacharromyces cerevisiae*. Mol Cell Biol 1990; 10:2294-2301.
23. Georgatsou E, Alexandraki D. Two distinctly regulated genes are required for ferric reduction, the first step of iron uptake in *Saccharomyces cerevisiae*. Mol Cell Biol 1994; 14:3065-3073.
24. Askwith C, Eide D, Van Ho A, Bernard PS, Li L, Davis-Kaplan S, Sipe DM, Kaplan J. The *FET3* gene of *S. cerevisiae* encodes a multicopper oxidase required for ferrous iron uptake. Cell 1994; 76:403-410.
25. Harris ED. Iron-copper interactions: some new revelations. Nutr Rev 1994; 52:311-5.

26. Harris ED. The iron-copper connection: the link to ceruloplasmin grows stronger. Nutr Rev 1995; 53:170-173.

27. Dancis A, Yuan DS, Haile D, Askwith C, Eide D, Moehle C, Kaplan J, Klausner RD. Molecular characterization of a copper transport protein in *S. cerevisiae*: and unexpected role for copper in iron transport. Cell 1994; 76:393-402.

28. Stearman R, Yuan DS, Yuko Y-I, Klausner RD, Dancis A. A permease-oxidase complex involved in high-affinity iron uptake in yeast. Science 1996; 271:1552-1557.

29. Dix DR, Bridgham JT, Broderius MA, Byersdorfer CA, Eide DJ. The *FET4* gene encodes the low affinity Fe(II) transport protein of *Saccharomyces cerevisiae*. J Biol Chem 1994; 269:26092-26099.

30. Dancis A, Haile D, Yuan DS, Klausner RD. The *Saccharomyces cerevisiae* copper transport protein (Ctr1p). Biochemical characterization, regulation by copper, and physiological role in copper uptake. J Biol Chem 1994; 269:25660-25667.

31. Kampfenkel K, Kushnir S, Babiychuk E, Inze D, Van Montagu M. Molecular characterization of a putative *Arabidopsis thaliana* copper transporter and its yeast homologue. J Biol Chem 1995; 270:28479-28486.

32. Knight SAB, Labbe S, Kwon LF, Kosman DJ, Thiele DJ. A widespread transposable element masks expression of a yeast copper transport gene. Genes & Dev 1996; (in press).

33. Yamaguchi-Iwai Y, Stearman R, Dancis A, Klausner RD. Iron-regulated DNA binding by the AFT1 protein controls the iron regulon in yeast. EMBO J 1996; 15: 3377-3384.

34. Bull PC, Thomas GR, Rommens JM, Forbes JR, Cox DW. The Wilson disease gene is a putative copper transporting P-type ATPase similar to the Menkes gene. Nature Genetics 1993; 5:327-337.

35. Bull PC, Cox DW. Wilson disease and Menkes disease: new handles on heavy-metal transport. Trends Genet 1994; 10:246-251.

36. Yamaguchi Y, Heiny ME, Shimizu N, Aoki T, Gitlin JD. Expression of the Wilson disease is deficient in the Long-Evans Cinnamon rat. Biochem J 1994; 301:1-4.

37. Sawaki M, Enomoto K, Hattori A, Tsuzuki N, Mori M. Role of copper accumulation and metallothionein induction in spontaneous liver cancer development in LEC rats. Carcinogenesis 1994; 15:1833-1837.

38. Masuda R, Yoshida MC, Sasaki M, Dempo K, Mori M. High susceptibility to hepatocellular carcinoma development in LEC rats with hereditary hepatitis. Jpn J Cancer Res 1988; 79:825-835.

39. Togashi Y, Li Y, Kang J-H, Takeichi N, Fujioka Y, Nagashima K, Kobayashi H. D-penicillamine prevents the development of hepatitis in Long-Evans Cinnamon rats with abnormal copper metabolism. Hepatology 1992; 15:82-87.

40. Kang J-H, Togashi Y, Kasai H, Hosokawa M, Takeichi N. Prevention of spontaneous hepatocellular carcinoma in Long-Evans Cinnamon rats with hereditary hepatitis by the administration of D-penicillamine. Hepatology 1993; 18:614-620.

41. Yuan D, Stearman A, Dancis A, Dunn T, Beeler T, Klausner RD. The *S. cerevisiae* Menkes/Wilson disease homologue provides copper to an extracellular ceruloplasmin-like oxidase required for iron uptake. Proc Natl Acad Sci USA 1995; 92:2632-2636.

42. Fu D, Beeler TJ, Dunn TM. Sequence, mapping and disruption of *CCC2*, a gene that cross-complements the Ca^{2+}-sensitive phenotype of *csg1* mutants and encodes a P-type ATPase belonging to the Cu^{2+}-ATPase subfamily. Yeast 1995; 11:283-292.

43. Kagi JHR, Schaffer A. Biochemistry of metallothionein. Biochemistry 1988; 27:8509-8515.

44. Kagi JHR. Overview of Metallothionein. In: Riordan JF, Vallee BL, eds. Methods in Enzymology. San Diego, CA: Academic Press, Inc, 1991:613-626

45. Tamai KT, Gralla EB, Ellerby LM, Valentine JS, Thiele DJ. Yeast and mammalian metallothioneins functionally substitute for yeast copper-zinc superoxide dismutase. Proc Natl Acad Sci USA 1993; 90:8013-8017.

46. Culotta VC, Howard WR, Liu XF. *CRS5* encodes a metallothionein-like proetin in *Saccharomyces cerevisiae*. J Biol Chem 1995; 269:25295-25302.

47. Hamer DH Thiele DJ, Lemontt JE. Function and autoregulation of yeast copperthionein. Science 1985; 228:685-690.

48. Thiele DJ, Hamer DH. Tandemly duplicated upstream control sequences mediate copper-induced transcription of the *Saccharomyces cerevisiae* copper-metallothionein gene. Mol Cell Biol 1986; 6:1158-1163.

49. Welch J, Fogel S, Buchman C, Karin M. The *CUP2* gene product regulates the expression of the *CUP1* gene, coding for yeast metallothionein. EMBO J 1989; 8:255-260.

50. Thiele DJ. *ACE1* regulates expression of the *Saccharomyces cerevisiae* metallothionein gene. Mol Cell Biol 1988; 8:2745-2752.

51. Zhu Z, Szczypka MS, Thiele DJ. Transcriptional regulation and function of yeast metallothionein genes. In: Sarkar B, ed. Genetic Response to Metals. New York, NY: Marcel Dekker, Inc., 1995.

52. Thorvaldsen JL, Sewell AK, Tanner AM, Peltier JM, Pickering GN, George GN, Winge DR. Mixed Cu^+ and Zn^{2+} coordination in the DNA binding domain of the AMT1 transcription factor from *Candida glabrata*. Biochemistry 1994; 33:9566-9577.

53. Koch KA, Thiele DJ. Autoactivation by a *Candida glabrata* copper metalloregulatory transcription factor requires critical minor groove interactions. Mol Cell Biol 1996; 16:724-734.

54. Dobi A, Dameron S, Hu S, Hamer D, Winge DR. Distinct regions of Cu(I)-ACE1 contact two spatially resolved DNA major groove sites. J Biol Chem 1995; 270:10171-10178.

55. Gralla EB, Thiele DJ, Silar P, Valentine JS. ACE1, a copper-dependent transcription factor, activates expression of the yeast copper, zinc superoxide dismutase gene. Proc Natl Acad Sci USA 1991; 88:8558-8562.

56. Fridovich I. Superoxide radical and superoxide dismutases. Ann Rev Biochem 1995; 64:97-112.

57. Bilinski T, Krawiek Z, Liczmanski A, Litwinska J. Is hydroxyl radical generated by the Fenton reaction in vivo? Biochem Biophys Res Commun 1985; 130:533-539.

58. Galiazzo F, Labbe-Rois R. Regulation of Cu,Zn- and Mn-superoxide dismutase transcription in *Saccharomyces cerevisiae*. FEBS Lett 1993; 315:197-200.

59. Westerbeek-Marres CAMW, Moore MM, Autor AP. Regulation of manganese super-oxide dismutase in *Saccharomyces cerevisiae*. Eur J Biochem 1988; 174:611-620.

60. Benov LT, Fridovich I. *Escherichia coli* expresses a copper—and zinc-containing superoxide dismutase. J Biol Chem 1994; 269:25310-25314.

61. Marres CAM, Van Loon APGM, Oudshoorn P, Van Steeg H, Grivell LA, Slater EC. Nucleotide sequence analysis of the nuclear gene coding for manganese superoxide dismutase of yeast mitochondria, a gene previously assumed to code for the Rieske iron-sulphur protein. Eur J Biochem 1985; 147:153-161.

62. Church SL, Grant JW, Ridnour LA, Oberley LW, Swanson PE, Meltzer PS, Trent JM. Increased manganese superoxide dismutase expression suppresses the malignant phenotype of human melanoma cells. Proc Natl Acad Sci USA 1993; 90:3113-3117.

63. Balasundaram D, Tabor CW, Tabor H. Oxygen toxicity in a polyamine-depleted *spe2Δ* mutant of *Saccharomyces cerevisiae*. Proc Natl Acad Sci USA 1993; 90:4693-4697.

64. Gralla EB, Valentine JS. Null mutants of *Saccharomyces cerevisiae* Cu,Zn-superoxide dismutase: characterization and spontaneous mutation rates. J Bacteriol 1991; 173:5918-5920.

65. Liu XF, Elashvili I, Gralla EB, Valentine JS, Lapinskas P, Culotta VC. Yeast lacking superoxide dismutase. Isolation of genetic suppressors. J Biol Chem 1992; 267:18298-18302.

66. Van Loon A, Pesold-Hurt B, Schatz G. A yeast mutant lacking mitochondrial manganese-superoxide dismutase is hypersensitive to oxygen. Proc Natl Acad Sci USA 1986; 83:3820-3824.

67. Guidot DM, McCord JM, Wright RM, Repine JE. Absence of electron transport (Rho°) restores growth of a manganese-superoxide dismutase-deficient *Saccharomyces cerevisiae* in hyperoxia. J Biol Chem 1993; 268:26699-26703.

68. Balzan R, Bannister WH, Hunter GJ, Bannister JV. *Escherichia coli* iron superoxide dismutase targeted to the mitochondria of yeast cells protects the cells against oxidative stress. Proc Natl Acad Sci USA 1995; 92:4219-4123.

69. Zhu D, Scandalios JG. Expression of the maize MnSod (*Sod3*) gene in MnSOD-deficient yeast

rescues the mutant yeast under oxidative stress. Genetics 1992; 131: 803-809.

70. Liu XF, Culotta VC. The requirement for yeast superoxide dismutase is bypassed through mutations in *BSD2*, a novel metal homeostasis gene. Mol Cell Biol 1994; 14:7037-7045.

71. Lapinskas PJ, Cunningham KW, Liu XF, Fink GR, Culotta VC. Mutations in *PMR1* suppress oxidative damage in yeast cells lacking superoxide dismutase. Mol Cell Biol 1995; 15:1382-1388.

72. Rudolph HK, Antebi A, Fink GR, Buckley CM, Dorman TE, LeVitre J, Davidow LS, Mao JI, Moir DT. The yeast secretory pathway is perturbed by mutations in *PMR1*, a member of a Ca²⁺-ATPase family. Cell 1989; 58:133-145.

73. Antebi BN, Fink GR. The yeast Ca²⁺-ATPase homologue, *PMR1*, is required for normal Golgi function and localizes in a novel Golgi-like distribution. Mol Biol Cell 1992; 3:633-654.

74. Cunningham KW, Fink GR. Calcineurin-dependent growth control in *Saccharomyces cerevisiae* mutants lacking *PCM1*, a homolog of plasma membrane Ca⁺ ATPases. J Cell Biol 1994; 124:351-363.

75. Archibald FS, Fridovich I. The scavenging of superoxide radical by manganous complexes in vitro. Arch Biochem Biophys 1982; 214:452-463.

76. Liu XF, Cullota VC, The copper and cadmium homeostasis gene, *BSD2*, is localized to the endoplasmic reticulum membrane in yeast. 35th Annual Meeting of the Society of Toxicology, Anaheim CA (Abstract in *The Toxicologist* 1996; 30(1):1521).

77. Thiele DJ. Metal-regulated transcription in eukaryotes. Nucl Acids Res 1992; 20:1183-1191.

78. Templeton DM, Cherian MG. Toxicological significance of metallothionein. In: Riordan JF, Vallee BL, eds. Methods in Enzymology. San Diego, CA: Academic Press, Inc., 1991:11-24

79. Schwarz MA, Lazo JS, Yalowich JC, Allen WP, Whitmore M, Bergonia HA, Tzeng E, Billiar TR, Robbins PD, Jack R, Lancaster J, Pitt BR. Metallothionein protects against the cytotoxic and DNA-damaging effects of

nitric oxide. Proc Natl Acad Sci USA 1995; 92:4452-4456.

80. Bremner I. Nutritional and physiological significance of metallothionein. In: Riordan JF, Vallee BL, eds. Methods in Enzymology. San Diego, CA: Academic Press, Inc., 1991:25-35

81. Felix K, Lengfelder E, Hartmann H-J, Weser U. A pulse radiolytic study on the reaction of hydroyl and superoxide radicals with yeast Cu(I)-thionein. Biochim Biophys Acta 1993; 1203:104-108.

82. Tamai KT, Liu X, Silar P, Sosinowski T, Thiele DJ. Heat shock transcription factor activates yeast metallothionein gene expression in response to heat and glucose starvation via distinct signalling pathways. Mol Cell Biol 1994; 14:8155-8165.

83. Liu XD, Thiele DJ. Oxidative stress induced heat shock factor phosphorylation and HSF—dependent activation of yeast metallothionein gene transcription. Genes Dev 1996; 10:592-603.

84. Lin SJ, Culotta VC. The *ATX1* gene of *Saccharomyces cerevisiae* encodes a small metal homeostasis factor that protects cells against reactive oxygen toxicity. Proc Natl Acad Sci USA 1995; 92:3784-3788.

85. Kim K, Kim IH, Lee KY, Rhee SG, Stadtman ER. The isolation and purification of a specific "protector" protein which inhibits enzyme inactivation by a thiol/Fe(III)/O2 mixed—function oxidation system. J Biol Chem 1988; 263:4704-4711.

86. Rhee SG, Kim KH, Chae HZ, Yim MB, Uchida K, Netto LE, Stadtman ER. Antioxidant defense mechanisms: a new thiol-specific antioxidant enzyme. Ann NY Acad Sci 1994; 738:86-92.

87. Chae HZ, Robison K, Poole LB, Church G, Storz G, Rhee SG. Cloning and sequencing of thiol-specific antioxidant from mammalian brain: alkyl hydroperoxide reductase and thiol-specific antioxidant define a large family of antioxidant enzymes. Proc Natl Acad Sci USA 1994; 91:7017-7021.

88. Chae HZ, Kim IH, Kim K, Rhee SG. Cloning, sequencing, and mutation of thiol-specific antioxidant gene of *Saccharomyces cerevisiae*. J Biol Chem 1993; 268:16815-16821.

89. Kim IH, Kim K, Rhee SG. Induction of an antioxidant protein of *Saccharomyces cerevisiae* by O2, Fe3+, or 2-mercaptoethanol. Proc Natl Acad Sci USA 1989; 86:6018-6022.

90. Chae HZ, Uhm TB, Rhee SG. Dimerization of thiol-specific antioxidant and the essential role of cysteine 47. Proc Natl Acad Sci USA 1994; 91:7022-7026.

91. Yim MB, Chae HZ, Rhee SG, Chock PB, Stadtman ER. On the protective mechanism of the thiol-specific antioxidant enzyme against the oxidative damage of bio-macromolecules. J Biol Chem 1994; 269:1621-1626.

92. Ruis H, Hamilton B. Regulation of yeast catalase genes. In: Scandalios JG, ed. Molecular Biology of Free Radical Scavenging Systems. Cold Spring Harbor, NY: Cold Spring Harbor Laboratory Press, 1992: 153-172.

93. Mathews CK, van Holde KE. Biochemistry. Redwood City, CA: The Benjamin/Cunnings Publishing Co., 1990.

94. Wieser R, Adam G, Wagner A, Schuller C, Marchler G, Ruis H, Krawiec Z, Bilinski T. Heat shock factor-independent heat control of transcription of the *CTT1* gene encoding the cytosolic catalase T of *Saccharomyces cerevisiae*. J Biol Chem 1991; 266:12406-12411.

95. Healy AM, Mariethoz E, Pizurki L, Polla BS. Heat shock proteins in cellular defense mechanisms and immunity. Ann NY Acad Sci 1992; 663:319-330.

96. Hamilton KK, Lee K, Doetsch PW. Detection and characterization of eukaryotic enzymes that recognize oxidative DNA damage. Methods Enzymol 1994; 234:33-44.

97. Kastan MB. Experimental models of human carcinogenesis. Nature Genet 1993; 5:207-208.

98. Hutchinson F. Chemical changes induced in DNA by ionizing radiation. Prog Nucleic Acid Res Mol Biol 1985; 32:115-154.

99. Teoule R. Radiation-induced DNA damage and its repair. Int J Radiat Biol 1987; 51:573-589.

100. Ward JF. DNA damage rpoduced by ionizing radiation in mammalian cells: identities, mechanisms of formation, and repairability. Prog Nucleic Acid Res Mol Biol 1988; 35:95-125.

101. Brennan RJ, Swoboda BE, Schiestl RH. Oxidative mutagens induce intrachromosomal recombination in yeast. Mutat Res 1994; 308:159-167.

102. Prakash S, Sung P, Prakash L. DNA repair genes and proteins of *Saccharomyces cerevisiae*. Annu Rev Genet 1993; 27:33-70.

103. Breen AP, Murphy JA. Reactions of oxyl radicals with DNA. Free Radic Biol Med 1995; 18:1033-1077.

104. Kolodner RD. Mismatch repair: mechanisms and relationship to cancer susceptibility. Trends Biochem Sci 1995; 20:397-401.

105. Loeb LA, Preston BD. Mutagenesis by apurinic/apyrimidinic sites. Annu Rev Genet 1986; 20:201-30.

106. Cross CE, Halliwell B, Borish ET, Pryor WA, Ames BN, Saul RL, McCord JM, Harman D. Oxygen radicals and human disease [clinical conference]. Ann Intern Med 1987; 107:526-545.

107. Ramotar D, Popoff SC, Gralla EB, Demple B. Cellular role of yeast Apn1 apurinic endonuclease/3'-diesterase: repair of oxidative and alkylation DNA damage and control of spontaneous mutation. Mol Cell Biol 1991; 11:4537-4544.

108. Kunz BA, Henson ES, Roche H, Ramotar D, Nunoshiba T, Demple B. Specificity of the mutator caused by deletion of the yeast structural gene (*APN1*) for the major apurinic endonuclease. Proc Natl Acad Sci USA 1994; 91:8165-8169.

109. Wilson D3, Bennett RA, Marquis JC, Ansari P, Demple B. Trans-complementation by human apurinic endonuclease (Ape) of hypersensitivity to DNA damage and spontaneous mutator phenotype in apn1-yeast. Nucleic Acids Res 1995; 23:5027-5033.

110. Umezawa H, Maeda K, Takeuchi T, Akami Y. New antibiotics bleomycin A and B. J Antibiot Ser 1966; A19.

111. Umezawa H, ed. Anticancer Agents Based on Natural Product Models. London: Academic Press, 1980:147-166.

112. Masson J-Y, Ramotar D. The *Saccharomyces cerevisiae IMP2* gene encodes a transcriptional activator that mediates protection against DNA damage caused by bleomycin and other oxidants. Mol Cell Biol 1996; 16:2091-2100.

113. Keszenman DJ, A. SV, Nunes E. Effects of bleomycin on growth kinetics and survival of *Saccaromyces cerevisiae*: a model of repair pathways. J Bacteriol 1992; 174:3125-3132.

114. Donnini C, Lodi T, Ferrero I, Puglisi PP. *IMP2*, a nuclear gene controlling the mitochondrial dependence of galactose, maltose and raffinose utilization in *Saccharomyces cerevisiae*. Yeast 1992; 8:83-93.

115. Farr SB, Kogoma T. Oxidative stress responses in *Escherichia coli* and *Salmonella typhimurium*. Microbiol Rev 1991; 55:561-585.

116. Jamieson DJ. *Saccharomyces cerevisiae* has distinct adaptive responses to both hydrogen peroxide and menadione. J Bacteriol 1992; 174:6678-6681.

117. Flattery-O'Brien J, Collinson LP, Dawes IW. *Saccharomyces cerevisiae* has an inducible response to menadione which differs from that to hydrogen peroxide. J Gen Microbiol 1993; 139:501-507.

118. Collinson LP, Dawes IW. Inducibility of the response of yeast cells to peroxide stress. J Gen Microbiol 1992; 138:329-335.

119. Jamieson DJ, Rivers SL, Stephen DW. Analysis of *Saccharomyces cerevisiae* proteins induced by peroxide and superoxide stress. Microbiology 1994; 140:3277-3283.

120. Davies JM, Lowry CV, Davies KJ. Transient adaptation to oxidative stress in yeast. Arch Biochem Biophys 1995; 317:1-6.

121. Izawa S, Inoue Y, Kimura A. Oxidative stress response in yeast: effect of glutathione on adaptation to hydrogen peroxide stress in *Saccharomyces cerevisiae*. FEBS Lett 1995; 368:73-76.

122. Harshman KD, Moye-Rowley WS, Parker CS. Transcriptional activation by the SV40 AP-1 recognition element in yeast is mediated by a factor similar to AP-1 that is distinct from GCN4. Cell 1988; 53:321-330.

123. Hussain M, Lenard J. Characterization of *PDR4*, a *Saccharomyces cerevisiae* gene that confers pleiotropic drug resistance in high-copy number: identity with *YAP1*, encoding a transcriptional activator [corrected] [published erratum appears in Gene 1991 Oct 30; 107(1):175]. Gene 1991; 101: 149-152.

124. Hertle K, Haase E, Brendel M. The *SNQ3* gene of *Saccharomyces cerevisiae* confers hyper-

125. Schnell N, Entian KD. Identification and characterization of a *Saccharomyces cerevisiae* gene (*PAR1*) conferring resistance to iron chelators. Eur J Biochem 1991; 200: 487-493.

126. Moye-Rowley WS, Harshman KD, Parker CS. Yeast *YAP1* encodes a novel form of the jun family of transcriptional activator proteins. Genes Dev 1989; 3:283-292.

127. Wu A, Wemmie JA, Edgington NP, Goebl M, Guevara JL, Moye-Rowley WS. Yeast bZip proteins mediate pleiotropic drug and metal resistance. J Biol Chem 1993; 268:18850-18858.

128. Bossier P, Fernandes L, Rocha D, Rodrigues-Pousada C. Overexpression of *YAP2*, coding for a new yAP protein, and *YAP1* in *Saccharomyces cerevisiae* alleviates growth inhibition caused by 1,10-phenanthroline. J Biol Chem 1993; 268:23640-23645.

129. Nolan GP. NF-AT-AP-1 and Rel-bZip: hybrid vigor and binding under the influence. Cell 1994; 77:795-798.

130. Struhl K. The DNA-binding domains of jun oncoprotein and the yeast GCN4 transcriptional activator protein are functionally homolgous. Cell 1987; 50:841-846.

131. Stephen DW, Rivers SL, Jamieson DJ. The role of the *YAP1* and *YAP2* genes in the regulation of the adaptive oxidative stress responses of *Saccharomyces cerevisiae*. Mol Microbiol 1995; 16:415-423.

132. Wemmie JA, Szczypka MS, Thiele DJ, Moye-Rowley WS. Cadmium tolerance mediated by the yeast AP-1 protein requires the presence of an ATP-binding cassette transporter-encoding gene, *YCF1*. J Biol Chem 1994; 269:32592-32597.

133. Wemmie JA, Wu AL, Harshman KD, Parker CS, Moye-Rowley WS. Transcriptional activation mediated by the yeast AP-1 protein is required for normal cadmium tolerance. J Biol Chem 1994; 269:14690-14697.

134. Hirata D, Yano K, Miyakawa T. Stress-induced transcriptional activation mediated by *YAP1* and *YAP2* genes that encode the Jun family of transcriptional activators in *Sac-*

charomyces cerevisiae. Mol Gen Genet 1994; 242:250-256.

135. Schnell N, Krems B, Entian KD. The *PAR1* (*YAP1/SNQ3*) gene of *Saccharomyces cerevisiae*, a c-jun homologue, is involved in oxygen metabolism. Curr Genet 1992; 21:269-273.

136. Szczypka M, Wemmie JA, Moye-Rowley WS, Thiele DJ. A yeast metal resistance protein similar to human cystic fibrosis transmembrane conductance regulator (CFTR) and multidrug resistance-associated protein. J Biol Chem 1994; 269:2285322857.

137. Li Z-S, Szczypka M, Lu YP, Thiele DJ, Rea PA. The yeast cadmium factor protein (YCF1) is a vacuolar glutathione S-conjugate pump. J Biol Chem 1996; 271:6509-6517.

138. Wu AL, Moye-Rowley WS. *GSH1*, which encodes gamma-glutamylcysteine synthetase, is a target gene for yAP-1 transcriptional regulation. Mol Cell Biol 1994; 14:5832-5839.

139. Kuge S, Jones N. *YAP1* dependent activation of *TRX2* is essential for the response of *Saccharomyces cerevisiae* to oxidative stress by hydroperoxides. EMBO J 1994; 13:655-664.

140. Ohtake Y, Satou A, Yabuuchi S. Isolation and characterization of glutathione biosynthesis-deficient mutants of *Saccharomyces cerevisiae.* Agric Biol Chem 1990; 54:3145-3150.

141. Ohtake Y, Yabuuchi S. Molecular cloning of the gamma-glutamylcysteine synthase gene of *Saccharomyces cerevisiae.* Yeast 1991; 7:953-961.

142. Grant CM, Collinson LP, Roe JH, Dawes IW. Yeast glutathione reductase is required for protection against oxidative stress and is a target for yAP-1 transcriptional regulation. Mol Microbiol 1996; 21:171-179.

143. Zitomer RS, Lowry CV. Regulation of gene expression by oxygen in *Saccharomyces cerevisiae.* Microbiol Rev 1992; 56:1-11.

144. Zhang L, Guarente L. Heme binds to a short sequence that serves a regulatory function in diverse proteins. EMBO J 1995; 14: 313-320.

145. Lapinskas P, Ruis H, Culotta V. Regulation of *Saccharomyces cerevisiae* catalase gene expression by copper. Curr Genet 1993; 24:388-393.

146. Marchler G, Schuller C, Adam G, Ruis H. A *Saccharomyces cerevisiae* UAS element controlled by protein kinase A activates transcription in response to a variety of stress conditions. EMBO J 1993; 12:1997-2003.

147. Schuller C, Brewster JL, Alexander MR, Gustin MC, Ruis H. The HOG pathway controls osmotic regulation of transcription via the stress response element (STRE) of the *Saccharomyces cerevisiae CTT1* gene. EMBO J 1994; 13:4382-4389.

148. Martinez-Pastor MT, Marchler G, Schuller C, Marchler-Bauer A, Ruis H, Estruch F. The *Saccharomyces cerevisiae* zinc finger proteins Msn2p and Msn4p are required for transcriptional induction through the stress-response element (STRE). EMBO J 1996; 9:2227-2235.

149. Estruch F, Carlson M. Two homologous zinc finger genes identified by multicopy suppression in a SNF1 protein kinase mutant of *Saccharomyces cerevisiae.* Mol Cell Biol 1993; 13:3872-3881.

150. Jungmann J, Reins HA, Lee J, Romeo A, Hassett R, Kosman D, Jentsch S. MAC1, a nuclear regulatory protein related to Cu-dependent transcription factors is involved in Cu/Fe utilization and stress resistance in yeast. EMBO J 1993; 12:5051-5056.

151. Hassett R, Kosman DJ. Evidence for Cu(II) reduction as a component of copper uptake by *Saccharomyces cerevisiae.* J Biol Chem 1995; 270:128-134.

152. Krems B, Charizanis C, Entian KD. Mutants of *Saccharomyces cerevisiae* sensitive to oxidative and osmotic stress. Curr Genet 1995; 27:427-434.

153. Krems B, Charizanis C, Entian KD. The response regulator-like protein *Pos9/Skn7* of *Saccharomyces cerevisiae* is involved in oxidative stress resistance. Curr Genet 1996; 29:327-334.

154. Brown JL, North S, Bussey H. *SKN7*, a yeast multicopy suppressor of a mutation affecting cell wall b-glucan assembly, encodes for a product with domains homologous to prokaryotic two-component regulators and to heat shock transcription factors. J Bacteriol 1993; 175:6908-6915.

155. Ota IM, Varshavsky A. A yeast protein similar to bacterial two-component regulators. Science 1993; 262:566-569.

156. Maeda T, Wurgler-Murphy SM, Saito H. A two-component system that regulates an

osmosensing MAP kinase cascade in yeast. Nature 1994; 369:242-245.

157. Morimoto RI, Sarge KD, Abravaya K. Transcriptional regulation of heat shock genes. J Biol Chem 1992; 267:21987-21990.

158. Morimoto RI. Cells in stress: transcriptional regulation of heat shock genes. Science 1993; 259:1409-1410.

159. Flint DH, Tuminello JF, Emptage M. The inactivation of Fe-S cluster containing hydro-lyases by superoxide. J Biol Chem 1993; 268:22369-22376.

160. Rouault TA, Klausner RD. Iron-sulfur clusters as biosensors of oxidants and iron. Trends Biochem Sci 1996; 21:174-177.

161. Wiedau-Pazos M, Goto JJ, Rabizadeh S, Gralla EB, Roe JA, Lee MK, Valentine JS, Bredesen DE. Altered reactivity of superoxide dismutase in familial amyotrophic lateral sclerosis. Science 1996; 271:515-518.

162. Rabizadeh S, Gralla EB, Borchelt DR, Gwinn R, Valentine JS, Sisodia S, Wong P, Lee M, Hahn H, Bredesen DE. Mutations associated with amyotrophic lateral sclero-

sis convert superoxide dismutase from an antiapoptotic gene to a proapoptotic gene: studies in yeast and neural cells. Proc Natl Acad Sci USA 1995; 92:3024-3028.

163. Gurney ME, Pu H, Chiu AY, Dal Canto MC, Polchow CY, Alexander DD, Caliendo J, Hentati A, Kwon YW, Deng HX, Chen W, Zhai P, Sufit RL, Siddique T. Motor neuron degeneration in mice that express a human Cu,Zn superoxide dismutase mutation [published erratum appears in Science 1995 Jul 14; 269:149]. Science 1994; 264:1772-1775.

164. Brown JRH. Amyotrophic lateral sclerosis: recent insights from genetics and transgenic mice. Cell 1995; 80:687-692.

165. Borchelt DR, Lee MK, Slunt HS, Guarnieri M, Xu ZS, Wong PC, Brown R Jr., Price DL, Sisodia SS, Cleveland DW. Superoxide dismutase 1 with mutations linked to familial amyotrophic lateral sclerosis possesses significant activity. Proc Natl Acad Sci USA 1994; 91:8292-8296.

CHAPTER 7

GENERAL STRESS RESPONSE: IN SEARCH OF A COMMON DENOMINATOR

Marco Siderius and Willem H. Mager

1. GENERAL STRESS

Cells must be able to cope with newly encountered situations in order to maintain optimal growth rate since the natural environment of unicellular organisms does not provide constant growth conditions. Yeast cells growing in such a variable environment will be challenged by changes in, e.g., temperature, osmolarity, pH, the presence of reactive oxygen intermediates, the availability of nutrients or a combination of these agents. In order to continue growing under the above mentioned variable circumstances cells have to sense the changes in the environment, respond to them and integrate this information into the mechanism that determines the growth potential of the cell. Indeed, a wide variety of cellular responses to different stress situations have been described in *Saccharomyces cerevisiae* (see previous chapters). These responses encompass the immediate production of compounds such as trehalose[1,2] which at an early stage protect cellular components from the detrimental effects of the stress.[2-4] Also signaling pathways are immediately activated, which at a later stage results in the induced expression of a wide variety of genes whose products confer protection to the cells.[5-7] All above mentioned protective responses to stress lead to a 'fine-tuned' web of specific reactions to certain stresses in addition to molecular events that can be designated global or general stress response. This chapter deals with the presumed common denominator of different types of stress responses.

2. ACQUIRED STRESS RESISTANCE AND CROSS PROTECTION

An intrinsic aspect of the stress response of yeast cells is the phenomenon of acquired stress resistance: cells can withstand a severe stress

Yeast Stress Responses, edited by Stefan Hohmann and Willem H. Mager.

condition more easily when they are previously exposed to a mild form of (the same) stress. For instance, a short pretreatment of cells with 0.7 M NaCl leads to an increase in the number of surviving cells when they are subsequently exposed to 1.4 M NaCl.[8,9] The same has been described for cells exposed to a severe heat shock when first treated with a mild, non-lethal rise in temperature[10] or in the case of hydrogen peroxide pretreatment prior to an oxidative stress challenge.[11] Thus mild stress conditions most likely trigger the pertinent cellular responses in order to get cells prepared to cope with a severe stress.

The finding of cross protection events was intriguing (see Fig. 7.1). This phenomenon was considered the first evidence for the occurrence of a general stress response. For instance, pretreatment of cells with a mild osmotic shock conferred resistance to heat shock[8,9] and exposure of yeast to high ethanol concentrations or weak acids conferred thermotolerance.[10] On the other hand, although the cross protection phenomenon is indicative of a shared response route, this is only partly true since in some instances stress tolerance acquisition does not occur. For example, a mild heat shock does not result in increased osmotolerance.[8] When Varela et al analyzed uptake and

incorporation of [^{35}S]methionine as a measure of stress resistance, preconditioning with elevated temperature was consistently not found to protect cells to osmotic shock.[9] On the other hand, Schüller et al showed that osmoprotection by a pretreatment with elevated temperature did occur.[12] An explanation for these seemingly contradictory data could be that the type of experiments performed are not readily comparable. Trollmo et al tested cross protection by pretreatment of exponentially growing cells with mild heat shock or osmoshock followed by growth on salt-containing (1.5 M NaCl) plates.[8] Schüller et al, however, exposed heat-pretreated cells to a severe osmoshock (3 M NaCl for 8 hours) after which the fraction of surviving cells was determined after growth on YPD.[12] The ability of cells to grow on salt-containing media was measured in the first type of experiments. Survival from osmoshock was determined in the second experiment (although these cells also experienced a hypo-osmotic shock due to plating onto YPD). These experimental differences—together perhaps with strain-dependent differences—make the data hard to compare and show the importance of properly defining what one wants to determine.

Fig. 7.1. Different stress conditions evoke common molecular responses which may lead to the acquisition of general stress resistance: cross-protection.

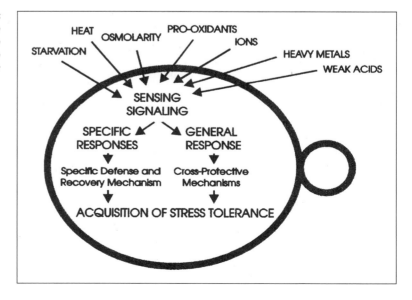

Other examples of nonreciprocal cross protection phenomena have also been described. Treatment of yeast cells with H_2O_2 does not evoke resistance to the superoxide generating drug menadione[13] whereas the opposite treatment does induce resistance. Similarly, preexposure to heat shock results in ethanol tolerance acquisition but the converse does not hold true.[14] Therefore parts of the stress response of yeast cells appears to be shared and lead to a certain level of cross protection. In addition stress-specific responses also clearly account for the full stress response necessary for survival under adverse conditions.

3. GROWTH CONDITIONS INFLUENCE STRESS RESPONSE

Tolerance of *S. cerevisiae* to a variety of detrimental conditions appeared to be correlated with the growth conditions since cells approaching stationary phase are more resistant to heat and other stresses than exponentially growing cells.[15,16] Cells reaching stationary phase accumulate trehalose, shut down part of the metabolism and induce the expression of several stress-responsive genes.[15] Similar molecular events occur upon starving yeast cells for nitrogen sources. The correlation between the specific changes operating when cells are starved for an essential nutrient and the acquired resistance to a variety of stresses again indicates that part of the responses of *S. cerevisiae* to stress in general is shared. Moreover suggests that stress events are controlled by mechanisms that also regulate growth. This observation is substantiated by work of Gross and Watson who showed that cells grown on acetate are more tolerant to heat shock while accumulating less trehalose than glucose grown cells.[17] In fact in some respects stress control and growth control may be considered as two sides of the same coin (see further). In the remainder of this chapter we will discuss the genes whose expression is induced in the general stress response, signal transduction routes involved in the sensing of stress and the integration of incoming information as well as the charac-

terization of *cis*-acting elements and *trans*-acting factors that serve as the ultimate targets of stress-induced signaling.

4. GENERAL STRESS-RESPONSIVE GENES

A common rationale for the stress-responsiveness of a gene is that stress-induced changes in its level of expression are beneficial to the cell. Although in this chapter we will mainly focus on changes at the level of gene transcription, one should bear in mind that this is not the full story. Changes in translational activities or post-translational modifications influencing stress-associated enzymatic activities also contribute to the actual response of *S. cerevisiae* under stress conditions. Part of the response of yeast to the detrimental effect of different stress conditions is likely to be caused by common mechanisms such as denaturation of proteins, disordering of membranes, DNA damage or metabolic disturbances.[6,18-20] Stress-responsive genes that are part of the general stress response in yeast are thus presumed to code for proteins that exert functions which are directly or indirectly necessary to cope with this type of damage under the various stress conditions. Table 7.1 contains a compilation of some typical 'general stress-responsive' genes; some of them will be discussed below.

- Since heat shock is the most extensively studied stress condition in yeast, heat shock protein genes (*HSP*) are among the best characterized stress-responsive genes.[6] The finding that some Hsps are also expressed when cells encounter other types of stress is generally considered a reflection of their important function as chaperones (in particular Hsps70) implicated in the repair of damaged proteins generated during the pertinent stress condition. The putative protease function of Hsp104 could be involved in disaggregation of clotted damaged proteins.[21] Therefore induced synthesis of this Hsp under a variety of stress conditions makes sense.

- Another gene expressed under different stress conditions is the polyubiquitin gene *UBI4*[22] which is involved in the nonlysosomal proteolysis of proteins.[23] By expressing this gene, yeast is able to get rid of damaged proteins that might be toxic to the cell.
- The *DDR2* gene is induced either by DNA damage or heat shock,[24] possibly to avoid the general adverse effects of stress on cellular DNA.
- Increase in trehalose synthesis belongs to the immediate responses to most if not all stress challenges. The initial rise in trehalose concentration is due to post-translational enzyme activation.[19] In addition, expression of the genes involved in trehalose biosynthesis are also targets of the stress response since cell transcription of the trehalose genes is increased upon imposing all kinds of stress to the yeast.
- Finally, most likely to avoid detrimental effects of reactive oxygen intermediates—which also contribute to the damaging effects of stress[25]—the gene encoding catalase T in *Saccharomyces cerevisiae* is induced upon heat shock, high osmotic shock and oxidative stress.[12,19,26] Activating genes like the ones mentioned above upon various stress situations therefore probably aims at overcoming the general damage on proteins, lipids and DNA or the consequence thereof.

5. NEGATIVE REGULATION OF TRANSCRIPTION OF GENERAL STRESS-RESPONSIVE GENES

As described above, a particular set of stress-responsive genes is expressed upon exposure to a variety of stress challenges. The term 'general stress response' would apply if expression of this group of genes is regulated in a coordinate fashion via a general stress-responsive route and mediated by specific *trans*-acting factors and *cis*-acting promoter sequences. A first clue for the existence of such a general pathway arises from the notion that the stress-responsive genes described above were shown to be negatively regulated by the Ras-cAMP signaling pathway.[15,27-29] The general stress response of yeast is sometimes referred to as metabolic stress since different stress conditions have in common that they all affect cellular metabolism.[7] Indeed, the Ras-cAMP signaling pathway—known to be involved in controlling metabolism (for extensive review see ref. 30)—is likely to play a part in regulating the general stress response (Fig. 7.2). A first indication for involvement of the Ras-cAMP pathway in the general stress response is the fact that *S. cerevisiae* mutants having high protein kinase A activity are highly sensitive to heat shock and starvation. Conversely, mutants with low protein kinase A activity show the opposite characteristics.[19,31] Furthermore, the specific regulation of stress-responsive genes upon ap-

Table 7.1. Examples of general stress-responsive genes in S. cerevisiae

Gene	Function	Ref.
HSP104	ATPase driven protein refolding	21
HSP26	unknown; chaperone?	
HSP12	unknown; chaperone?	29
UBI4	polyubiquitin	22
CTT1	catalase	33
DDR2	DNA damage repair?	24
TPS2	trehalose phosphate phosphatase	61
GAC1	regulatory subunit protein phosphatase type 1	49
CYC7	iso-2-cytochrome c	37

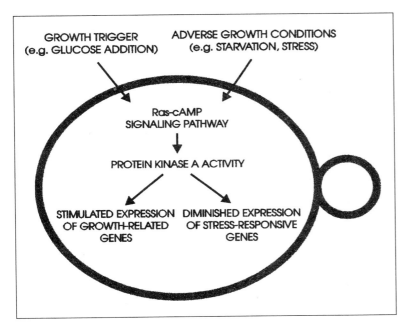

Fig. 7.2. Growth control and stress control represent counteracting molecular events. Favorite growth conditions which may lead to elevated growth potential activate the Ras-cAMP pathway and hence stimulate the activity of Protein kinase A. Adverse growth conditions (in general: stress) result in inactivation of the Ras-cAMP pathway. Protein kinase A activity in its turn has opposite effects on the transcription of growth-stimulated vs. stress-responsive genes.

proach to stationary phase is also indicative of a regulatory role of the Ras-cAMP pathway. Functional analysis of promoters of the *SSA3*, *CTT1*, *DDR2*, *HSP12* and *CYC7* genes consistently revealed sequences involved in the negative control of gene expression by high protein kinase A activity (see below). These findings may reflect a general metabolic control mechanism for stress-responsive genes in yeast mediated by the Ras-cAMP pathway (see further).

Bissinger et al[32] observed regulation by the Ras-cAMP pathway using fusions of the *CTT1* promoter to the bacterial *lacZ* gene. Interestingly, a couple of mutants generated in this study—*ctn1-1* and *ctn 5-1*—that showed an increased β-galactosidase and catalase T activity under both basal and nutrient-limited conditions were found to accumulate glycogen and trehalose which are typical indications for entrance into stationary phase.[15] The *ctn1-1* strain appeared to contain a mutation in the *CDC25* gene and the *ctn 1-5* strain in *RAS2*. The first demarcation of the *cis*-acting element conferring the modulatory activity of the Ras-cAMP pathway to the stress-responsive genes was performed using a *lacZ*-fusion of the promoter of the *HSP70* homologue *SSA3*. Boorstein and

Craig described a 35 bp UAS_{pds} which serves to modulate the induction of *SSA3* gene expression after a diauxic shift (i.e., at lowered protein kinase A activity).[28] Induction of expression after diauxic shift appeared to be dependent on the presence of both UAS_{pds} and a HSE (heat shock factor binding site) but the reported modulation of transcription activity by cAMP was governed by the UAS_{pds}. Interestingly, the heat shock induction of the *SSA3* promoter was also modulated synergistically by both elements. This finding demonstrates that the genes expressed in the general stress response may also be modulated by other stress-signaling pathways, thus ensuring both a general and a specific response for every stress-responsive gene when a specific stress is imposed to cells. Regulation of expression of the *CTT1-lacZ* fusion gene under nitrogen starvation, following heme induction or by altering protein kinase A activity is mediated by different parts of the promoter.[33] This suggests that net expression of *CTT1* is the result of the input of various signals, one of which might be the general stress signal.

LacZ-promoter fusion studies on two different stress-responsive genes (*CTT1*,[26] *DDR2*[24]) and *gus*-promoter fusions of the *HSP12* gene[34] further narrowed down the stress/protein kinase A responsive elements in the respective promoters. Mutational analysis of the *CTT1* promoter revealed a 13 bp element that was able to drive induction of expression upon nitrogen starvation and repression by high protein kinase A activity.[26] Moreover, the same region appeared to be involved in regulation of the response to a shift in temperature (23°C to 37°C) as had been described earlier.[35] The element—designated STRE for general stress responsive element—was subsequently used in an artificial promoter fused to the *lacZ* reporter.[26] Multiple elements in either one or the other orientation turned out to be more effective in activating gene expression than the single element. However, when the degree of induction of the wild type promoter was compared with that of a single element this difference was less pronounced (factor of ten versus eight times induced). The effect of the number of stress-responsive elements present in stress-induced promoters

on the net expression levels was also described by Gounalaki and Thireos[36] in their analysis of the stress-induced expression of the trehalose-6-phosphatase gene *TPS2*. The artificial *lacZ*-fusion construct bearing multiple elements was also tested in the response to NaCl, sorbitol and H_2O_2 shock and in all cases appeared to mediate a response. The studies of Kobayashi and McEntee[24] on the *DDR2* and of Varela et al[34] with the *HSP12* gene clearly revealed a pentanucleotide element (CCCCT) capable of mediating this stress response. This CCCCT-element is further designated STRE. Figure 7.3 shows a schematic representation of a number of STRE-containing promoter regions. Very recently the STREs previously detected in the promoter sequence of the gene for iso-2-cytochrome c CYC7 were demonstrated to be stress-responsive.[37] These STREs probably mediate both induction by a sudden temperature shift, by osmotic stress and by low cAMP or approach to stationary phase and

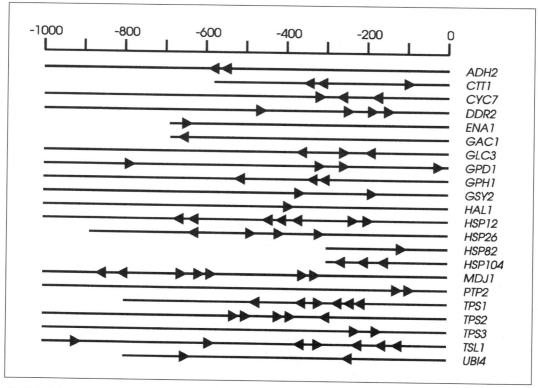

Fig. 7.3. Schematic representation of STRE-containing promoter regions. Numbers indicate the nucleotide position relative to ATG. Arrowheads indicate positions and orientation of the CCCCT motif.

hence represent prototypical general stress-responsive elements. Also notable in the case of *CYC7*, the promoter region contains in addition to STREs other regulatory elements such as a Hap1-site.[37] The UAS$_{pds}$ from the *SSA3* promoter has a slightly different sequence (CCCT or TCCCT). It remains to be determined whether this difference reflects a slight variation of the STRE sequence or a distinct functional role, e.g., as a target of a different set of input signals.

In addition to the involvement of the Ras-cAMP pathway there appears to be another clue for a link between regulation of metabolism and stress control. Stress-responsive genes such as *HSP12* are known to be repressed by glucose.[29] The underlying mechanism so far, however, is elusive. Glucose repression has been shown to involve the regulatory role of Snf1p, an AMP-dependent protein kinase homologue.[38] According to present models (see chapter 1) Snf1p inhibits Mig1p which, jointly with other factors, functions in repression of transcription of glucose-repressed genes. In addition, Snf1p is supposed to affect transcription in a Mig1p-independent fashion, possibly via interaction with the Polymerase II associated mediator complex.[39] Snf1p in turn is inactivated through Hxk2p upon addition of glucose to nonfermenting yeast cells.[39] As such protein kinase A may be considered as counteracting Snf1p. There is no evidence for the presence of Mig1-sites in the *HSP12* promoter though some STREs display a similar structure.[34] On the other hand, STRE-mediated transcription activation was also observed in *snf1*-mutant strains.[40] As will be discussed in the next section, several factors revealed to be implicated in general stress-regulated transcription show a functional link with Snf1p. It remains to be investigated whether these intriguing findings reflect the involvement of the SNF1 pathway in regulating the general stress response.

The data discussed above taken together indicate that the STRE is involved in the transcriptional induction of several stress-responsive genes and it mediates the effect of a metabolic signaling pathway (Ras-cAMP). These characteristics are in agreement with the requirements for a general stress response in yeast as predicted from the physiological stress response studies. Having discussed the Ras-cAMP signaling pathway as a negative regulator of the transcription of the general stress-responsive genes, we now turn to the characterization of signaling pathways that induce or de-repress expression of these genes.

6. POSITIVE REGULATION OF STRE-MEDIATED TRANSCRIPTIONAL ACTIVATION

Considering the STRE-mediated negative regulation of the expression of stress-responsive genes by the Ras-cAMP pathway, triggering of gene expression via STRE upon stress exposure most likely occurs in conjunction with the additional promoter elements. Of course this raises the question as to whether a common signaling pathway might trigger the stress-regulated gene transcription via STRE or whether multiple pathways exert their effects via this element. The fact that the High Osmolarity Glycerol (HOG) pathway appeared to transduce the specific signal evoked by high osmolarity to the expression of the appropriate genes via the STRE is a first indication that multiple signaling pathways exist (Fig. 7.4).

As described in chapter 4, the HOG signaling pathway is yet another example of the use of a MAP kinase signaling module in *S. cerevisiae* (and other eukaryotes) to monitor and respond to extracellular changes.[41-43] Briefly, the input of the high osmolarity signal into the HOG pathway (Fig. 7.4; see also Fig. 5.7) comes from two different osmosensor systems in the plasma membrane. First, a 'two-component' system (Sln1p/Ssk1p) has been reported that upon sensing high osmolarity stimulates (by lifting the blockade of the phosphorylated Ssk1p) the redundant pair of MAPKK kinases Ssk2p/Ssk22p that further activate Pbs2p.[44,45] Alternatively, the input can

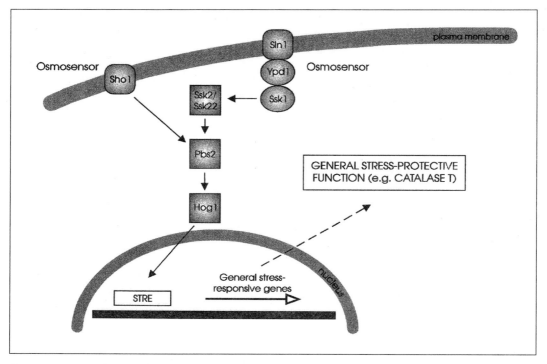

Fig. 7.4. HOG pathway as a positive regulator of STRE-mediated transcriptional activation. For further explanation see chapter 4.

derive from Sho1p, a SH3 domain containing putative transmembrane osmosensor that most likely directly interacts with Pbs2p.[45] Stimulation of the HOG pathway leads to the induced expression of genes involved in glycerol metabolism[46-48] but also of the genes that are considered to be part of the general stress response in yeast like *CTT1*,[12] *HSP12*[34] and *CYC7*.[37]

The influence of the HOG pathway on transcription of the general stress-responsive genes has been most extensively studied using the *CTT1* promoter and the artificial STRE-containing promoter derived from *CTT1* which harbors 7 copies of the STRE sequence fused to the *lacZ* reporter gene[12] as well as the *HSP12* gene promoter.[34] Schüller et al showed that the *CTT1* mRNA levels and β-galactosidase activity (for the artificial STRE promoter-*lacZ* fusion) are reduced after a mild hyperosmotic shock in *hog1Δ* and *pbs2Δ* strains when compared to wild type.[12] Although both deletions also affect the induction by weak acid stress, oxidative chal-

lenge and ethanol treatment on *CTT1* mRNA levels and β-galactosidase activity, this does not seem to be mediated directly by Hog1p since in contrast with osmostress conditions this MAP kinase is not phosphorylated under these stress circumstances. Point mutations in Hog1p preventing the activating phosphorylation of the MAP kinase also abolished the rise in catalase T activity upon a mild hyperosmotic challenge. The *HOG1* gene is also clearly necessary for full induction of *HSP12* mRNA levels upon mild hyperosmotic shock since under these circumstances this messenger only slightly increases in *hog1Δ* and *pbs2Δ* strains, whereas overexpression of *HOG1* results in overinduction of *HSP12* mRNA levels.[34] Likewise, osmostress activation of the *CYC7* promoter was recently found to be dependent of Hog1p.[37] It thus seems likely that the HOG pathway mediates its effect via STRE elements in the *CTT1*, *HSP12* and *CYC7* promoters. On the other hand, it is clear that other stress-induced signaling pathways also regulate stress-induced gene expression via STRE in a Hog1-independent fashion. This would mean that the *hog1* deletion only has an effect on stress-induced gene expression when cells sense high osmolarity. Furthermore, it may imply that there is a difference between STRE elements in promoters containing multiple STRE elements. There is evidence indeed for both. Schüller et al[12] showed that when testing mRNA levels of other STRE-containing genes (*DDR2, HSP104, GAC1* and *UBI4*) upon application of different stress conditions, indeed the *hog1* deletion only affected osmoshock response. The results were not entirely straightforward because the *DDR2* response to 5 mM sorbate and pH 2.8 was also affected by *hog1Δ* and the *GAC1* induction upon high osmolarity was not so ·clearly Hog1-dependent. In agreement with the latter finding for *GAC1*—and also for *UBI4*[12]—osmostress induction of *CYC7* is not entirely lost in a *hog1Δ* background but it is strongly delayed.[37] Apart from the conclusion that non-Hog1p-mediated signaling could also be responsible for high

osmolarity-induced expression of genes, another conclusion could be that the mere presence of STRE elements alone is not sufficient for Hog1p-induced induction of expression.

A possible explanation for the above described phenomenon could be that different signaling pathways driving stress-induced gene expression via STRE elements could only utilize STRE elements when they act together with other (as yet unknown) regulatory promoter elements. If this is true, there could be a difference in transcription regulation between individual STRE elements in promoters with multiple STREs. An indication for this was obtained from the studies of Varela et al[34] who showed that inactivating one STRE element in the *HSP12* promoter by point mutations did not always result in the same reduction of salt-inducibility of the *HSP12* mRNA (Fig. 7.5).

Expression of the *GPD1* gene—involved in the specific high osmolarity response—is also stimulated via Hog1p activity.[46] Indeed, in the *GPD1* promoter

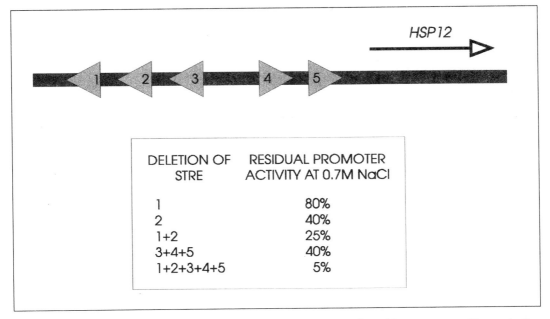

DELETION OF STRE	RESIDUAL PROMOTER ACTIVITY AT 0.7M NaCl
1	80%
2	40%
1+2	25%
3+4+5	40%
1+2+3+4+5	5%

Fig. 7.5. Differential effects of individual STRE-motif disruptions in the HSP12 promoter. The activation potential of the various HSP12-promoter constructs has been tested by measuring transcription of a gus-reporter gene.[34]

STRE elements area also present although it is not clear whether they function in high osmolarity-induced expression (P. Eriksson, personal communication). In addition, this gene so far has not been found to be responsive to other stress conditions. Therefore studies of the *GPD1* promoter may help in defining STRE-independent Hog1-specific inducible elements (see chapter 4).

Evidence for additional (signaling) components implicated in STRE-mediated transcriptional activation recently came from investigation of the general stress induction of *CYC7*.[37] These studies revealed a possible role in the general stress response of the previously identified nuclear factor Rox3p. *Rox3Δ* strains are temperature sensitive and in this genetic background both heat and osmostress activation of *CYC7* is virtually absent. *ROX3* expression was previously reported to be increased during anaerobiosis and the recent experiments extended this finding to other stress conditions. Subsequent deletion analysis of the *ROX3* promoter elucidated a novel putatively stress-responsive *cis*-element: GA_{10} GGAA.[37] A single AT bp deletion was found to abolish the transcriptional activation governed by this element. Notably, similar elements occur in the *HSP12* promoter but their biological role in this context is as yet elusive.

A search for high copy number suppressors of the *rox3ts* allele resulted in the isolation of *RTS1* encoding a protein homologous to the B' subunit of the serine-threonine phosphatase PP2A.[37] A subsequent *RTS1* deletion strongly affected the stress response of *CYC7* while basal levels were found to be slightly increased. Overexpression of *RTS1* in the *rox3ts* strain led to greatly elevated stress-activated transcript levels which may explain the original suppression phenotype. Epistatic analyses suggested that Rox3p acts downstream from Rts1p. On the basis of these results the authors presented a model in which both Rox3p and Rts1p are involved in mediating multiple stress signals. The finding of a phosphatase as a putative compo-

nent of general stress signal transduction is intriguing, since it may affect the stress response by antagonizing the action of (a) kinase(s). This fits in present models that triggering of a signaling pathway involves the activation of kinases while desensitization may occur by the action of (constitutively active) phosphatases. So far, however, the effects of *ROX3* and *RTS1* deletions have only been examined for *CYC7* and we have to await analyses of other general stress-responsive genes before being able to interpret these interesting results in a reliable fashion.

7. *TRANS*-ACTING FACTORS IN THE GENERAL STRESS RESPONSE

An obvious intriguing part of the signaling pathways mediating the general stress response is played by the *trans*-acting factors binding to the promoters of the different stress-responsive genes. Apart from the possibility that multiple signaling pathways generate various gene expression responses mediated by STREs that could respond depending on their flanking promoter sequences, *trans*-acting factors are a means of diverting input signals to increase 'fine-tuning' of the response. Characterization of these *trans*-acting factors will be necessary to understand how triggering of signal transduction pathways after sensing a specific stress may lead to the enhanced expression of genes that make up the specific response as well as to the induction of the general stress-responsive genes. Furthermore, the effects of the metabolic state of the cell (for instance reflected by the activity of the Ras-cAMP pathway) have input on the net induction of stress-responsive gene expression and might also converge to these *trans*-acting factors.

The first evidence for a protein binding to STRE came from Kobayashi and McEntee[24] who used a labeled STRE-containing oligonucleotide probes from the *DDR2* promoter region and detected binding to a 140 kDa protein (see further). A second indication for the presence of a factor involved in STRE-mediated gene expression came from the work of Gounalaki and Thireos.[36] Overexpression of the yeast AP-1 homologue *YAP1* yields a pleiotropic drug resistance phenotype. In order to understand this phenotype the authors searched for revertants and the trehalose-6-phosphatase gene (*TPS2*) was isolated from their screening. The cycloheximide and heat shock-induced expression of this gene was shown to be diminished in a *yap1Δ* mutant. Also, the activity of the artificial STRE-driven promoter based on the STRE-containing oligonucleotide from the *DDR2* promoter as used by these authors appeared to be diminished in the *yap1Δ* mutant.[24] Unfortunately, in vitro binding of Yap1p to the artificial promoter construct could not be detected. Thus there is a possibility that the *YAP1* effects on STRE-driven transcription are of an indirect nature. Yap1p has previously been demonstrated to serve as a transcriptional activator of several oxidative stress-induced genes.[20] An obvious candidate gene for Yap1-stimulated transcription via STRE elements therefore would be *CTT1*. However, although Yap1p activity appears to be stimulated by H_2O_2[50], catalase T activity and the β-galactosidase activity encoded by the artificial STRE-*lacZ* fusion construct are hardly induced by H_2O_2.[26,43] The induction of the *HSP12* gene upon encountering high-osmotic stress or heat stress was shown to be unaffected by the simultaneous deletion of both *YAP1* and the related *YAP2*.[34] Therefore it may be concluded that Yap1p and Yap2p are not involved in all STRE-mediated gene expression, although we cannot rule out that some STRE elements with flanking regions like those present in the *TPS2* promoter are Yap1-regulated.

The possibility that Rox3p may act as a *trans*-acting factor mediating general stress-responsive activation of transcription has been discussed above. Interestingly, Rox3p has been found to be part of the mediator complex which is associated with RNA Polymerase II.[51] This mediator complex is known to contain several proteins required for transcriptional activation.

It is striking that Rox3p is one of these components since this might reflect its role in stress-specific regulation of transcription. Further studies are needed to test this appealing hypothesis.

Recently the transcriptional activators Msn2p and Msn4p[52] were shown to be involved in STRE-driven gene expression since the *msn2 msn4* double deletion strain did not show any significant induction of transcription of the *HSP26*, *HSP12*, *CTT1* and *DDR2* genes after carbon-source starvation, heat shock treatment, high osmotic challenge or treatment with sorbate or ethanol.[40] Some residual induction of *HSP12* transcription is still present under high-osmolarity stress, which is perhaps indicative of a STRE sequence that is not Msn2p/Msn4p driven or an alternative way of inducing *HSP12* expression under these circumstances. Msn2p exerts its regulatory activity of STRE driven transcription by physically interacting with the C_4T element as evidenced by the recent isolation of *MSN2* in a screening of a λgt11 yeast DNA library with C_4T containing DNA as a probe.[53] Msn2p proved to represent the activity binding to STRE previously identified by the same group as a 140 kDa protein (predicted molecular mass of Msn2p: 78 kDa).[24,53] Expression of the *SSA3* gene with the slightly different stress-responsive element (CCCT or TCCCT) is not affected by the *msn2 msn4* double deletion, indicating that this element may be regulated in a different way. The double mutant is only stress-sensitive when severe stress conditions are encountered (T = 45°C, 7 hours at 3 M NaCl or 1 hour in 5 mM H_2O_2). The lack of a phenotype at mild stress conditions of the double deletion mutant is indicative for the presence of additional components (genes or proteins) that could provide the first-line protection to milder stress challenges.

Constitutive expression of the *MSN2* and *MSN4* genes resulted in increased thermotolerance and resistance to starvation. Notably, *MSN2* and *MSN4* have previously been identified as suppressors of the *snf1ts* mutation which causes a growth defect at glucose starvation conditions. Notably, the same holds true for Rox3p discussed above, which also has been isolated as a suppressor of *snf1*.[51] This set of data emphasizes the link between nutrient availability and stress response and lends support to the idea that the general stress response may serve to monitor and react to global metabolic disturbances. Consistent with this observation, the stress-responsive genes *CTT1* and *HSP26* are even expressed under nonstress conditions, albeit at low levels. Overexpression of both *MSN2* and *MSN4* in cells with the artificial STRE-*lacZ* reporter gene shows a slight induction of β-galactosidase activity (factor of four for *MSN2* and two for *MSN4*).[40] It is likely that stress-induced regulation occurs most probably through post-translational activation of the factors since more than 10-fold induction is observed under

stress conditions. To verify that the observed effects of *MSN2* and *MSN4* are mediated through the STRE elements in the promoters of the stress-responsive genes, gel retardation studies have been performed which point to Msn2p/Msn4p-binding to STRE elements in the *HSP12*, *CTT1* and *DDR2* promoters.[40] In conclusion, the *MSN2* and *MSN4* gene products could be mediators of STRE-driven gene expression. The available data are summarized in Figure 7.6.

8. CONCLUSION AND QUESTIONS

In this chapter we have reviewed the responses in *Saccharomyces cerevisiae* that are known under the term general stress response. The induction of several genes coding for proteins that could play a protective role under various stress conditions led

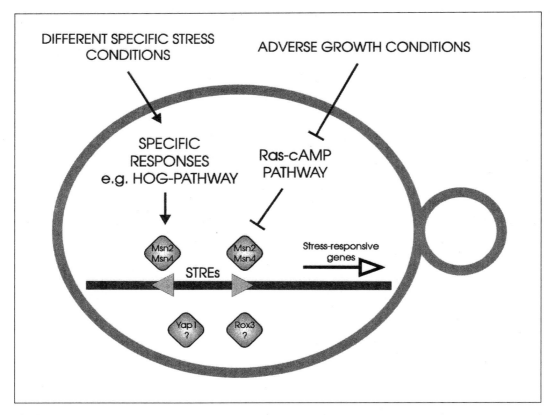

Fig. 7.6. Summary of signaling pathways, cis-acting elements and trans-acting factors implicated in the general stress response.

to the idea that expression of these genes may be mediated by a common pathway.

Indeed, the metabolic thermometer of the yeast cell i.e., the Ras-cAMP pathway, appears to have a general inhibitory effect on both stress resistance and expression of the stress responsive genes. A consequence of the involvement of the Ras-cAMP pathway in the general stress response is that all modulations of this pathway influence and complicate the stress response. The first examples of this aspect have already been described for *YAK1* and *SCH9* (genes influencing the protein kinase A activity in the cell) which have effect on heat-resistance,[54] as is the case with *SOK1* and *SOK2*.[55,56]

The only signaling pathway known so far that modulates expression of various stress-responsive genes in a positive manner is the HOG pathway. This MAP kinase signaling pathway includes the stress input 'receptors' for the high-osmolarity stress signal as well as the output in the form of altered gene expression. Of course the protein kinases that make up the HOG pathway—such as Pbs2p—could, apart from stimulating Hog1p, also have other substrates that are of importance in the early stress response. Furthermore, the activity of protein kinases in the HOG pathway could also be subject to modulation by other signaling pathways (e.g., ref. 57) in order to 'fine-tune' the response to the stress encountered. The characterization of other stress-sensing and stress-responsive signaling pathways as well as cross-talk between pathways will increase our understanding of the stress responses in yeast and help identifying transcription factors and *cis*-acting elements involved. Interestingly, a second 'two-component system' is proposed to be involved in a oxidative stress induced signaling pathway[58] which could be indicative of a new HOG pathway-like signaling route mediating the oxidative stress response (see chapter 6).

STRE sequences present in the promoters of the general stress-responsive genes appear to mediate at least HOG-induced gene expression. The Msn2p and Msn4p factors are so far the only elements known that mediate the transcription of all tested stress-responsive genes having STRE sequences in the promoter. The Yap1p factor could mediate expression of a subset of genes which may also hold for Rox3p. Interesting in this respect will be a further characterization of STRE sequences in the promoters of stress-responsive genes: how is the interplay between HOG and Ras-cAMP at STRE regulated? Do other signaling pathways operate which mediate transcription via STRE? If so, what is the importance of sequences flanking STRE?

Having described the cellular components that could mediate (part of) the cross protection phenomenon as a result of a general response to stress, the final question remains: what is the common denominator? The protective nature of trehalose, the increased levels of this disaccharide upon encountering stress conditions and the STRE-mediated expression of the trehalose genes suggest that trehalose could be a common factor in the general stress response. However, not all stress conditions encountered evoke a rise in trehalose concentration and acquisition of thermotolerance without any increase in trehalose content has been described,[19,59] indicating that it is not a common factor in the general stress response.

The general stress-responsive genes as described above encode proteins that could exert 'healing' functions in stressed cells. Expression of this set of genes is, however, not the same in every stress situation encountered: e.g., the *UBI4* gene is hardly expressed upon high osmolarity challenge[12] and *CTT1* is hardly expressed when *S. cerevisiae* encounters oxidative stress.[26] Probably this reflects the input of STRE-mediated expression by different signal transduction pathways. Therefore these pathways are not the shared feature we are looking for.

The only shared regulator in the general stress responses that has been identified is protein kinase A activity. Both the trehalose concentration[19] and STRE-mediated expression of general stress responsive genes is lowered by high protein kinase A activity. Furthermore, since high protein kinase A activity is indicative for optimal growth conditions[30] it could serve as a cellular switch mechanism between growth under optimal conditions and induction of the general stress response when encountering adverse circumstances. Transition to stationary phase and G_1 arrest—both of which are accompanied by a lowered protein kinase A activity—consistently result in the induction of expression of the general stress-responsive genes. How could the protein kinase A activity be envisaged as the metabolic switch of the general stress response? A hypothetical model of protein kinase A regulating the general stress response is depicted in figure 7.7. In cells growing under optimal growth conditions protein kinase A activity is high and activated by the RAS/Adenylate cyclase pathway and/or the Fermentable Growth Medium induced pathway (FGM).[30] Protein kinase A phosphorylates proteins that regulate growth in a positive way and the stress response in a negative way. We speculate that under adverse growth conditions protein kinase A activity may be lowered by disappearance of the input from cAMP or FGM and/or by signaling pathways induced by the encountered stress that inhibit or translocate protein kinase A catalytic subunits. Protein phosphatase activity in the cell will decrease the level of protein kinase A phosphorylation, thus contributing in switching the cell from growth to stress response. The signal transduction routes induced by encountering stress—like the HOG pathway—could then de-repress expression of a subset of general stress-responsive genes by stimulating expression

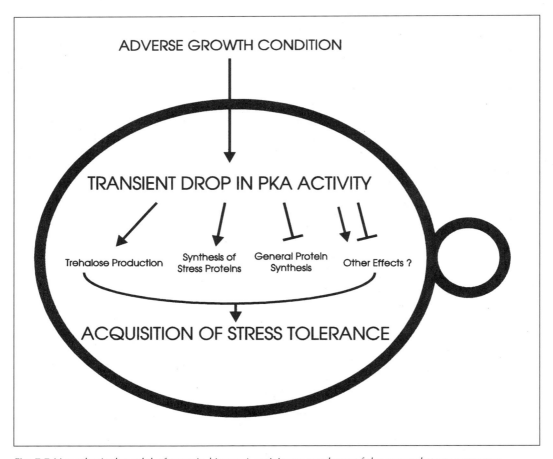

Fig. 7.7 Hypothetical model of protein kinase A activity as regulator of the general stress response.

via specific STRE elements. Furthermore, induction of these specific signal transduction pathways also leads to stress-specific responses like *GDP1* expression upon a high osmolarity challenge. Specificity in the stress response on the protein level might be generated by activation of specific phosphatases that counteract protein kinase A phosphorylation. For example, calcineurin is proposed to mediate protein kinase A-opposing effects on *ENA1* expression when cells are experiencing sodium stress.[60]

In the hypothetical model in Figure 7.7, the acquired resistance and cross protection is a consequence of the combined effect of the rise in trehalose content and the increased expression of general stress-responsive genes. In some cases preconditioning with one stress does not lead to cross protection, whereas encountering stress in the opposite order does.[8,14] This observation can be explained by assuming that both general and specific responses occur upon encountering a stress situation. Exposure of cells to a certain stress may provide enough protection to cope with the second challenge, but an essential component of the response might be lacking in the reverse situation. Assuming modulation of protein kinase A activity as the common denominator in the stress response can explain all phenomena described in this chapter. A very interesting line of future research would be to investigate if protein kinase A activity (or location of protein kinase A catalytic subunits) is altered when yeast cells are encountering adverse growth conditions.

ACKNOWLEDGMENTS

The authors thank Drs. Ellen de Groot for critically reading the manuscript.

REFERENCES

1. Mackenzie KF, Singh KK, Brown AD. Water stress plating hypersensitivity of yeasts: protective role of trehalose in *Saccharomyces cerevisiae*. J Gen Microbiol 1988; 134: 1661-1666.

2. Singh KK, Norton RS. Metabolic changes induced during adaptation of *Saccharomyces cerevisiae* to a water stress. Arch Microbiol 1991; 150:38-42.

3. Mager WH, Varela JCS. Osmostress response of the yeast *Saccharomyces cerevisiae*. Mol Microbiol 1993; 10:253-258.

4. Chowdhury S, Smith KW, Gustin MC. Osmotic stress and the yeast cytoskeleton: phenotype-specific suppression of an actin mutation. J Cell Biol 1992; 118:561-571.

5. Lindquist S. The heat shock response. Annu Rev Biochem 1986; 55:1151-1191.

6. Mager WH, Moradas-Ferreira PM. Stress response of yeast. Biochem J 1993; 290:1-13.

7. Mager WH, de Kruijff AJJ. Stress-induced transcriptional activation. Microbiol Rev 1995; 59:506-531.

8. Trollmo C, André L, Blomberg A, Adler L. Physiological overlap between osmotolerance and thermotolerance in *Saccharomyces cerevisiae*. FEMS Microbiol Lett 1988; 56:321-326.

9. Varela JCS, van Beekvelt CA, Planta RJ, Mager WH. Osmostress-induced changes in yeast gene expression. Mol Microbiol 1992; 6:2183-2190.

10. Coote PJ, Cole MB, Jones MV. Induction of increased thermotolerance in *Saccharomyces cerevisiae* may be triggered by a mechanism involving intracellular pH. J Gen Microbiol 1991; 137:1701-1708.

11. Davies JM, Lowry CV, Davies KJA. Transient adaptation to oxidative stress in yeast. Arch Biochem Biophys 1995; 317:1-6.

12. Schüller C, Brewster JL, Alexander MR, Gustin MC, Ruis H. The HOG pathway controls osmotic regulation of transcription via the stress-response element (STRE) of the *Saccharomyces cerevisiae* CTT1 gene. EMBO J 1994; 13:4382-4389.

13. Jamieson DJ. *Saccharomyces cerevisiae* has distinct adaptive responses to both hydrogen peroxide and menadione. J Bacteriol 1992; 174:6678-6681.

14. Piper PW. The heat shock and ethanol stress responses of yeast exhibit extensive similarity and functional overlap. FEMS Microbiol Lett 1995; 134:121-127.

15. Werner-Washburne M, Braun E, Johnston GC, Singer RA. Stationary phase in the yeast

Saccharomyces cerevisiae. Microbiol Rev 1993; 57:83-401.

16. Werner-Washburne M, Braun EL, Crawford ME, Peck VM. Stationary phase in *Saccharomyces cerevisiae*. Mol Microbiol 1996; 19:1159-1166.

17. Gross C, Watson K. Heat shock protein synthesis and trehalose accumulation are not required for induced thermotolerance in derepressed *Saccharomyces cerevisiae* cells. Biochem Biophys Res Comm 1996; 220:766-772.

18. Wolff SP, Garner A, Dean RT. Free radicals, lipids and protein degradation. Trends Biochem Sci 1986; 11:27-31.

19. Piper PW. Molecular events associated with the acquisition of heat tolerance by the yeast *Saccharomyces cerevisiae*. FEMS Microbiol Rev 1993; 11:339-356.

20. Moradas-Ferreira P, Costa V, Piper P, Mager WH. The molecular defences against reactive oxygen species in yeast. Mol Microbiol 1996; 19:651-658.

21. Parsell DA, Sanchez Y, Stitzel JD, Lindquist S. Hsp 104 is a highly conserved protein with two essential nucleotide-binding sites. Nature 1991; 353:270-273.

22. Finley D, Özkaynak E, Varshavsky A. The yeast polyubiquitin gene is essential for resistance to high temperatures, starvation and other stresses. Cell 1987; 48:1035-1040.

23. Jentsch S, Seufert W, Sommer T, Reins HA. Ubiquitin-conjugating enzymes: novel regulators of eukaryotic cells. Trends Biochem Sci 1990; 15:195-198.

24. Kobayashi N, McEntee K. Identification of cis and trans components of a novel heat shock stress regulatory pathway in *Saccharomyces cerevisiae*. Mol Cel Biol 1993; 13:248-256.

25. Davidson JF, Whyte B, Bissinger PH, Schiestl RH. Oxidative stress is involved in heat-induced cell death in *Saccharomyces cerevisiae*. Proc Natl Acad Sci USA 1996; 93:5116-5121.

26. Marchler G, Schüller C, Adam G, Ruis, HA *Saccharomyces cerevisiae* UAS element controlled by protein kinase A activates transcription in response to a variety of stress conditions. EMBO J 1993; 12:1997-200.

27. Tanaka K, Matsumoto K, Toh-e A. Dual regulation of the expression of the polyubiquitin gene by cyclic AMP and heat

shock in yeast. EMBO J 1988; 7:495-502.

28. Boorstein W, Craig EA. Regulation of a yeast HSP70 gene by cAMP responsive transcriptional control element. EMBO J 1990; 9:2543-2553.

29. Praekelt UM, Meacock PA. HSP12, a new small heat shock gene of *Saccharomyces cerevisiae*: analysis of structure, regulation and function. Mol Gen Genet 1990; 223:97-106.

30. Thevelein JM. Signal transduction in yeast. Yeast 1994; 10:1753-1790.

31. Shin DY, Matsumoto K, Iida H, Uno I, Ishikawa T. Heat shock response of *Saccharomyces cerevisiae* mutants altered in cAMP-dependent protein phosphorylation. Mol Cell Biol 1987; 7:244-250.

32. Bissinger P, Wieser R, Hamilton B, Ruis H. Control of *Saccharomyces cerevisiae* catalase T gene (CTT1) expression by nutrient supply via the RAS-cAMP pathway. Mol Cell Biol 1989; 9:1309-1315.

33. Belazzi T, Wagner A, Wieser R, Schanz M, Adam G, Hartig A, Ruis, H. Negative regulation of transcription of the *Saccharomyces cerevisiae* catalase T (CTT1) gene by cAMP is mediated by a positive control element. EMBO J 1991; 10:585-592.

34. Varela JCS, Praekelt UM, Meacock PA, Planta RJ, Mager, WH. The *Saccharomyces cerevisiae* HSP12 gene is activated by the high-osmolarity glycerol pathway and negatively regulated by the protein kinase A. Mol Cell Biol 1995; 15:6232-6245.

35. Wieser R, Adam G, Wagner A, Schüller C, Marchler G, Ruis H, Krawiec Z, Bilinski T. Heat-shock factor-independent heat control of transcription of the CTT1 gene encoding the cytosolic catalase T of *Saccharomyces cerevisiae*. J Biol Chem 1991; 266:12406-12411.

36. Gounalaki N, Thireos G. Yap1, a yeast transcriptional activator that mediates multidrug resistance, regulates the metabolic stress response. EMBO J 1994; 13:4036-4041.

37. Evangelista CC, Rodriguez Torres AM, Limbach MP, Zitomer RS. Rox3 and RTS1 function in the global stress response pathway in bakers yeast. Genetics 1996; 142:1083-1093.

38. Celenza J, Carlson M. A yeast gene that is essential for release from glucose repression

encodes a protein kinase. Science 1986; 233:1175-1180.

39. Trumbly R. Glucose repression in the yeast *Saccharomyces cerevisiae*. Mol Microbiol 1992; 6:15-21.

40. Martinez-Pastor MT, Marchler G, Schüller C, Marchler-Bauer A, Ruis H, Estruch, F. The *Saccharomyces cerevisiae* zinc finger proteins Msn2p and Msn4p are required for transcriptional induction through the stress-response element (STRE). EMBO J 1996; 15:101-109.

41. Herskowitz I. MAP kinase pathways in yeast: for mating and more. Cell 1995; 80:187-197.

42. Levin DE, Errede B. The proliferation of MAP kinase signalling pathways in yeast. Curr Op Cell Biol 1995; 7:197-202.

43. Ruis H, Schüller C. Stress signalling in yeast. Bioessays 1995; 17:959-965.

44. Maeda T, Wurgler-Murphy SM, Saito H. A two-component system that regulates and osmosensing MAP kinase cascade in yeast. Nature 1994; 369:242-245.

45. Maeda T, Takekawa M, Saito H. Activation of yeast PBS2 MAPKK by MAPKKKs or by binding of an SH3-containing osmosensor. Science 1995; 269:554-58.

46. Albertyn J, Hohmann S, Thevelein JM, Prior, BA. GPD1, which encodes glycerol-3-phosphate dehydrogenase, is essential for growth under osmotic stress in *Saccharomyces cerevisiae*, and its expression is regulated by the high-osmolarity glycerol response pathway. Mol Cell Biol 1994; 14:4135-4144.

47. Norbeck J, Påhlman A-K, Akhtar N, Blomberg A, Adler L. Purification and characterization of two isoenzymes of DL-glycerol-3-phosphatase from *Saccharomyces cerevisiae*. J Biol Chem 1996; 271: 13875-13881.

48. Hirayama T, Maeda T, Saito H, Shinozaki K. Cloning and characterization of seven cDNAs for hyperosmolarity-responsive (HOR) genes of *Saccharomyces cerevisiae*. Mol Gen Genet 1995; 249:127-138.

49. Franscois JM, Thompson-Jaeger S, Skroch J, Zellenka U, Spevak W, Tatchell K. GAC1 may encode a regulatory subunit for protein phosphatase type I in *Saccharomyces cerevisiae*. EMBO J 1992; 11.

50. Kuge S, Jones N. YAP1 dependent activation of TRX2 is essential for the response of *Saccharomyces cerevisiae* to oxidative stress by hydroperoxides. EMBO J 1994; 13:655-664.

51. Song W, Treich I, Qian N, Kuchin S, Carlson M. SSN genes that affect transcriptional repression in *Saccharomyces cerevisiae* encode SIN4, ROX3, and SRB protein associated with RNA polymerase II. Mol Cell Biol 1996; 16:115-120.

52. Estruch F, Carlson M. Two homologous zinc finger genes identified by multicopy suppression in a SNF1 protein kinase mutant of *Saccharomyces cerevisiae*. Mol Cell Biol 1993; 13:3872-3881.

53. Schmitt AP, Mcentee, K. Msn2p, a zinc finger DNA-binding protein is the transcriptional activator of the multistress response in *Saccharomyces cerevisiae*. Proc Natl Acad Sci USA 1996; 93:5777-5782.

54. Hartley AD, Ward M, Garrett, S. The Yak1 protein kinase of *Saccharomyces cerevisiae* moderates thermotolerance and inhibits growth by an Sch9 protein kinase-independent mechanism. Genetics 1994; 136: 465-474.

55. Ward M, Garrett S. Suppression of a cyclic AMP-dependent protein kinase defect by overexpression of SOK1, a yeast gene exhibiting sequence similarity to a developmentally regulated mouse gene. Mol Cell Biol 1994; 14:5619-5627.

56. Ward M, Gimeno CJ, Fink GR, Garrett S. SOK2 may regulate cyclic AMP-dependent protein kinase-stimulated growth and pseudohyphal development by repressing transcription. Mol Cell Biol 1995; 15:6854-6863.

57. Jiang B, Ram AFJ, Sheraton J, Klis FM, Bussey H. Regulation of cell wall beta-glucan assembly: PTC1 negatively affects PBS2 action in a pathway that includes modulation of EXG1 transcription. Mol Gen Genet 1995; 248:260-269.

58. Krems B, Charizanis C, Entian K-D. The response regulator-like protein Pos9/Skn7 of *Saccharomyces cerevisiae* is involved in oxidative stress resistance. Curr Genet 1996; 29:327-334.

59. Van Dijck P, Colavizza D, Smet P, Thevelein JM. Differential importance of trehalose in stress resistance in fermenting and nonfermenting *Saccharomyces cerevisiae* cells. Appl Env Microbiol 1995; 61:109-115.

60. Hirata D, Harada S, Namba H, Miyakawa T. Adaptation to high salt stress in *Saccharomyces cerevisiae* is regulated by Ca^{2+}/calmodulin-dependent phosphoprotein phosphatase (calcineurin) and cAMP-dependent protein kinase. Mol Gen Genet 1995; 249:257-264.

61. De Vergilio C, Bürckert N, Bell W, Jenö P, Boller T, Wiemken A. Disruption of TPS2, the gene encoding the 100kDa subunit of the trehalose-6-phosphate synthase/phosphatase complex in *Saccharomyces cerevisiae*, causes accumulation of trehalose-6-phosphate and loss of trehalose-6-phosphatase activity. Eur J Biochem 1993; 212:315-323.

CHAPTER 8

YEAST STRESS RESPONSES: ACHIEVEMENTS, GOALS AND A LOOK BEYOND YEAST

Helmut Ruis

1. INTRODUCTION

1.1. GENERAL REMARKS

What should be the goal of a conclusions chapter in a multi-author book? When I was reading the individual chapters, trying to get an overview of their contents and attempting to get "take home messages" in areas less familiar to me, I realized that this is not always easy. It is almost impossible sometimes to cover in one chapter all the interesting questions and potential answers, as well as all the intriguing and sometimes apparently contradictory findings of a certain area. On the other hand, to combine this with sufficient emphasis on the most important findings and conclusions in order to inform a nonexpert of the most interesting messages a field currently has to offer and about the directions the research in this area will and should take in the near future probably goes beyond the scope of the individual chapters. Furthermore, it cannot be expected that treatment of individual aspects of the field covered in this book will also deal with more general questions that are not only relevant specifically to the yeast *Saccharomyces cerevisiae*, but to fungi in general or to higher plants or animals or even to humans. I will try to be helpful particularly to the nonexperts or to those who are not experts for all aspects covered in this book. My selection of topics may be somewhat arbitrary, however. Hopefully what I have to conclude about stress in yeast and about its relevance for other systems will, on one hand be correct and relevant in the eyes of the experts (my co-authors). On the other hand, I hope that it will be sufficiently exciting to stimulate further research. What I attempt to address could stimulate future investigations with the system covered in this volume—yeast—but also with other systems. I will therefore particularly emphasize cross-connections and valid generalizations.

I shall try to emphasize in this chapter the external conditions triggering stress responses in yeast. I will then try to summarize the cellular targets of these stress factors and the hierarchy of responses. The nature of stress sensors which I will discuss next is a very interesting topic, although little hard knowledge is available at present concerning this point and we will have to wait for future developments. Alternatively, and if we are impatient, we will have to work in this direction ourselves. An area where more progress has been made already, although there are undoubtedly more unanswered questions than definitive answers, is that of the stress signaling mechanisms which transmit the information concerning external stress factors via the primary sensors to final cellular targets. Finally, I will focus on one type of target, not because it is the only important one but because it plays a role in every subarea of the stress response field covered in this book: stress-activated transcription factors and the types of DNA control elements binding them specifically.

1.2. YEAST AS A MODEL SYSTEM IN STRESS RESPONSE STUDIES

Why write a monograph about stress responses in yeast? There are several good reasons for concentrating on this organism:

1. The yeast studied in most investigations—*S. cerevisiae*—is an extremely well-characterized eukaryote in genetic terms. Genetic approaches employed in this system are extremely powerful and many examples are presented in the various chapters of this volume, demonstrating the usefulness of genetics for any type of physiological investigation. The majority of the studies described in this volume could not be carried out without the availability of a genetic

system unique among eukaryotic organisms.

2. With good reason, the molecular genetic characterization of *S. cerevisiae* which has now culminated in the complete sequencing of its genome, is mentioned here separately. Molecular and classical genetics are based on different types of strategies and many examples have been presented here illustrating this fact. However, one of the greatest advantages of the molecular genetic characterization of yeasts (particularly *S. cerevisiae* and *Schizosaccharomyces pombe*) is that any type of approach derived from molecular genetics or molecular biology (e.g., the construction of gene knockouts or the use of reporter genes) creates the possibility to develop new strategies based on classical genetics. This combination of classical and molecular genetics has been important since transformation methods have been developed for *S. cerevisiae*.[1-3] However, its attractiveness has been tremendously increased by the availability of the complete *S. cerevisiae* genome sequence[4,5] and studies of stress phenomena will certainly profit greatly from the availability of this information in the future.

3. An important aspect concerns the relevance of information obtained by yeast molecular genetics for studies with other organisms, especially higher eukaryotes. The yeast sequencing project has demonstrated that a great number of yeast gene products are homologous to those of higher plants, mammals and man. Since more and more genes from higher eukaryotes are becoming available while their functional characterization is lacking, in many cases it becomes attractive to introduce such genes into yeast mutants lacking the respective yeast gene and to test for functional complementation (see, e.g., the studies carried out on *HOG1* homologues from mammalian cells[6,7]). Furthermore, it will often be helpful to study genes of higher eukaryotes

more carefully in a direction suggested by data obtained with their yeast homologues. There is no doubt that these general remarks are also valid in the stress field and a number of examples for this have been presented in previous chapters of this volume (see, for example, HSF covered in chapter 3, the channel proteins of the MIP family discussed in chapter 4 or the cases of familial amyotrophic lateral sclerosis (FALS), a human genetic disease associated with superoxide dismutase mutations, covered in chapter 6).

4. There is a tremendous amount of biochemical information available on yeasts. As shown in various chapters, this information is sometimes obviously and sometimes surprisingly relevant for the stress field.

5. Yeasts have been utilized for biotechnology since ancient times. Stress in yeast cells cannot always be avoided in industial applications. While this might pose problems for the process in many instances, stress effects could sometimes be used for practical purposes, particularly if their effects on the physiology of yeast are well understood (e.g., to increase the viability of yeast produced).

In summary, studies on stress phenomena using yeast as a model system are not only interesting for understanding the organism itself, but should be relevant for the more complex higher eukaryotic systems and will become more attractive in areas as different as pure biology, biotechnology, agriculture or human medicine.

2. STRESS RESPONSE MECHANISMS IN YEAST

2.1. STRESS CONDITIONS AFFECTING YEAST CELLS

After dealing with these general aspects I will discuss more specific aspects and pose a rather trivial-looking question: which external factors are stressful to yeast cells or to other organisms—particularly microorganisms—and what makes them stressful to cells? Before going into details one should perhaps emphasize that, for example, as discussed in chapter 2, stress conditions are a very common event for yeast cells in nature although many of us try to avoid them in the research laboratory or in many biotechnological applications. As emphasized in chapter 2, most microorganisms including yeasts spend the majority of their time in stationary phase. The stress factors discussed extensively in the individual chapters of this volume provide a list which is relevant for yeasts but also for other (prokaryotic) microorganisms which share at least some important aspects of their physiology with yeasts. Generally these factors will also cause stress to higher eukaryotes or at least to some specialized cell types of such organisms. The factors emphasized in this book are nutrient depletion, heat shock, osmotic or salt stress and oxidative stress (Fig. 8.1). Of course, these classifications should not be misleading and the situation is actually more complex: nutrient depletion is a very broad phenomenon dealing, e.g., with depletion of carbon, nitrogen, sulfur or phosphorus sources. Mechanisms relevant to these phenomena overlap but are not identical. In the case of salt stress one should discriminate between an osmotic component and a true salt component as well as between effects of cations and anions. As outlined in chapter 5, anion toxicity does not appear to play a major role in *S. cerevisiae* but does so in some other organisms. Oxidative stress factors trigger a number of overlapping but not identical responses. Some of the responses are caused predominantly by superoxide anions, some by hydrogen peroxide or indirectly by various different heavy metal ions. Of course there are also other physiological conditions stressful to microorganisms not covered in separate chapters of this book because they are either less important in the case of yeast or have been used less in experimental investigations. Some stress factors important to yeast, such as osmotic stress for example, may be less important in, for example, animals. However, as emphasized in this case in chapter

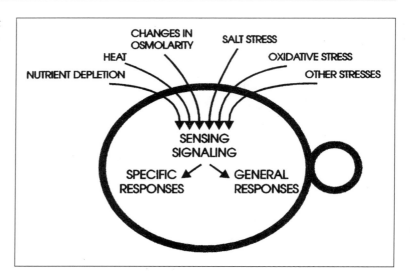

Fig. 8.1. Stress and the yeast Saccharomyces cerevisiae.

4, this does not apply in the case of some specialized cells (kidney cells are exposed to dramatic changes in osmolarity).

As emphasized in Figure 8.1, these factors may induce general responses common to all or most of them (see particularly chapter 7) or specific responses only caused by one type of stress. Which effect on cells do these factors have in common? Is there a general way of defining conditions causing stress to cells of microorganisms? One may address this question in reverse by asking for ideal living conditions for microorganisms. What is "yeast paradise"? Of course this "paradise" should offer optimal conditions for the survival of a population of yeast cells (not of an individual cell). Quite empirically, but also in agreement with our understanding of rules that apply in evolutionary selection, such conditions will allow the rapid formation of the greatest possible number of viable cells. This will give the larger cell population a greater chance for survival of less favorable conditions than the more slowly growing smaller one (at least if these two types of population are equally sensitive to stress; see below). I do not fully differentiate here between cell growth and cell division although this distinction is rather important and these are two coordinated but distinct phenomena.

Taking this into consideration, every deviation from ideal growth conditions should create stress to yeasts and other microorganisms. All conditions causing suboptimal growth (e.g., heat, nutrient starvation, suboptimal carbon or nitrogen sources, high osmolarity) should be stressful to yeast cells (see, e.g., chapter 1 for details). If this effect on cell growth is a crucial aspect of yeast stress responses cells might register stress in general by the decrease of activity of one or several key factors controlling cell growth. Protein kinase A has been emphasized in this context in several chapters. If a decrease in protein kinase A activity simultaneously allows cells to activate their stress protection mechanisms (e.g., the STRE-dependent system;[8,9] also see chapter 7) allowing a larger percentage of nongrowing or slowly growing cells to survive, this system would make sense in terms of physiology. However, no direct control of protein kinase A activity by stress conditions with the exception of carbon source starvation has yet been unambiguously demonstrated. Furthermore, it should be emphasized that not all stress conditions appear to have the same type of effect on cell growth. As discussed in chapter 2, only stress by exhaustion of nutrients leads to an entry into stationary phase. Other stress conditions cause slower growth of yeast cultures. It is unclear in all cases how much of this is due to a direct regulation of protein kinase A, which then controls cell growth, and how much to indirect effects, which in the extreme case affect cell viability due to severe damage of some cellular components. However, it may not be justified to limit the factors to be taken into consideration with respect to both growth control and stress responses to protein kinase A. While this becomes clear from the contents of virtually all chapters of this volume, some examples taken from chapter 1 may suffice to illustrate this point: as discussed, other systems involved in carbon catabolite repression play an important role. Furthermore, nitrogen catabolite repression, the general control of amino acid biosynthesis and the stringent response have to be considered. In addition, cross pathway control mechanisms should be better understood and taken into consideration. While protein kinase A certainly acts on many signaling pathways, other pleiotropic regulators undoubtedly exist. Stress signaling mechanisms may therefore have a general basis differing from the one discussed above or they may have no such general basis at all. In the one case where stress signaling and sensing is perhaps best understood in molecular terms—high osmolarity signaling via the HOG pathway in *S. cerevisiae* (see mainly chapter 4 and notice that the situation appears different in *S. pombe*)—no evidence has yet been obtained for a direct mechanistic coupling of stress sensing and growth control. Therefore sensing and signaling mechanisms may differ for different types of stress.

2.2. TARGETS, LEVELS AND HIERARCHY OF STRESS EFFECTS

It is evident already from the broad definition of stress (everything that has a negative effect on cell growth) that there must be both multiple targets and levels of effects of stress factors. An incomplete summary of such targets is presented in Figure 8.2. At the level of molecules that may be damaged by various stress types and should therefore be protected by stress responses are DNA, certain stress-hypersensitive proteins and membrane lipids. All these classes of targets and some of the mechanisms for their protection have been

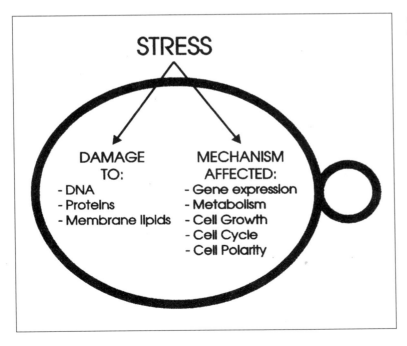

Fig. 8.2. Main targets of stress conditions in yeast.

studied most carefully in connection with oxidative stress but also to some extent with respect to salt stress (discussed in chapters 5 and 6). In the case of DNA, repair processes are obviously of functional importance in the stress response since only one or a few functional molecules that could be inactivated by stress are available per cell. Studies on DNA repair have been carried out extensively in yeast and illustrate the model character of this system particularly well.

Gene expression is influenced by stress conditions in rather complex ways. At least under some stress conditions the expression of the majority of genes is reduced. The exact reasons for this may be rather trivial (e.g., a stress-hypersensitive component of the general transcription machinery) but this is not well understood. Some of these aspects are discussed in chapter 2. We also do not understand why a certain class of genes whose expression is induced by stress via various transcription factors is apparently not or less affected by the general negative effect on gene expression. The reason for the expression of these genes escaping the general inhibition of transcription must be connected with the mode with which stress-activated transcription factors interact with the general transcription machinery. This is best illustrated by the expression of reporter genes which is negatively affected by stress via their normal control elements, but actually activated when driven by DNA elements binding stress-activated factors.

Although still incomplete, there is currently a better understanding of mechanisms triggering the specific induction of expression of stress genes. Many aspects of this phenomenon are discussed rather extensively in all chapters of this volume and will not be further discussed here. There are a number of cellular phenomena beyond gene expression which are affected by stress, such as many aspects of metabolism, cell growth, cell division and, e.g., cell polarity. Metabolic effects of stress might mechanistically be based on effects on the

activity or on the level of enzymes. Transporters (into the cell, out of the cell, to or from the vacuole) might also be affected by stress with similar consequences for metabolism as illustrated in various chapters, particularly in chapter 5 dealing with salt effects. Some of these effects on enzymes and transport systems may be indirect (e.g., via changes in gene expression), but some of them are among the early, direct stress effects (see Fig. 8.3). The complex biochemical effects of starvational stress are clearly illustrated in chapter 1 of this volume and other chapters demonstrate that complex biochemical changes are not only triggered by nutritional starvation. These biochemical events will include a modulation of the *activities* of enzymes, transporters and other protein molecules. In a later stage *levels* of many proteins will also vary. As a consequence of these events at the level of catalytic macromolecules, transport, biosynthesis and degradation of many substrates will be altered and levels of many small molecules will also be changed and new steady states will be reached.

Many other parameters will also be changed indirectly by a rather direct stress-triggered change in a few cellular parameters. Therefore cause and effect relationships have to be distinguished from mere correlations. This is not always easy. Genetic effects (e.g., mutant phenotypes, epistatic relationships) may be extremely helpful in truly establishing cause and effect relationships. However, there are pleiotropic mutant phenotypes and it is well documented that some of these phenotypes may be rather remote consequences of the true molecular alterations caused by a mutation. This demonstrates that in the stress field—like in all areas dealing with complex living cells—we have to search for primary effects. As is illustrated, e.g., in chapter 1, more sophisticated biochemical investigations might be helpful in this respect. Kinetics of effects observed may yield important information. There are numerous examples of stress effects discussed in this

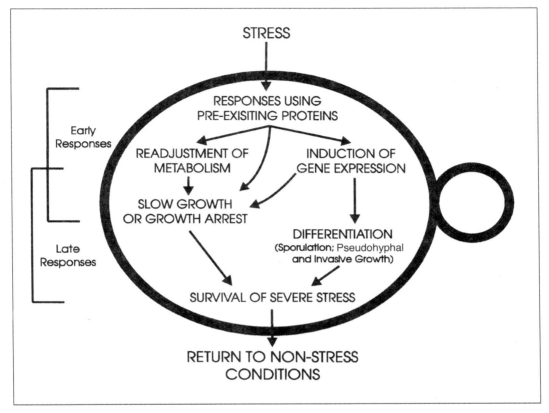

Fig. 8.3. Hierarchy of yeast stress responses. Early responses, as defined, are independent of protein synthesis. The overlap of early and late responses indicated means that the responses in the overlapping region may occur early or late or, in some cases, early and late.

volume where temporal relationships of effects observed are not clear at all.

Of course one has to bear in mind that for some purposes, like in biotechnology (or, e.g., in the case of human cancer), not only early and perhaps primary effects are relevant. One should also remember that primary consequences of, e.g., a mutation will in many cases only be discovered if one searches for them rather directly using the information gained from the nature of the gene products affected, their control, their subcellular location, etc. There are numerous examples in the history of yeast genetics where mutations were known (and named) by a phenotype that had little to do with the direct function of the gene product. Some of these examples have been discussed in various parts of this book. Furthermore, some rather interesting effects

observed experimentally may depend on some less obvious experimental details such as culture density or the use of liquid or solid media (chapter 2). This means that in the stress field—like in many other areas of yeast biology—we only truly understand a certain phenomenon if we have already studied many of its molecular details. In many cases these details cannot be uncovered without a multidisciplinary approach investigating very different facets of a system.

After having discussed which factors create stress to yeast cells and at which levels these effects are principally exerted, I shall outline certain possible generalizations concerning the hierarchy of stress responses. These generalizations have been very schematically depicted in Figure 8.3. When exposed to stress conditions of various kinds cells in general will react with a cascade of responses which might be classified as early, primary or immediate and secondary or late. While "early" appears to implicate a time scale, it is difficult or impossible to define this time scale in absolute terms. However, in technical terms early responses should happen independent of protein synthesis. Furthermore, one might generalize that the term "early responses" should be limited to events taking place within one round of mitosis. This means that the absolute time scale for "early" will differ depending on how fast cells will grow. Obviously this cannot apply at all for cells in stationary phase, attainment of which is a late response in itself. Time scales involved under these conditions have been discussed in chapter 2. Some of the early responses will certainly happen at the level of metabolism (chapter 1) and will involve the modulation of the activities of preexisting enzymes. The biological function of such events should in the first place be an almost immediate protection against stress. This protection will usually only be functional in the case of mild stress. Again, the limit between mild stress and severe stress is hard to define, but operationally one might define mild stress as conditions still allowing cell growth, e.g., the formation of visible colonies on plates. This is not possible anymore in the case of severe stress. We should not forget that, at least in unicellular microorganisms, the immediate main strategy of previously nonstressed cells to sudden stress would be to outgrow it by sheer number. The factors needed for any constitutive stress responses beyond a certain level (selected for by evolution) would mean an unnecessary constitutive burden to yeast growth and therefore a selective disadvantage.

In a realistic natural situation stress conditions occur suddenly and usually last for some time. There is also a certain probability that stress will become more severe with time. Therefore cells should be prepared for this situation as fast as possible and at least some changes in the induc-

tion of gene transcription and an adjustment of cell growth to the new situation will be among the early events. The second function of these early events will be to prepare cells for processes undoubtedly requiring the synthesis of new proteins; in this respect two different strategies will apply: to grow more slowly under conditions of mild stress and to grow not at all but to survive under severe stress. As discussed above, slow growth could be an immediate consequence of stress, triggering the establishment of stress resistance. The two strategies (to grow slowly or to try to survive without growing at all) should have a different molecular basis. This is exemplified by the formation of protective proteins under the control of HSF supporting growth under mild stress and of different proteins under control of STRE-binding transcription factors providing protection against severe stress. This dualism is discussed in chapter 3 and later in more general terms in this chapter.

Late responses which should make cells more resistant to severe stress and should help them to survive such regimes may—depending on the nature and the extent of stress—occur in slowly growing cells or in cells which have switched to a different, more economic growth pattern (e.g., pseudohyphal or invasive growth during nutrient starvation). In some other cases cells may at least transiently stop growing. They may start to grow again after this early response or they may enter stationary phase (a G_0-like state; the question how far the stationary phase of yeast cells corresponds to the G_0 state of higher eukaryotic cells has been discussed in chapter 2). Under some metabolic conditions cells will sporulate provided they are diploid. It might be advantageous to use different strategies simultaneously, in particular when the cells cannot anticipate how the environmental conditions will develop further.

As has been strongly emphasized in chapter 1 of this book it will also be vital for cells living in nature to return to nonstress conditions in a rapid and ordered way ("returning to the feast"; see chapter 1). It should be noted that the ability to switch quickly to a proper response to both stress and nonstress conditions should offer a tremendous selective advantage to cells living under natural conditions. As emphasized already in chapter 3, for example, heat shock proteins acting as chaperones and other proteins acting synergistically with them will on the one side, help to renature damaged proteins and on the other side, remove them if renaturation is not possible. In this respect the Hsp104 molecule appears to play a very central role as a catalyst of protein disaggregation or reactivation.[10,11] Proteolytic activities requiring ubiquitin and the proteasome should have important functions as long as stress conditions last as well as in the return to more favorable conditions.[12,13]

2.3. STRESS SENSORS

A question discussed in a number of chapters of this volume concerns the nature of the immediate stress sensors. This question is obviously important, but only little is known about stress sensors in any eukaryotic system. It is therefore not surprising that there is a lot of speculation concerning this point. An enzyme like protein kinase A should obviously not be a stress sensor in spite of its potential central role in stress responses discussed above. If protein kinase A plays a general role it would be in the transmission of stress signals to transcription factors and/or other protein targets but not in sensing. Sensors of external stress could act at the cell surface and could be located in the plasma membrane or they could be intracellular constituents. One case where stress sensors have been identified in yeast are the components of the HOG pathway Sln1p and Sho1p. They are at least extremely likely to play a role as osmosensors.[14,15] However, not even in this case do we currently understand in molecular details how these molecules act in osmosensing. Two component systems which have been considered to be limited to prokaryotes[16] have

now also been detected in eukaryotes (like Sln1p-Ssk1p of the HOG pathway). Therefore they might play a more general role in stress sensing in yeasts and other eukaryotes. As discussed in chapter 6, a gene (*POS9=SKN7*[17]) that is involved in the response to oxidative stress according to genetic evidence is a putative response regulator protein and hence a putative part of a two component system. However, in the absence of any further information on the functional role and the mechanism of action of this protein, any generalization concerning the role of two component systems in stress sensing remains entirely speculative. Beyond the information available on the HOG pathway plasma membrane transporters have been suggested to play a role as stress sensors (see in chapters 1, 4 and 5, for example).

The general problem of sensing mechanisms is discussed more systematically in chapter 3. The evidence for a role of denatured proteins, of intracellular pH, of plasma membrane fluidity, of the proton motive force at the plasma membrane and a role of the plasma membrane H^+-ATPase (including a potentially quite important regulatory function of Hsp30) as signals and/or sensors are critically compared. Intracellular proteins specifically responsive to a certain type of stress also play a role as sensors. Ace1p is rather well understood as a copper sensor even in relevant molecular details.[18] The protein acts as a transcription activator when a conformational change is induced by copper ions, which allows Ace1p to bind to its DNA consensus sequence. The role of metal homeostasis in sensing heavy metal ions and oxidative stress and the potential use of metal sulfur clusters in sensing oxidative stress is also discussed in chapter 6. Obviously, transport systems through the plasma membrane and the vacuolar membrane (the vacuole acts as a sequestering compartment for a number of compounds) should function in homeostasis. Homeostasis appears to play a crucial role in stress sensing and/or signaling at least in some cases.

2.4. Signaling Mechanisms

There appears to be no doubt that signaling mechanisms of various complexity will transmit the stress signals sensed in different ways to intracellular targets. In some cases this appears to occur in a rather direct manner such as in the case of Ace1p discussed above. In other cases signal transduction pathways of various complexities are or may be involved in the transmission of stress signals. These pathways may or may not utilize second messengers like cAMP, calcium ions or others. Changes in concentration and in the intracellular distribution of heavy metal ions might also have a second messenger-like function. Their importance is illustrated by two severe human genetic diseases—Menkes syndrome and Wilsons disease—which are due to defects in copper ion distribution and are discussed in chapter 6.

The HOG MAP kinase pathway is an example of a rather well understood mechanism functioning in the transmission of increases in osmolarity. Different aspects of the function of this pathway are discussed in various chapters, most extensively in chapter 4. As emphasized there, the specificity of this pathway for high osmolarity stress (and perhaps low pH stress) seems to distinguish it from similar signaling pathways in mammalian cells and even in *S. pombe*. This is remarkable since at least some MAP kinases involved in these stress signaling pathways from higher eukaryotes can functionally replace Hog1p in *S. cerevisiae* at least to some extent.[6,7] Further studies will have to show what factors distinguish the *S. cerevisiae* system from those of other eukaryotes and whether other MAP kinases activated by other (perhaps multiple) stresses under physiologically relevant conditions exist in *S. cerevisiae*. It is obvious from the discussion of the HOG pathway in chapters 4 and 5 that it does play a crucial role in high osmolarity and high salt stress. However, in spite of extensive efforts by a number of laboratories not all aspects of its function such as the nature of downstream targets or its mode of action in the control of the *GPD1* gene have

been clarified (see chapter 4). Furthermore, other modes of high osmolarity signaling appear to exist and seem to play a role, especially in slower responses to increases in osmolarity.

The question arises from the very successful story of the HOG pathway as well as from findings made with mammalian systems whether MAP kinase cascades play a more general role in the transmission of stress signals. Whereas the findings made in other systems than yeast point to a more general role, the evidence obtained with the yeast *S. cerevisiae* does not yet justify such a conclusion for this organism. Furthermore, one should bear in mind that structural homologies between signal transduction components or transcription factors of different types of organisms do not necessarily reflect functional similarities. While there are many such examples, one case that should also be outlined also for other reasons should suffice here. As explained in virtually every textbook of biochemistry, the second messenger cAMP functions as a starvation stress signal in *Escherichia coli* and in mammalian liver. One is tempted to generalize the role of rising cAMP levels since this is undoubtedly valid in some other cases as well. As is illustrated in chapter 1 this generalization does not hold for *S. cerevisiae* where cAMP is certainly not a starvation signal. Probably cAMP as a second messenger has been "invented" quite early in evolution, but then it has been used by different organisms and different cell types to transmit different types of signals. In principle the same situation appears to apply in the case of MAP kinase modules.

Nevertheless, another such MAP kinase module potentially involved in yeast stress signaling should be mentioned here. As also discussed in chapter 3 and 4 the PKC-pathway is activated by heat shock and hypo-osmotic shock and has been implicated in nutrient sensing. However, the physiological role of this effect is not entirely clear. The phenotypes of PKC-pathway mutants observed may rather be a consequence of its function in the control of cell wall assembly than of a true role in sensing and transmitting changes to low osmolarity. It might be added in this context that no effect of mutations in the pathway on STRE-activated transcription—which is affected by many types of stress—has been detected (C. Schüller, H. Ruis, unpublished).

As discussed to some extent in chapter 4, the coexistence of different MAP kinase pathways even sharing some molecular components or possessing structurally similar components will make it necessary to ensure specificity of signaling. On the other hand, a certain well-controlled degree of cross talk between MAP-kinase pathways and of some of their components with other signaling pathways (e.g., with protein kinase A) will be biologically advantageous. It is not nearly entirely clear how cells solve the specificity problems, but the role of the specific scaffold protein Ste5p in maintaining specificity has been demonstrated in the case of the mating pheromone response pathway.[19,20] No such proteins have yet been identified for other MAP kinase pathways but may well exist.

We have to expand our view on stress signaling if one considers the exclusive availability of nonfermentable carbon sources as a stress factor for yeast cells. All proteins necessary for the derepression of systems involved in the utilization of nonfermentable carbon source in the absence of sugars will be involved in signaling this type of stress. The Snf1p protein kinase is discussed in chapter 1 as an important example for this type of signaling system. However, similar to the situation in the PKC-pathway, data concerning phenotypes of *snf1* mutants are not easy to interpret since these are extremely pleiotropic. Such data should therefore be interpreted with caution. A similar situation might exist in the case of protein phosphatases involved in signaling. Calcineurin (protein phosphatase 2B) is discussed in chapter 5 as an important example for this type of regulatory factor and the lack of specificity of action of this factor has been emphasized. There is no doubt that in vir-

tually every case where the activity of a protein is controlled via a protein kinase, one or several protein phosphatases will also play a role in regulation. The specificity could be derived from the substrate specificity of only one of the partners (kinase or phosphatase) and stress signals could be transmitted via one of these components, while the other one just counteracts its activity in a constitutive way or transmits another type of signal.

The situation is complicated by the fact that mutant data obtained might be misleading. In a mutant cell lacking a factor such as a protein phosphatase with limited substrate specificity, its activity may be replaced to some extent by another protein with overlapping specificity but not normally acting in its place in a wild type cell. The consequence of this might be a noninformative phenotype of the mutant. In this sense the (slight) overproduction of a factor with the potential to replace a functional protein might suppress the phenotype of a mutant lacking this protein, even though in wild type cells the two proteins do not share their substrates. If such data are not interpreted with due caution one might conclude for instance that a certain phosphatase exerts a function in a pathway where it actually has no role in wild type cells. The extent of specificity of this group of enzymes in undisturbed wild type cells is not sufficiently understood at the moment. Also, in the case of the protein kinase A pathway, reliable conclusions concerning its physiological function are not easy to draw from experiments with mutants deficient in a component or overproducing one. Cells lacking the cAMP-binding regulatory subunit of protein kinase A (*bcy1* mutants) exhibit many phenotypes and there seems no doubt that a number of them are a very indirect consequence of uncontrolled protein kinase A activity. Therefore conclusions based solely on *bcy1* mutant phenotypes should be confirmed by independent types of experimental evidence. The role of protein kinase A in stress signaling and in linking cellular stress control to growth control has been discussed above, and particularly in chapters 1 and 7. Some of the open questions concerning the function of protein kinase A in the general stress response will be discussed in more detail below.

2.5. STRESS-ACTIVATED TRANSCRIPTION FACTORS AND DNA ELEMENTS BINDING THEM

While it is clear from what has been discussed in this book and earlier in this chapter that stress factors will influence different types and different levels of cellular events, there is also no doubt that the control of the initiation of transcription of specific genes is one of the most crucial aspects in the yeast stress response field. Before some aspects of this phenomenon are discussed it should be emphasized that gene expression might be affected by stress also at various post-transcriptional levels. There is little if any hard information concerning this point. Induction of expression by single, specific stress factors exists in a variety of cases, and numerous data confirming this conclusion are presented in this volume. Beyond these specific effects there exist several DNA elements and transcription factors binding to them which are of more general importance insofar as they act in the regulation of transcription of multiple genes and are also activated by more than one stress condition. Although it is difficult to draw a limit between general and specific factors, it may not be too arbitrary to include three types of DNA elements in a group which appears to play a broader role in the induction of transcription by stress factors: AREs, HSEs and STREs. All three types of elements have been covered in various chapters of this volume, most systematically in chapter 6 in the case of AREs, in chapter 3 in the case of HSEs and in chapter 7 in the case of STREs. I will add a few summarizing remarks concerning the current knowledge of their mode of action and of their functional relevance:

1. **AREs** (AP-1-responsive elements) of yeast cells activate the expression of a number of yeast genes by binding tran-

scription factors homologous to mammalian AP-1 (Yap1p, Yap2p). As discussed in more detail in chapter 6, both transcription factors binding to ARE appear to have overlapping albeit not necessarily identical functions and play a role in the adaptation of cells to oxidative stress (H_2O_2). Beyond that these factors appear to play a role in the resistance to heavy metals, a fact that probably can be rationalized by the role of these metals in the generation of some types of oxidative stress. A number of target genes of these transcription factors have been identified. An intriguing finding concerning functions of Yap1p concerns its role in the regulation of STRE activity. As discussed in chapter 7, STRE-driven expression of a reporter gene is diminished in a *yap1* mutant. However, Yap1p does not bind to STREs and therefore this effect appears to be indirect.

2. **HSEs** (heat shock elements) and the heat shock transcription factor (HSF) appear to exist in virtually all types of eukaryotic cells. The yeast response to heat shock is predominantly mediated via HSEs-HSF as covered in chapter 3. Only some aspects of the function of HSF in the stress response will be emphasized again in the present context. While a great number of genes activated by heat shock appear to be controlled by HSF, evidence for the role of most of these genes in stress protection is lacking. The most notable exception is the *HSP104* gene whose expression is likely to be controlled via HSEs and STREs. Most remarkably, HSF itself is not required for the acquisition of tolerance to a severe heat shock[21] above 50°C while at least some HSF-controlled genes are needed for growth at 37-39°C.

Furthermore, yeast HSF is encoded by a single essential gene, i.e., it is required also in the absence of stress. The latter finding is most easily explained by the fact that HSF controls the expression of yeast chaperone genes, which are required also by nonstressed cells. At least some of the genes encoding this class of proteins require HSEs also for basal expression. It appears to be relevant in this respect that—in contrast to HSFs of higher eukaryotes—HSFs of the yeasts *S. cerevisiae* and *Kluyveromyces lactis* bind to HSEs constitutively and exhibit some activity as transcriptional trans-activators also in nonstressed yeast cells. From these and similar observations the picture is emerging that at least in yeast, HSF functions in nonstressed and moderately stressed cells while other factors are required for the protection of cells against severe stress.

Even this simplified picture of the functions of HSF would be incomplete without mentioning an aspect covered in chapter 6 of this volume. The *S. cerevisiae CUP1* gene encoding metallothionein, which plays a crucial role in the response to oxidative stress, has been demonstrated to be induced by growth on nonfermentable carbon sources or by a generator of superoxide anions. This induction has been demonstrated to be independent of Ace1p which mediates induction of this gene by copper ions. Instead, the induction requires HSF and a HSE in front of the *CUP1* gene.[22] This surprising finding may have its counterpart in the activation of the mammalian HSF proteins by generators of oxidative stress.[23] In yeast cells oxygen radicals would not only be generated by superoxide, but also by the respiratory growth on nonfermentable carbon sources (obligatory utilization of oxygen generating oxygen radicals as a byproduct).

3. STREs—systematically covered in chapter 7—were originally discovered as positive control elements which are under negative control by protein kinase A and as DNA elements mediating the HSF-independent heat shock induction of some genes (discussed in

chapter 3). The role of HSE-HSF in the stress response may be broader than it seems at the moment, as discussed above. However, there appears to be no doubt that the broad specificity of induction of STRE-dependent transcription, the involvement of protein kinase A in STRE control and the pattern of genes under control by this UAS element make it a much better candidate for a crucial component in the general stress response playing an important role in enabling cells to survive periods of extreme stress of different types.

The biological advantage gained by such a general response system providing protection against a number of severe stresses when induced by a mild version of one of these stress conditions may not be obvious. It should be clarified first that the general response system probably does not include all relevant stress responses. It rather appears to be a cellular system that plays a fairly central but not the sole role in the establishment of the inducible resistance of cells against a number of, but apparently not all, severe stress conditions. The capacity to outgrow stress without making cells particularly stress resistant is at least partly lost if growth conditions of yeast cells deteriorate because of any type of stress. Under natural conditions it will be quite likely that a limited population of cells is confronted with some other type of stress before growth conditions improve again. Therefore a system protecting cells against multiple stress types and probably against some combinations of stress factors damaging cells in a synergistic way is advantageous in evolutionary terms. Then one may ask how far this consideration is also valid for other types of organisms. A general stress response—perhaps at least in some cases with a (partly) different mechanistic basis—appears to exist in many classes of organisms. The selective advantage outlined above in a speculative manner should exist for unicellular organisms. The details of the arguments in favor of such a mechanism would have to differ in the case of multicellular organisms because of the increased complexity of these organisms and of their mode of interaction with their environment.

Our knowledge about transcription factors binding to STREs and mediating stress signals is very recent[24,25] and therefore a number of important questions in connection with this system remain unanswered at the moment. Msn2p and Msn4p are Cys$_2$His$_2$ zinc finger transcription factors binding specifically to STREs. Based on existing genetic data, Msn2p directly or indirectly plays an important role in the response to many or all stresses mediated by STREs and in the general response. Msn4p may or may not have exactly the same specificity with respect to stress factors activating it, but it seems to have at least functions overlapping with those of Msn2p. However, while, e.g., the heat response mediated by STREs is completely dependent on Msn2p and/or Msn4p, this is clearly not the case, e.g., when STRE-dependent transcription is activated by increases in osmolarity. This might be explained by the existence of more transcription factors similar to Msn2/4p perhaps displaying different activation patterns. Alternatively, entirely different activation mechanisms may operate in the case of some stresses (high osmolarity) which, however, according to available experimental data would have to involve STREs and at least an important indirect role for Msn2/4p.

The interplay between growth control mediated by protein kinase A and stress control via STREs is discussed extensively in earlier publications,[8,26] in chapter 7 and in an earlier part of this chapter. One crucial mechanistic question shall be emphasized further. As outlined in Figure 8.4A, at least some stress conditions might activate transcriptional responses to stress by (indirectly) inhibiting protein kinase A and its negative effect on STRE-mediated activity. This mechanism is definitely not valid for all stress factors activating STRE-dependent transcription, as is illustrated by the mode of action of the HOG pathway on STREs.[26] Furthermore, direct evidence

for such a mechanism is lacking except for the control of STRE activity by carbon sources availability via the Ras-cAMP pathway. Remarkably, cAMP levels are also decreased under salt stress,[27] and this may play some role in the control of STRE activity if this element is activated via high salt concentrations. Other types of stresses might also affect cAMP levels but this has not yet been investigated thoroughly. In other instances the FGM pathway, discussed mainly in chapter 1, or similar yet unknown signaling pathways might transmit cAMP-independent stress signals to protein kinase A, but direct evidence for this is currently lacking. On the other hand, as illustrated in Figure 8.4B, protein kinase A might not be directly involved in the transmission of at least some stress signals. As high osmolarity stress in the case of the HOG pathway other stresses might control Msn2p or other STRE-binding transcription factors directly and protein kinase A could act as a modulator of their activity. In this case the activity or cellular location of protein kinase A would not be altered by stress conditions. Of course some stress conditions might transmit their effect via protein kinase A, other stress effects might converge at STRE-binding factors or factors interacting with STRE-binding proteins and bypass protein kinase A. It should be emphasized in this connection that a direct modulation of the activity of Msn2/4p is still hypothetical and remains to be demonstrated in molecular details in the case of protein kinase A as well as in that of Hog1p. As visualized in Figure 8.4, the most important question concerning the mechanism of the general response that is still open at the moment but might be solved soon is which stress conditions affect the cellular location or activity of protein kinase A. On the other hand the role of Msn2p-like proteins as transcription factors functioning in the general response has undoubtedly been recognized recently.

Finally, one might summarize that stress responses in a simple eukaryotic organism like the yeast S. cerevisiae have

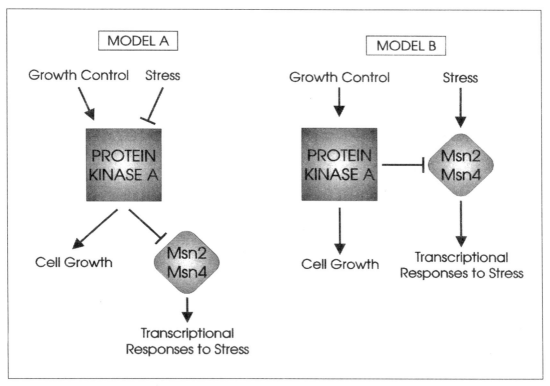

Fig. 8.4. The role of protein kinase A in the response to stresses mediated by STREs. A: A model implying a direct role of protein kinase A in the transmission of stress signals. B: An alternative model implying protein kinase A as a modulator of stress signals. In both cases, negative effects on Msn2p have not yet been demonstrated to occur by direct phosphorylation or interaction with protein kinase A. Other STRE-binding transcription factors or other proteins mediating protein kinase A effects may be involved.

turned out to be tremendously complex. Studies presented in this volume appear to indicate that a respectable part of the gene products that can be produced by this organism are at least to some extent involved in stress responses. While this is fascinating, it is also discomforting to some extent since we are currently in a situation where some fundamental questions appear to be partly solved, but where one could get the impression that there is no limit to the complexity of stress responses. On the one hand this latter impression is almost certainly misleading. New genes involved in the stress response will show up less frequently in the future. The small size of the S. cerevisiae genome that has now been sequenced completely sets a clear limit to further developments, although the number of possible relevant interactions between different gene products and factors modulating their activity is still tremendous. On the other hand, the impression remains valid that a significant percentage of the genome of an organism like S. cerevisiae functions in stress responses. This perhaps shows that in an environment more or less hostile to all organisms (in nature) microorganisms can only survive in the long run if they are pessimistic[26] and are ready for as many difficulties as possible which nature may suddenly present to them. Therefore, the complexity of stress responses is likely to be not only apparent to some extent but real. This teaches us that these responses are biologically more important than scientists would have expected some time ago.

Acknowledgments

I am extremely grateful to Christoph Schüller for critically reading and discussing this manuscript with me. Work in my own laboratory in the yeast stress field was supported by grants from the Fonds zur Förderung der Wissenschaftlichen Forschung, Vienna.

References

1. Hinnen A, Hicks JB, Fink GR. Transformation of yeast. Proc Natl Acad Sci USA 1978; 75:1929-1933.
2. Beggs J. Transformation of yeast by a replicating hybrid plasmid. Nature 1978; 275:104-109.
3. Rothstein RJ. One-step disruption in yeast. Methods Enzymol 1983; 101:202-210.
4. Oliver SG. From DNA sequence to biological function. Nature 1996; 379:597-600.
5. Johnston M. Genome sequencing: the complete code of a eukaryotic cell. Current Biol 1996; 6:500-503.
6. Galcheva-Gargova Z, Derijard B, Wu I-H, Davis RJ. An osmosensing signal transduction pathway in mammalian cells. Science 1994; 265:806-808.
7. Han J, Lee J-D, Bibbs L, Ulevitch RJ. A MAP kinase targeted by endotoxin and hyperosmolarity in mammalian cells. Science 1994; 265:808-811.
8. Ruis H, Schüller C. Stress signaling in yeast. BioEssays 1995; 17:959-965.
9. Mager WH, de Kruijff AJJ. Stress-induced transcription activation. Microbiol Rev 1995; 59:506-531.
10. Sanchez Y, Taulien J, Borkovich KA, Lindquist S. Hsp104 is required for tolerance to many forms of stress. EMBO J 1992; 11:2357-2364.
11. Lindquist S, Kim G. Heat shock protein 104 expression is sufficient for thermotolerance in yeast. Proc Natl Acad Sci USA 1996; 93:5301-5306.
12. Finley D, Chau V. Ubiquitination. Annu Rev Cell Biol 1991; 7:25-69.
13. Heinemeyer W, Kleinschmidt JA, Saidowsky J, Escher CH, Wolf D. Proteinase yscE, the yeast proteasome/multicatalytic-multifunctional proteinase. EMBO J 1991; 10:555-562.
14. Maeda T, Wurgler-Murphy SM, Saito H. A two-component system that regulates an osmosensing cascade in yeast. Nature 1994; 369:242-245.
15. Maeda T, Takekawa M, Saito H. Activation of yeast PBS2 MAPKK by MAPKKKs or by binding of an SH3-containing osmosensor. Science 1995; 269:554-558.
16. Bourret RB, Borkovich KA, Simon MI. Signal transduction pathways involving protein phosphorylation in prokaryotes. Annu Rev Biochem 1991; 60:401-441.
17. Krems B, Charizanis C, Entian KD. The response regulator-like protein Pos9/Skn7 of *Saccharomyces cerevisiae* is involved in oxidative stress resistance. Curr Genet 1996; 29:327-334.
18. Zhu Z, Szczypka MS, Thiele DJ. Transcriptional regulation and function of yeast metallothionein genes. In: Sarkar B, ed. Genetic Response to Metals. New York: Marcel Dekker Inc., 1995.
19. Elion EA. Ste5: a meeting place for MAP kinases and their associates. Trends Cell Biol 1995; 5:322-327.
20. Faux MC, Scott JD. Molecular glue: kinase anchoring and scaffold proteins. Cell 1996; 85:9-12.
21. Smith BJ, Yaffe MP. Uncoupling thermotolerance from the induction of heat shock proteins. Proc Natl Acad Sci USA 1991; 88:11091-11094.
22. Liu XD, Thiele DJ. Oxidative stress induced heat shock factor phosphorylation and HSF-dependent activation of yeast metallothionein gene transcription. Genes Dev 1996; 10:592-603.
23. Morimoto RI. Cells in stress: transcriptional regulation of heat shock genes. Science 1993; 259:1409-1410.
24. Martinez-Pastor MT, Marchler G, Schüller C, Marchler-Bauer A, Ruis H, Estruch F. The *Saccharomyces cerevisiae* zinc finger proteins Msn2p and Msn4p are required for transcriptional induction through the stress-response element (STRE). EMBO J 1996; 15:2227-2235.
25. Schmitt AP, McEntee K. Msn2p, a zinc finger DNA-binding protein, is the transcriptional activator of the multistress response in *Saccharomyces cerevisiae*. Proc Natl Acad Sci USA 1996; 93:5777-5782.

26. Schüller C, Brewster JL, Alexander MR, Gustin MC, Ruis H. The HOG pathway controls osmotic regulation of transcription via the stress response element (STRE) of the *Saccharomyces cerevisiae CTT1* gene. EMBO J 1994; 4382-4389.

27. Marquez JA, Serrano R. Multiple transduction pathways regulate the sodium-extrusion gene *PMR2/ENA1* during salt stress in yeast. FEBS Lett 1996; 382:89-92.

INDEX